Children and Pollution

Children and Pollution: Why Scientists Disagree

Colleen F. Moore

OXFORD
UNIVERSITY PRESS

2009

OXFORD
UNIVERSITY PRESS

Oxford University Press, Inc., publishes works that further
Oxford University's objective of excellence
in research, scholarship, and education.

Oxford New York
Auckland Cape Town Dar es Salaam Hong Kong Karachi
Kuala Lumpur Madrid Melbourne Mexico City Nairobi
New Delhi Shanghai Taipei Toronto

With offices in
Argentina Austria Brazil Chile Czech Republic France Greece
Guatemala Hungary Italy Japan Poland Portugal Singapore
South Korea Switzerland Thailand Turkey Ukraine Vietnam

Copyright © 2009 by Oxford University Press, Inc.

Published by Oxford University Press, Inc.
198 Madison Avenue, New York, New York 10016
www.oup.com

Oxford is a registered trademark of Oxford University Press

Library of Congress Cataloging-in-Publication Data
Moore, Colleen F., 1950—
Children and pollution: why scientists disagree / Colleen F. Moore.
     p. cm.
Includes bibliographical references and index.
ISBN 978-0-19-538666-0
1. Pediatric toxicology.   2. Environmental toxicology.
3. Pollution—Psychological aspects.   I. Title.
RA1225.M667 2009
618.92'98—dc22
2008041656

Originally published as *Silent Scourge: Children, Pollution,
and Why Scientists Disagree*

9 8 7 6 5 4 3 2 1
Printed in the United States of America
on acid-free paper

To the victims of pollution poisoning disasters all over the world.
May we develop the wisdom to prevent further needless suffering.
And to my graduate school mentors: Carol S. Dweck,
Michael H. Birnbaum, and Don E. Dulany

Acknowledgments

While I wrote this book I was fortunate to receive encouragement, love, and support from my swimming pals, friends, and family members. Extra-special thanks are due to three scholars who read and commented on drafts of the first edition for me: Michael Kaschak, Nazan Aksan, and Kate Zirbel, all of whom gave me important encouragement as well as insightful feedback. I am indebted to Michael Kaschak in particular for encouraging me to highlight the ethical issues in the material. I also received very helpful reviews of the first edition on behalf of Oxford University Press from Joseph Jacobson and Kristin Shrader-Frechette and from an anonymous reviewer that helped me to improve the final product considerably. I want to express my thanks to the scientists who conducted the original research that I have summarized in this book. Without their work, we would all be ignorant about the effects of pollution on children's development. Undoubtedly some of these scientists will take issue with my interpretations of their work, but those differences are part of the processes of healthy science. I am grateful to the students in my Psychology of Environmental Issues lab groups and classes over the last several years for their help in grappling with some of the issues that formed the background for this book. I also wish to thank Pat Klitzke and Jane Fox-Anderson, not only for their impeccable professional work but also for their friendship. I am grateful to the Cartography Laboratory of the Geography Department for artwork, and to the College of Letters and Science, University of Wisconsin–Madison, for sabbatical leave during which I undertook this project initially, and during which I prepared the second edition. Finally, I thank Joan Bossert and Sarah Harrington at Oxford University Press for giving the material new life in the second edition.

Contents

Prologue

Controversies swirl around environmental issues. Future generations will be affected dramatically by decisions our society makes today, just as all of us now are being affected by the environmental policies made in the last several generations. It is our responsibility to think about the world we are leaving for future generations and about how our decisions as individuals and as a culture impact the future. For most of us, those thoughts start at home with our own children, nieces, nephews, and grandchildren. I think that the public should know as much as possible about the effects of pollution on children's development, why the experts argue about those effects, and how different decision-making criteria and ethical considerations are inextricably inter-twined in both science and policy.

This book reviews some of the best science on how pollution affects our children's behavioral health and examines why scientific experts often dis-agree about the best science. The term behavioral health refers to intellectual functioning, behavior, and emotional states. Most news coverage of environ-mental issues is heavily slanted toward diabetes, cancer, asthma, or other diagnosable diseases, or toward broad issues such as global warming and deforestation. Those issues are very important. But how pollution affects our daily behavioral functioning is equally important and seriously neglected.

One of the main points of this book is that we do not need a "body count" of deaths or cancer cases in order to conclude that pollution has serious devel-opmental effects. Most children are exposed to all of the types of pollution that I cover: lead, mercury, PCBs, pesticides, noise, chemical wastes, and radioac-tivity. These types of pollution are usually silent and insidious. The effects of these pollutants on children's well-being are revealed by carefully constructed psychological assessments of memory, attention, learning, motor skills, intel-ligence, personality, emotion, and other characteristics. Sometimes these effects are serious enough that the child needs special education or therapy.

The behavioral effects of pollution also have large dollar costs, in addition to the emotional toll taken on families and society of raising and educating a child with special needs.

There are simple things anyone can do to reduce exposure without panicking or spending extra money. These tips are collected in the last chapter "Protect Your Family, Protect Our Planet." Many of the things we can do to cut our pollutant exposures will also help reduce global warming. Family health and the health of our planet are linked.

But personal protective measures are not enough. Because virtually all children are exposed to and affected by the kinds of pollution discussed in this book, the effects of low-level exposure to these pollutants need to be taken seriously by the government agencies that regulate them. In a democratic society citizens should voice their concerns. This book will enable you to ask the right questions about how decisions to regulate any pollutant are made.

Most of the pollutants I cover were thought to be safe until a mass poisoning incident of some type occurred. Some pollution disasters or crises were serious enough to cause people to evacuate their homes. Evacuation and relocation in themselves create stress and social disruption. Those stresses and disruptions can affect people's lives long after a pollution crisis is over. On top of that, people in the midst of a pollution crisis suffer from the fear of exposure in addition to any direct effects of the pollution itself. These disasters remind us how much damage uncontrolled pollution can cause, and why we have environmental regulations.

Readers of this book will learn about the specific kinds of pollution I cover and how they affect children. But I use the effects of these particular kinds of pollution on our children's behavioral health to address general issues that apply to most of today's environmental troubles. This is important not only for preventing future pollution disasters but also for preventing the kinds of widespread insidious negative effects that are occurring at present.

Questions about Pollution

Has There Been Adequate Monitoring?

Without vigilant research it is too easy to say, "According to what we know, this type of pollution is safe." A statement like this is often true only because adequate research has not been done. As use of a new technology grows, exposure to any pollution from it increases. Broad exposure of most of the population to any new kind of pollution raises the question of what the pollution is doing to people, especially children, and also what it is doing to ecosystems. In the case of lead in gasoline and paint, tests on the effects of low-level exposure to the public were not systematically done for approximately 50 years. When the research finally was done, it showed serious deleterious effects of lead on children's cognitive functioning and behavior. And we still do not know how lead might be contributing to the occurrence of dementia later in life. Research on the behavioral effects of pesticide exposures is just starting, although many of the current pesticides have been in use since the end of World War II. PCBs were used globally in a huge variety of industrial processes and consumer products. When studies were done after 40 years of use, it was found that PCBs had permeated virtually every ecosystem on the globe. A similar situation is now occurring with flame retardants known as PBDEs.

Have Effects Other Than Frank Poisoning Been Studied?

A disease view of the effects of pollution says that a person has or does not have a disease as a result of exposure. In the disease viewpoint there is a search for a threshold of exposure that results in the "illness." Instead, many pollutants show a continuous dose-response effect. Higher doses yield more serious negative effects, and lower doses yield smaller negative effects. As you will see, the research on lead exposure in children questions the disease and threshold view of pollution. I do not mean to imply that there are never thresholds for the effects of pollutants. But the threshold will depend on what outcome is measured.[1] Obviously, the threshold of lead, PCB, or mercury exposure for death is much higher than the threshold for altered behavioral development in children. Controversies often erupt around the issue of whether there is a threshold, and also around whether small negative effects are serious enough to warrant government regulation. Asking about dose-response effects rather than thresholds for poisoning can change how these controversies are played out.

How Does the Timing of Exposure Influence the Effects?

Prenatal exposure to mercury, PCBs, noise, pesticides, or ionizing radiation can be a hazard to the fetus and later child development. This is true even when the pregnant mother does not show noticeable adverse effects herself. Because some kinds of pollution (for example, lead, PCBs, some pesticides, some radioactive isotopes) are stored in our bodies for very long periods of time, prenatal effects imply that women need to avoid exposure even before they become pregnant. And unfortunately, the behavioral effects of fetal exposure to toxins are often irreversible because they can affect the way the brain forms its basic structures and functions. But not all exposures to pollution are irreversible, and not all kinds of pollution are more toxic prenatally than postnatally. Pollution that has irreversible effects on development needs regulation, even if the effects are not poisoning. On the other hand, when the effects of pollution are reversible, we might be able to take charge and protect our families by reducing our exposure—unless the pollutant is so widespread that it is unavoidable.

Who Is Exposed Most Heavily, How, and Why?

Exposure to pollution is not an equal-opportunity event in America, or in any part of the world. When lead was a gasoline additive, simply living in a large city meant that you and your children would be exposed to lead whether you wanted to be or not. Those with limited economic resources who live in substandard housing in America today have children who are much more highly exposed to lead than most children of well-to-do families. With PCBs and mercury spread over the entire globe, people living in subsistence economies that depend on fish and other animals are often more highly exposed than people in industrialized economies. And those who make their living in agriculture are exposed to pesticides more heavily than the rest of us. Who is exposed often affects how society regulates a pollutant. I raise this issue because I am concerned about fairness and social equity, which I believe are important foundations of any society, but especially a democratic one.

The Role of Decision Standards in Scientific Controversies and Environmental Conflicts

Disagreements among scientists are embedded in all environmental controversies. It would be easy to write a one-sided book about how bad pollution is for children. But good scientists often disagree about pollutants' effects on children. In this book I describe some arguments between scientists over the effects of pollution on children's behavioral health. In one dispute a researcher was charged with scientific misconduct, and in another a scientist was threatened with losing his job. Yet another scientist was removed from an Environmental Protection Agency (EPA) committee she had chaired, and another had her mail opened and was not allowed to apply for research funding. These extraordinary attacks on scientists were embedded in disputes over research methods and interpretations of results. They illustrate how normal scientific conflicts can escalate when the research impacts either public policy or corporate profit.

Different points of view are embedded in all scientific conflicts, whether routine or extreme. Scientists differ in what decision standards should be used for drawing a conclusion that a pollutant is harmful. They adopt different starting assumptions and theories for the research. These decision standards and assumptions are very important in determining whether a type of pollution will be regulated by government. One of my goals is to demystify how scientific research is used in government risk assessments so that the average citizen can participate in debates over environmental issues. For each pollutant, I show why the scientists disagree, how they attempt to resolve their differences, and how policy makers deal with conflicting scientific conclusions in regulating pollution.

Science is a human enterprise, not a method to find cut-and-dried facts. Judgment calls are involved in both research and in the risk assessments done by government agencies. Most environmental decisions are driven by the research findings and the risk assessments. This can leave the average citizen out of the process. There is also a growing cynicism about scientific experts: each side in an environmental controversy finds its own scientific researcher to present one side as a hired gun. Litigants do engage scientists to be their hired guns, but the scientists have usually taken opposite sides prior to being involved in a lawsuit. If you understand a bit of the whys behind scientists' debates among themselves, you will see that scientists can take opposite sides and still be both scientific and honest.

One key to understanding how scientists can take sides and still be scientific is to realize that scientific uncertainty is never eliminated even in the best science. Although science is often presented in the mass media as if it draws conclusions about facts with certainty, all scientific conclusions involve uncertainty and nuances of interpretation. The scientific process should provide us with a measure of the degree of uncertainty—in other words, the likelihood that a particular scientific conclusion is incorrect. Throughout the book I have highlighted the sources of scientific uncertainty and how the scientists' disagreements are intertwined with the uncertainties. Scientific uncertainty is an important aspect of that debate because it relates to the central question of environmental policy: how much evidence of harm is enough to require regulation of a pollutant?

A second key is to understand that the conclusions a scientist draws from research data, especially conclusions that pertain to public policy, involve at least a little bit of the researcher's own values. Some of these values pertain to beliefs about what is required in order to draw a scientific conclusion (values having to do with what counts as knowledge). But a scientist's personal beliefs and ethics are likely the motivation for undertaking research on a particular topic in the first place. And the cultural and historical context in which the scientist works form the background for the whole research enterprise.[2] The values that can influence a scientist's conclusions run the full spectrum from simple assumptions about methods, to commitment to a prevailing theoretical perspective that may prevent open-minded consideration of alternatives, to more blatant bias, perhaps in favor of a financial sponsor or a personal political commitment. While all of us can agree that explicit bias does not belong in science, even a simple assumption about which methods and theory are best can tilt a piece of research toward one conclusion or another. I show this throughout the book.

Thomas Kuhn's (1970) famous treatise on scientific revolutions showed how a predominant theory can filter out research findings that contradict that theory. He also discussed why researchers are reluctant to discard preferred theories. The majority of contemporary philosophers of science agree that there are no universal rules in science for deciding which theory gives the best explanation for a phenomenon. Because of this aspect of the normal processes of science, the proper interpretation of a piece of research can become the subject of debate among scientists, with the scientists forming camps that support different viewpoints. These kinds of controversies are common in all science, not just in science dealing with environmental issues.

Other views of science assert that adherence to a particular theory is driven partly by the degree to which the theory is consistent with cultural norms, beliefs, and values.[3] One example that illustrates how the cultural and historical context can influence the way research is carried out and interpreted is Stephen Jay Gould's 1981 book, "The Mismeasure of Man." Gould describes how racial bigotry was incorporated in taking and interpreting anthropological and psychological measurements—a notion we find abhorrent today. Another example is Ruth Bleier's indictment of biological research on sex differences as having been conducted and interpreted in a manner that would perpetuate cultural myths and the subordination of women. Scientists try to avoid biases such as those illustrated in Gould's and Bleier's books. But they may be blinded by their own skewed viewpoints because their scientific community and culture is biased.

Even though science is a human enterprise subject to human foibles, it is still the best approach we have for finding out whether and how pollutants affect people and the earth's ecosystems. Good scientists strive to be as objective as possible from the start of data collection to the final conclusion. Research that is published in scientific journals is reviewed by other scientists. As reviewers, scientists do their best to assess the adequacy of the methods and the logic that connects the results to the conclusions. While this peer review process is also imperfect, it serves to enforce the standards of the scientific community and the goal of objectivity in research.

Many environmental controversies are couched in terms of whether the best science provides enough evidence to justify regulation of a pollutant. This is an ethical issue. Although the best science is based on data collected and interpreted by scientists who are striving to be rational and objective, the conclusion about whether to regulate a pollutant requires scientists to decide how much evidence is enough to justify regulation. The decision ultimately has implications for whether people will suffer harm from a pollutant and whether someone will have to pay to clean it up. For these reasons, I regard most environmental controversies as primarily value and ethical differences. Because environmental policy decisions affect all of us and are based on scientific conclusions that involve many interpretations, scientific conclusions as they apply to environmental policy can and should be questioned. After reading this book you will know several kinds of questions to ask in order to ferret out the value issues that are embedded in any expert's conclusions.

Why Alcohol, Tobacco, Violence, and Poverty Are Omitted

If I had a magic wand to improve the lives of children in America today, I would eliminate poverty, violence, alcohol abuse, and tobacco addiction before any of the pollutants I cover in this book. But none of these are what we think of as environmental pollution, so it is someone else's job to write about them. All have serious deleterious effects on children's lives. I touch on the developmental effects of tobacco in Chapter 4 on pesticides. The effects of poverty are addressed in social justice issues that are included in each chapter. The fact that I do not focus on violence, alcohol abuse, tobacco addiction, and poverty does not imply they are unimportant.

This Book Will Not Diagnose a Child

Almost one of every seven children in the United States—approximately 15%—receives special education services of some kind. The special education rate is twice that for minority students in some school districts. The pollutants I cover in this book have the potential to increase the number of special education students because they affect intellectual, emotional, and behavioral development. We all want fewer special education students. The research evidence shows that decreasing children's exposure to certain pollutants will help decrease special education needs and costs in the United States and the rest of the world.[4] Some pollutants are associated with a particular pattern of developmental changes in children.[5] However, the cause of a given type of special education need in a particular child usually cannot be pinned down. For example, reading disability can be caused by many things. Living in a high noise environment is just one possible cause of reading disability. After exposure to a pollutant has done its deed, it is virtually impossible to say with certainty that a particular pollutant was responsible for a child's problems.

Why Worry?

My goal is not to scare my readers. I intend to provoke deep thought about environmental pollution and its effects on the psychological well-being of current and future generations. But the topic can be frightening, especially for parents. One reaction to the information here might be excessive worry about every little detail of life and whether it is exposing your child to an unwanted pollutant. Another reaction might be to shrug and dismiss the information, because after all, you cannot worry about everything. I hope my readers will find a middle ground.

One way to think about the hazards of pollution to your family is to think about car safety seats for children. The use of child safety seats reflects not worry, but normal prudence. Most parents are not afraid to strap a child into the car seat and go for a drive. Thirty years ago you would not have used a safety seat. At that time children had a higher chance of being seriously injured or killed in auto collisions than today. Government policy in America changed that. Child safety seats are now required. I think of the information in this book as analogous to car safety seats. There are simple things you can do to protect your children from daily exposure to certain pollutants that will reduce the chance of harm. But public policy is as important as personal action. Just as the mandate for safety seats for kids saves lives and prevents injuries, public policies to prevent pollution at the source can improve every-one's quality of life.

Point of View

Like other scientists, I also bring my own point of view, personal values, and cultural background to the interpretation of the research. My professional training is in both developmental psychology and the psychology of judgment and decision making. I have done my best to give a fair and accurate overview of the research on how pollution affects children's behavior and quality of life. My review is selective rather than comprehensive. I have presented studies that I consider to be the best science, along with the major issues and argu-ments they have spawned.

Aside from my professional training, my view is rooted in my childhood experiences in southern California, growing up at a time when the air pollu-tion there was even more serious than it is now. As children walking home from a summer afternoon at the swimming pool, my friend Barry and I would try to trick each other into taking a deep breath. Why? It hurt to take a deep breath after playing outdoors in the smog! The effects of the smog went beyond temporary lung pain. Living in the smog took the joy out of physical exercise.

When the view of the mountains was obliterated by smog, the world felt dingy and distorted. But at the same time that our chests ached every afternoon, the state of California was building more freeways all over the state, exacerbating the air pollution problem. These experiences are part of why I think that not enough has been done by either our governments or ourselves as individuals to reduce pollution and its effects.

I began to question whether the *psychological* effects of pollution go beyond loss of enjoyment. What about pollution you cannot feel so clearly or immediately as the deep smog-ache in the lungs? As a professor at the University of Wisconsin, I decided to teach a course about the psychological effects of pollution on people. When I first began teaching that course, I was surprised to find that a large amount of scientific research had been done on some topics starting in the 1970s—in particular, on lead exposure and children's school performance, as well as on how noise affects children's reading. But the topics covered in this book are still not found in most textbooks on developmental psychology, educational psychology, pediatrics, nursing, or related fields. The more I learned, the more it seemed clear to me that the topics in this book were important to bring to the broad attention of both the public and professionals such as health care providers, teachers, psychologists, and social workers. I have tried to write the book in an accessible style. I hope it will be useful not only for developmental psychologists but also for people without a specialized background in behavioral science.

I also believe that government regulation of pollution has not given sufficient weight to behavioral effects, especially for children. Such neglect appears to me to be partly a value judgment based on the perception that a physical disease such as cancer is somehow more "real" than a psychological problem such as failing school performance. I think the policy neglect originates partly in ignorance about the negative effects of pollution on children's development and overall psychological well-being. I hope this book will help to correct that ignorance.

Toward a New Dialogue on Environmental Issues

We hear public officials such as the head of the U.S. Environmental Protection Agency state that environmental policy should be based on the best science. I agree. But such an assertion is overly simplistic as stated. Science by itself can never be the basis of public policy because science is not supposed to take ethical or moral positions. But, public policy inevitably involves ethics. People take ethical and moral positions, and scientists are people. But the ethical and moral positions of scientists should not receive more weight in policy than the ethical and moral positions of others, at least not in a democratic society.

Most environmental policy controversies are not really about whether the science is good enough. The controversies are about whose values and ethics will take priority in policy making. The ethical arguments are veiled by finger pointing over the details of the science. Researchers who finger the inadequacies of each other's research are usually applying different decision standards that are based on their own beliefs and values. They are often arguing about how strong the evidence needs to be in order to conclude that a pollutant is harmful. This is a question of values, both epistemological values (i.e., what is required in order for us to say we have knowledge about something) and ethical values.

If we view environmental controversies as rooted in the decision standard for how much evidence is needed in order to conclude that a pollutant is harmful, then we can redirect the discussion squarely onto the morals and ethics behind those chosen decision standards. The best science about pollution will still be central, but it should not be permissible to use science to hide value judgments. In an ideal world, the participants in environmental controversies, including scientists, would be clear in distinguishing between the research results and their own ethical positions and value-based conclusions about what public policies follow from the results. I hope that this book will help move environmental debate toward a discussion of values and ethics in which *all* may participate, both those who regard themselves as "green" and those who feel that environmental threats are overstated.

Children and Pollution

CHAPTER 1

Lead and the Roots of Environmental Controversies

I n 2007 the popular toy, Thomas the Train, was recalled by its man-ufacturer because of a lead paint hazard. What are the hazards of lead for children? This has been studied extensively in the last 30 years. Exposure to lead lowers children's IQ test scores and raises the likelihood of restless and inattentive behavior, diagnosis with attention deficit hyperactivity disorder (ADHD), and juvenile delinquency. Lead exposure in the United States has dropped dramatically since lead was phased out of paint and gasoline. But lead exposure is still a problem for young children.

Shortly after one of my friends had a baby, his in-laws visited. The in-laws spent two weeks helping the new parents fix up the house, which was built in the 1920s. They painted the baby room, the stairs, and the living room. When the in-laws sanded the woodwork to remove the old paint, lead dust permeated the air. When the infant's pediatrician tested the child's blood lead, it was almost double the current cutoff for high lead (10 µg /dl). The infant had absorbed lead by breathing it from the air. This scenario shows that environmental policy decisions have long-term repercussions for public health. Let's look at the history of lead to see why an infant could be lead poisoned in 2008 by paint dust from home renovations.

The history of lead in the United States brings us face to face with this question: what decision standard do we use to regulate a new technology that has uncertain risks and benefits? Lead also provides a striking example of how scientists clash over environmental issues, with experts on opposite sides drawing different conclusions. A major part of the scientific conflict is rooted in the decision standards and assumptions on which the research is based. Those decision standards and assumptions involve implicit value judgments that affect the conclusions. In turn, the conclusions scientists draw ultimately affect the pollution prevention and cleanup decisions of our governments.

Lead also tells a tale of grave social injustice: African American children in low-income families in the United States have the highest exposure to lead of any segment of the American population. But scientists argue about whether lead exposure is a cause of lower IQ test scores or behavior problems or whether it just appears that lead exposure has negative effects on children because lead exposure is higher in socially disadvantaged circumstances. The conclusions a researcher draws about whether lead causes problems in children depend critically on the scientist's assumptions about IQ test scores and what influences them. So, en route to learning about how lead affects children's development, this chapter also involves decision standards in science and shows how unquestioned, status quo viewpoints within science influence scientists' interpretations of their research results in very important ways.

A Brief History of Lead Exposure and Regulation in the United States

Alice Hamilton, MD, Social Reformer

Lead has been known to be poisonous since the Roman period, when Pliny the Elder wrote about it. But factory workers after the industrial revolution were exposed to higher levels of lead in greater numbers than before. Alice Hamilton, MD, was an early crusader in the battles for worker safety. Figure 1.1 shows the postage stamp issued in her honor. Among her other accomplishments, she was the first woman to be a professor at Harvard

FIGURE 1–1. Postage stamp honoring Dr. Alice Hamilton.

University. Before she became a professor, starting in 1910, she systematically studied lead exposure and its effects in industrial workers.[1]

Dr. Hamilton and her team asked to enter the production facilities of companies where they suspected lead exposure in the workers. In her auto-biography, *Exploring the Dangerous Trades*, she described the reaction of Edward Cornish, who later became president of the National Lead Company. Mr. Cornish denied that his men were "leaded." He issued this challenge to Dr. Hamilton: if she could prove to him that the work environment was damaging the health of his workers, he would do whatever she said to prevent that lead poisoning, including hiring doctors.

After surveying local hospitals for cases of "plumbism" (lead poisoning), Dr. Hamilton presented Mr. Cornish with records of 22 verified cases of his own workers who had been lead poisoned (see Table 1.1 for symptoms of lead poisoning). Under Cornish's leadership, the Sangamon Street works of the National Lead Company in Chicago took unprecedented steps to prevent workers' exposure to lead dust and fumes. The company had its engineers invent methods for confining dust and fumes and then used the same meth-ods at other plants in the Chicago area. Dr. Hamilton wrote, "I have met many admirable men in industry throughout these thirty-two years, but my warm-est gratitude and admiration goes to Edward Cornish."[2]

Mr. Cornish, acting on the advice of Alice Hamilton, was clearly a pioneer in worker safety. But the advent of tetraethyl lead in gasoline posed other chal-lenges to industry leaders as well as public health physicians like Dr. Hamilton.

1926: Tetraethyl Lead in Gasoline Gets OK from Surgeon General

The battle over lead in gasoline began prior to any specific federal legal author-ity to regulate pollution. In some ways this struggle established precedents

TABLE 1–1. Effects of acute lead pensioning

Fatigue
Discolored (blue-gray) teeth and game
Jaundice
Colic
Numbness
Trembling and lack of motor control (palsy)
Muscle cramps and contractions
Pain in the extremities
Paralysis and loss of sensation
Hallucinations
Kidney failure
Seizures and coma
Death

for how pollution is regulated. There were two clear sides. Some argued that exposure to lead in gasoline was not enough above background levels to be harmful. Other scientists argued that lead from gasoline was a major pollutant and a public health hazard.[3]

In February 1925 Dr. Hamilton, along with other doctors, wrote to the Surgeon General suggesting an impartial investigation into the hazards of tetraethyl lead in gasoline. The public health doctors were extremely alarmed by the number of lead poisonings and deaths that were happening in the plants producing the tetraethyl lead. In October 1924 at the Standard Oil Company's tetraethyl lead labs in New Jersey, 35 of 49 workers became severely lead poisoned, with palsy (muscle tremors) and hallucinations. Five workers died. Other workers handling tetraethyl lead had died in the plant in Dayton, Ohio, and at the DuPont plant in New Jersey. In May 1925 the *New York Times* reported that 300 workers at the DuPont plant had been poisoned and that 8 had died. The *Times* said the workers called the tetraethyl lead facility the "House of Butterflies" because those who worked there, even just repairing the equipment, were often poisoned to the point of hallucinations.[4]

As a result of the outcry from public health doctors like Alice Hamilton, the Surgeon General held a conference in May 1925. At the conference at least three people spoke strongly against adding lead to gasoline: Dr. Hamilton, Professor Yandell Henderson of Yale University, and Dr. David Edsall, dean of the Harvard Medical School. Hamilton said, "You may control conditions within a factory, but how are you going to control the whole country?" Henderson expressed his belief that lead was a public hazard on the same scale as a serious infectious disease. He said that lead combustion in the engines of automobiles would cause lead to fall from the air into every major city in the country: "Conditions would grow worse so gradually and development of lead poisoning will come on...insidiously...before the public and the government awaken to the situation."[5] The argument in favor of the use of tetraethyl lead was that it increased engine efficiency and would extend scarce gasoline supplies. Its supporters called its invention a "gift from God."[6]

The majority of those who attended the conference concluded that tetraethyl lead should not be banned unless there was *proof* that it was a public health hazard. Research was obviously needed, so a blue ribbon committee of public health experts and industry representatives was appointed by the Surgeon General to carry out the research. The committee issued a report seven months after the conference. The report concluded that "there are at present no good grounds for prohibiting use of ethyl gasoline...provided that its distribution and use are controlled by proper regulation."[7] But the committee also issued a warning that the hazard from lead could grow over time and that as the number of automobiles in the country increased, it would be very important to study the possible effects of lead on a continuing basis. The committee suggested that funding from Congress be requested to conduct such studies.[8] Unfortunately, such follow-up assessments were not conducted.

A One-Sided Decision Standard: Benefits Supersede Risks?

An outcome of the Surgeon General's conference that turned out to be more important than the blue ribbon committee was what some call the "Kehoe Paradigm" for environmental decision making.[9] Dr. Robert Kehoe was a physician on the faculty at the University of Cincinnati who worked as a consultant with the lead industry and became director of the Kettering Laboratory. The Kettering Laboratory was funded by the lead industry and conducted research on lead and its effects.

At the Surgeon General's conference Kehoe proposed that any decision about tetraethyl lead be based on facts already known: "If it is shown . . . that an actual hazard exists in the handling of ethyl gasoline, that an actual hazard exists from exhaust gases from motors, that an actual danger to the public is had as a result of treatment of gasoline with lead, the distribution of gasoline with lead in it will be discontinued."[10] This decision standard requires evidence analogous to a "smoking gun" in a murder case before action is taken to protect the public. In contrast, the medical director of a New York City hospital, Dr. Touart, stated what is now called the "Precautionary Principle": "It seems to me that perhaps the attitude should be taken that this ethyl gasoline is under suspicion and therefore should be withheld from public consumption until it is conclusively shown that it is not poisonous."[11] Notice that the Surgeon General's special committee adopted the smoking gun standard. They recommended that *until* there was definitive evidence of harm, tetraethyl lead should be allowed to go forward.

Now let's restate the two positions and analyze them a bit.

Position 1 (an early version of the Precautionary Principle): Tetraethyl lead should be banned, until it is shown to be safe, on the basis of the known toxicity of lead to humans, deaths and illnesses among factory workers, the likely wide exposure of human populations to lead as a result of burning it in gasoline, and the likely harm that would result from widespread environmental contamination.

Position 2 (the Kehoe Paradigm or smoking gun principle): Tetraethyl lead should be allowed unless and until it is shown definitely to be a health hazard in the general population. There are benefits of its use. There is uncertainty about both the amount of exposure that would occur from burning it in gasoline and the effects of that exposure.

Position 1 has three elements: the known toxicity of the substance in high doses, the likely exposure of large populations to uncertain doses, and the potential damage to large populations. Position 2 also has three elements: the known benefits of the new technology, the unknown amount of exposure of the general population, and the unknown effects of exposure on the general population.

These two viewpoints implicitly weigh the unknowns of the future in dramatically different ways. Both positions contain strong *implicit value judgments* of the benefits of technological innovation in comparison with the uncertainties of negative health impacts. The precautionary approach says that uncertain harm should be prevented. The smoking gun principle says that a potential benefit should be grasped. It gives no importance to uncertain future harm. An important point to realize is that benefits of new technologies are often assumed to be a certainty, as stated in Position 2. Future benefits are not usually subjected to the same scientific scrutiny as are possibilities of future harm. But most importantly, the two positions differ not in their *scientific* evaluations but in the *personal moral values* they express. However, the difference in personal values is a bit veiled. Kehoe once objected that those opposed to the use of tetraethyl lead wanted *proof* that it was safe but that the opponents did not say what would constitute proof. On the other hand, the Kehoe Paradigm or smoking gun principle (Position 2) also requires proof of harm without providing a standard. These are starkly different decision standards.

We see these two contrasting viewpoints echoed in current controversies over many new technologies, including cell phones, genetically engineered crops, and nanotechnologies. And we see the arguments deflected from the basic value differences and instead placed on the science of estimating the amount of harm at different exposures. Each side in an environmental controversy will find an expert to provide an interpretation of the scientific evidence that is favorable to its own side. I hope to show you why scientists can disagree about the scientific evidence without being dishonest. They are usually applying different decision standards. But as you will see, the scientific controversies surrounding the effects of low-level exposure to lead are particularly fierce, and they continue right up to the present time.

1965: Clair Patterson Challenges the Scientific Status Quo

Prior to approximately 1965, the lead industry funded most of the research on lead and lead poisoning through the Kettering Laboratory. Dr. Kehoe described the Kettering Laboratory as "the only source of new information on the subject, and its conclusions have wide influence in this country and abroad in shaping the point of view and activities with respect to this question."[12]

Kehoe apparently thought lead in gasoline was not dangerous. He measured lead in workers who did not directly handle tetraethyl lead, and he found that they had rather high lead concentrations, though they were not visibly poisoned. He assumed that lead was *normally* present in humans, without realizing that these workers had been exposed by inhaling fumes. Later he studied people in a remote area of Mexico and found that they too had relatively high levels of lead in their blood. But these people too had been exposed to lead because they used tableware made with lead glazes. Kehoe reasoned that because symptoms of lead poisoning were not apparent in either the workers or the Mexican villagers, humans must have adapted to having a certain level of lead in their bodies. He concluded that lead exposure was *natural*. He thought there was a kind of equilibrium level for lead absorption and that when higher lead exposure occurred, more lead was eliminated from the body. Poisoning would therefore be rare. In short, he argued that there was a clear threshold for lead poisoning and that the human body had a natural defense against reaching that threshold.[13]

It was not until 1965 that Kehoe's evidence and arguments were seriously challenged. Dr. Clair ("Pat") Patterson, a geochemist at the California Institute of Technology, was the scientist who did it. He was studying the age of the earth. He began studying lead pollution because it was important to eliminate lead's impact on his laboratory measurements. Working in the 1950s in southern California, where smog was a health threat, Patterson found that lead in the air, especially from leaded gasoline exhaust, was the root of lead contamination in his lab. Careful attention to lead pollution in the air allowed Patterson to make a scientific breakthrough in 1955. The earth was more than a billion years older than scientists had thought. Previous researchers had conducted their measurements in lead-polluted air.[14]

In 1965 Patterson published a paper in a scientific journal in which he distinguished between "natural" and "contaminated" sources of lead in the environment and in people. Because his conclusion conflicted with the widely accepted viewpoint of Kehoe, and because Patterson was a geochemist and not a medical doctor, Patterson's article was met with ridicule, anger, and skepticism. Patterson had challenged a central point in Kehoe's case for the safety of lead: that a high background level of lead exposure is "natural." The ensuing controversy resulted in loss of Patterson's research funding. Some members of the Board of Trustees of the California Institute of Technology requested that he be fired. But Patterson continued to pursue his research. He analyzed deep ice from polar regions.He estimated lead contamination in pre-industrial humans by using bones from museums. He showed that lead levels in Americans were approximately 500 to 1,000 times that found in ancient humans (presently Americans have approximately 100–200 times ancient exposure). Patterson had shown that Kehoe was flat wrong about what "normal" was.[15]

In 1966 both Kehoe and Patterson testified in U.S. Senate hearings on the first Clean Air Act. Patterson bluntly pointed out that some of Kehoe's data on lead in the air in major U.S. cities were erroneous. More importantly, Patterson also pointed out the need to separate the scientific research itself from the *use* of scientific research to advocate a particular position: "It is clear, from the history of development of the lead pollution problem in the United States that responsible and regulatory persons and organizations concerned in this matter have failed to distinguish between *scientific activity* and the *utilization* of observations for a material purpose. [Such utilization] is not science....It is the defense and promotion of industrial activity. This utilization is not done objectively. It is done subjectively" (emphasis added).[16]

The conflict between Patterson and the status quo position of Kehoe shows the importance of distinguishing between scientific results themselves and the uses of scientific results to advocate certain public policies. In the expert testimony of a scientist, one ought to be able to find the line between the findings themselves and inferences and conclusions relevant to policy that are drawn from the findings. Regulatory decisions are political and value laden by their very nature. I will leave to the reader's judgment the question of whether the regulatory process follows the legacy of Kehoe's Paradigm of insistence on a smoking gun of evidence of harm prior to regulation, especially for new technologies.

Lead Exposure, Government Regulation, and Child Development

Lead in the environment comes not just from leaded gas but also from paint and other sources. Let's take a brief walk through time and consider some other key events with lead. At first acute poisoning was thought to be the only negative effect of lead. But after long-term effects were found, it sometimes took decades for regulations to be enacted. A time line of key industrial, government, and medical events relevant to lead in paint and gasoline is given in Table 1.2.

In 1909, before World War I, three European countries restricted the indoor use of paint containing white-lead pigment. This restriction was based on the dangers of lead paint to workers. Restrictions were imposed by other countries between 1922 and 1931. In 1922 the League of Nations Third International Labor Conference recommended banning the use of white-lead paint indoors. Meanwhile, in the United States, Dutch Boy white-lead paint was heavily advertised by its manufacturer, the National Lead Company, for use indoors, including around children. But by the mid-1920s many medical studies had been published documenting lead poisoning in both factory workers and children who were exposed to lead at home. In 1933, partly as a result of these studies, Robert Kehoe, still working for the Lead Industry Association, endorsed efforts to remove lead from children's environments. But in 1938 the Lead Industry Association seemingly ignored his advice and launched a campaign to promote the use of white-lead paint inside low-cost housing and institutions such as schools and hospitals.[17]

TABLE 1–2. Time line of events in the modern history of lead

1904–1927	Numerous medical articles published documenting lead planning in both children and workers
1909	France, Belgium, and Austria restrict use of lead print indoors
1910	Dr. Alice Hamilton convinces Edward Cornish of the National Lead Company to protect workers' health by controlling dust and fumes in the factory
1920–1929	Dutch Boy paint promotes the use of lead paint, including around children, and issues "paint booklets" for dealers to give away to children
1922	Third International Labor Conference of the League of Nations recommends banning lead in indoor paint
1922–1931	Great Britain, Sweden, Poland, Czechoslovakia, Spain, Yugoslavia, Tunisia and Greece restrict use of lead paint indoors
1925	Surgeon General's conference on leaded gasoline
1928	Lead Industries Association (LIA) formed
1933	Dr. Robert Kehoe endorses eliminating lead from children's environments
1938	LIA launches campaign promoting lead paint inside low-cost housing, schools, and hospitals
1939	National Paint, Varnish and Lacquer Association suggests that members voluntarily label toxic materials
1943	Time magazine covers research of Byers & Lord linking lead poisoning to later poor school performance and behavior problems
1945	California enacts lead paint labeling regulations
1949	Maryland enacts lead paint labeling regulations, shortly repealed
1952	LIA publishes book, "Lead in Modern Industry," stating that lead paint has "practically no undesirable qualities"
1954–55	Research shows lasting aftereffects of childhood lead poisoning
1954	AMA recommends that lead paint he labeled "poisonous" and not for interior use or around children
1954	New York City enacts regulation limiting lead to 1% in paint
1955	ANSI adopts voluntary limit of 1% lead in interior paint
1955	New York Daily News reports that 10 Brooklyn children died of lead poisoning during the preceding year
1955	Dr. Clair "Pat" Patterson publishes research revising the estimated age of the earth; results required eliminating lead pollution from his lab
1956	Parade magazine and CBS TV cover childhood lead poisoning
1957	City of Baltimore screens homes for lead paint
1959	Ethyl Corporation seeks permission to raise lead to 4 cc/gallon in gasoline
1959	Surgeon General convenes a committee to evaluate the health effects of atmospheric lead
1961	City of Baltimore Health Department suggests removal of lead paint from housing in areas of the city with high rate of lead poisoning
1965	Dr. Clair Patterson publishes paper questioning "natural" level of lead in the environment and people

(continued)

TABLE 1–2. (Continued)

1966	Senate hearings on Clean Air Act include testimony from Robert Kehoe and Clair Patterson
1970	Congress Passes Lead Paint Poisoning Prevention Act. Used of Lead paint inside federally funded housing is prohibited; federal funds are allocated to study health impacts of lead on children. Lead in paint to be phased out.
1970	Congress passes the Clean Air Act. EPA fails to write rules regulating lead in air.
1970	Surgeon General calls for early identification of children with lead exposure 40 μg/dl
1970	To meet air pollution regulation for hydrocarbons, General Motor announces it will begin installing catalytic converters that require lead-free gas
1971	New York City Health Department tests 76 kinds of paint and finds 8 with lead between 2% and 10%
1972	EPA issues rules requiring each gas station to have one lead-free pump
1973	Natural Resources Defense Council sues EPA for failing to issue air pollution regulation on lead. Appeal Court upholds suit.
1973	EPA issues first air pollution regulation for lead in gasoline. Ethyl Corp. and Dupont sue to prevent enforcement. EPA wins in Supreme Court appeal.
1976	California Air Resources Board sets standard for lead in air at maximum of 1.5 μg per cubic mater.
1976	Natural Resources Defense Council wins another suit against EPA over failure to set an air pollution standard for lead
1977	EPA endroses 1.5 μg/cubic meter standard for lead air pollution
1978	Lead in paint prohibited
1979	OSHA limits lead dust in workplaces to 40 μg per cubic meter, and workers blood lead to 50 μg/dl
1980	LIA petitions the EPA to rescind lead air pollution regulations
1982	EPA administrator Ann Gorsuch promises privately not to enforce lead regulation in New Mexico. Bad publicity ensues.
1982	EPA committee appointed to review health effects relevant to air polution regulation. Claire Ernhart and Herbert Needleman clash as members of the committee.
1986	EPA issues draft of health effects of air pollution
1986	New York Times reports that Ethyl Corporation representative said the EPA lead rules were influenced by "rabid environmentalists"
1991	CDC sets 10 μg/dl blood lead as "level of concern" for children and issues strategic plan to end childhood lead poisoning
1992	Title X of the Housing and Community Development Act directs HUD to make recommendations on lead paint hazard reduction and financing

Sources: Compiled from Berney, 1993; Flegal 1998; Markowitz & Rosen, 2000; Needleman, 1998a, 1998b, 2000; Nriagu 1998; Rosner & Markowitz, 1985

Lead poisoning in the 1930s was thought to be an acute disease from which children would make a full recovery if the sources of lead exposure were eliminated.[18] This fit Kehoe's theory of a threshold of exposure in order for illness to occur. According to the threshold idea, once a child's lead was brought back below that level, the child would recover. "Normal" lead levels were considered to be anything below about 70 μg/dl[19] unless accompanied by symptoms. Lead poisoning was diagnosed in small children by the presence of a known source of lead exposure (either from chewing paint off toys, cribs, and windowsills or from breathing lead paint fumes) and symptoms such as abnormal red blood cells, excretion of lead in abnormal quantities in the urine and stools, anemia or noticeable paleness, loss of appetite and vomiting, weakness and uncoordinated muscle movements, partial paralysis, colic (intestinal pain) or serious constipation, hypertension (high blood pressure), headache, seizures, and encephalopathy with swelling of the brain.[20]

The view that childhood lead poisoning was only an acute illness changed in 1943. Randolph Byers, MD, and Elizabeth Lord, PhD, are credited with being the first researchers to follow lead poisoned children to see how they were doing later in school and overall. They tracked 20 lead poisoned children who had been discharged from a Boston hospital as "cured." They gave the children IQ tests and also looked at their school progress and any reports of difficult behavior problems. Byers and Lord found that only one of the children could really be considered to be developing normally and succeeding in school. Five children were borderline mentally retarded (IQ scores below 85). Some of the other children's IQ test performance declined over the years. Three children had been expelled from schools for serious misbehavior, and two others showed marked restlessness and inattention.

Byers and Lord concluded that lead poisoning had disrupted the normal processes of development. They ended their article with this conclusion: "Lead poisoning is a serious disease developing from entirely man-made hazards, which should be controlled by appropriate legislation."[21]

Time magazine reported Byers and Lord's work, intensifying publicity of the toxic potentials of lead in children.[22] Lead poisoning was no longer only an acute illness; it was now recognized to have deleterious longer-term effects. Other follow-up studies also showed that many lead poisoned children continued to have problems later.[23]

In 1945, based on the mounting evidence of hazards from lead paint, California enacted lead paint labeling regulations. Maryland enacted labeling regulations in 1949, but they were subsequently repealed. In 1954 New York City enacted a regulation limiting lead in interior paint to 1%. That same year, the American Medical Association recommended labels on lead paint that would read: "WARNING: This paint contains an amount of lead which may be poisonous and should not be used to paint children's toys or furniture or interior surfaces in dwelling units which might be chewed by children."

This warning was not adopted verbatim anywhere in the country.[24] Instead, paint companies adopted warnings that did not use the term "poisonous." In the midst of these events, in 1952 the Lead Industries Association published a booklet on the value of lead, stating that white-lead paint has "practically no undesirable qualities."[25]

Negative publicity about childhood lead poisoning grew. In 1956, both *Parade* magazine (the Sunday newspaper insert) and CBS TV produced feature stories on lead poisoning in children. In 1955 the American National Standards Association (an industry standards organization) had adopted a *voluntary* limit of 1% lead in paint. But in 1971, when the New York City Health Department tested paints off the shelf, 10% of the brands contained anywhere from 2% to 10% lead, showing that voluntary limits had failed.

Finally, in 1970, during the Nixon administration, Congress passed the Lead Poisoning Prevention Act. It provided money to screen children for lead exposure. Also, the Consumer Products Safety Commission ruled that use of all lead paint must be phased out by the end of February 1978. The phaseout of leaded gasoline followed later under the Clean Air Act. It was a long road from the Surgeon General's meeting in 1925 to the effective regulation of lead in the United States in 1978.[26] And although the regulations limiting lead in children's products such as jewelry, toys, and furniture are clear, enforcement is conducted primarily by voluntary product recalls. Product recalls only occur *after* the toys have been sold to the public and potential exposure has already occurred (see Chapter 8, "Protect Your Family, Protect Our Planet").

The Behavioral Scientists Get Involved

In the 1970s research began in earnest on the intellectual and behavioral effects of childhood lead exposure. These studies were spurred partly by the funds Congress allocated for children's lead screening. Studies in the 1970s and early 1980s provided evidence that lead had negative effects on children at lower concentrations than previously thought: values ranging from about 10 to 60 µg/dl. These levels were formerly considered to be normal, or below the poisoning threshold. In 1979, Herbert Needleman, MD, published a study of approximately 150 children, almost all middle-class white children from suburbs around Boston. He found that as lead increased, children's performance on IQ tests declined and teacher ratings of attention also declined. The children's blood lead ranged from about 12 to 54 µg/dl.[27] These lead levels are quite high by today's standards but were below the threshold for poisoning at the time. In 1974 Claire Ernhart, PhD, and a graduate student, Joseph Perino, published a study based on 80 African American preschool children from Queens, New York City. They also found that lower IQ test scores were related to higher blood lead levels (but below 60 µg/dl) even after controlling for parent IQ score and birth weight.[28] Additional research in the 1970s yielded similar findings.

Conflicts Erupt Again

The controversy over lead that started in 1925 with the Surgeon General's conference and continued with the contentious Senate Hearing testimony of Patterson and Kehoe erupted again in the early 1980s. But now behavioral scientists were involved. In 1982–1983 Needleman and Ernhart both served on an EPA (Environmental Protection Agency) panel dealing with the dangers of lead in air. The two researchers disagreed violently about whether subpoisoning levels of lead exposure were harmful to children. Why the conflict?—hadn't both of them found negative effects of lead exposure in the 1970s? Yes. But Ernhart had done a follow-up study of the children at age 8–9 years. She concluded that lead exposure was no longer related to the children's IQ scores.[29] Herbert Needleman and his collaborators were also in the midst of long-term follow-up studies. The two scientists disagreed so strongly about the effects of lead exposure on children that the EPA appointed an independent committee to examine the details of their research.[30]

Allegations of biased science seemed to linger in the background of the disagreement. *Science* magazine reported that Ernhart's work was partially financially supported by the lead industry and also pointed out that Needleman's research was probably being gone over with a fine-tooth comb because of the lead industry's opposition to the EPA's cleanup proposals.[31] Ernhart[32] later pointed out that Needleman had potential to earn rather large fees as an expert witness. She said that her funding from the lead industry did not begin until 1983, so it could not have biased her 1981 results. Needleman[33] later explained that his own point of view on lead had been partly shaped by meeting some of the lead-exposed workers from the "House of Butterflies" when he had a summer job at DuPont.

The special committee appointed by the EPA decided that neither Ernhart's nor Needleman's study provided definitive evidence about the effects of subpoisoning lead exposure on children's IQ scores. In other words, neither study could be considered a smoking gun of evidence against lead. The EPA based its decisions to phase out lead from gasoline on other health effects.[34]

After the EPA meetings, Ernhart and Needleman published a series of commentaries critical of each other's work,[35] as well as reanalyses of their own earlier data. In 1985 Ernhart published a reanalysis of her 1974 and 1981 data and changed her 1974 conclusion.[36] In 1974 Perino and Ernhart had written, "While the effects of subclinical lead intoxication may not be noted in the individual cases seen in a pediatric clinic, analyses of group data indicate quite clearly that performance on an intelligence test is impaired."[37] The 1974 publication was based on Perino's PhD dissertation conducted under

Ernhart's supervision.[38] But in 1985 Ernhart and her colleagues concluded, "The reanalyses provide no reasonable support for an interpretation of lead effects in these data."[39] Perino was not a coauthor on either the 1985 reanalysis or the 1981 article.

What happened between 1974 and 1985 to make Ernhart change her mind about the effects of lead on children? The main issue was that after correction of minor errors and the elimination of data from one participant in the study, the statistics no longer met the standard cutoff for interpretation as "significant"—5 chances or less out of 100 that the results would occur as a fluke (a concept expressed mathematically as $p = .05$). Instead, the likelihood of getting the results purely by chance was between 7 and 9 chances in 100. Otherwise, the results reported in the 1985 reanalysis were virtually identical to those of 1974. This brings us back to the issue of what decision standards we use: the smoking gun principle advocated by Kehoe on behalf of the lead industry, or the Precautionary Principle advocated by the other public health doctors?

False Positive and False Negative Errors in Scientific Decision Making

Those readers less familiar with scientific research are probably surprised that the difference between 5 chances in 100 and 7 chances in 100 would make a researcher draw a totally different conclusion. If you were deciding whether to buy a lottery ticket, the difference between 5 and 7 in 100 probably wouldn't change your mind. Many other researchers would not have changed their minds the way Ernhart did. However, if Perino and Ernhart's original results had not met the cutoff of 5 chances in 100 or less of occurring by chance, they probably would not have been published. When research is not published, it does not affect policy or health practices.

The principles in this section on decision standards apply to most scientific research, not just behavioral science.[40] First, it is important for the public to realize that it is exceedingly rare for science to yield a totally definitive answer, or proof. If every child who came in contact with the teensiest amount of lead were killed, we wouldn't need scientific research to tell us that it was dangerous. But we live in a world in which events happen with uncertainty, so researchers rely on statistics to help them draw their conclusions.

When any scientist draws a conclusion such as "Substance X does 'significantly' harm children's development" or "Taking cholesterol lowering medication 'significantly' lowers the chance of having a heart attack," statistical analyses are behind it. Of course, some people taking cholesterol medication do have heart attacks, and some children are unharmed by substance X. The code word "significant" normally means that there are 5 chances in 100 or less that the conclusion "Substance X is harmful to children's development" is wrong. Saying that Substance X is harmful when it isn't is a *false positive error*. In the research on cholesterol medication, a false positive error would be concluding the medication helps when it does not. The "5 in 100 or lower" cutoff is a social convention among scientists that is essentially arbitrary. So when Ernhart changed her conclusion based on her reanalyses, she was following the social conventions of science, though more strictly than many others would have. The statistics said the chance of a false positive error was too large (it was more than 5 in 100) to allow her to conclude that lead harms children. So she concluded that "there was no reasonable support for lead effects in these data."[41]

The other kind of decision error is a *false negative error*. In environmental research, this is a conclusion that a substance is safe when it really is not. In medicine, an example of a false negative error is "The test provided no evidence that you have cancer," when you actually do have it. False positive and false negative errors have different consequences. The chance of making a false negative error is *not* preset by scientific convention. Science emphasizes having a *low preset chance of a false positive error*. This is so that when a scientist does draw a positive conclusion, we know there is only a small chance of it being wrong. That is what science is for: drawing conclusions we can rely on.

The standards in science are more lax for a conclusion that there is "no evidence of harm." In most research, the chance of a false negative error is much greater than 5 in 100. But studies that draw the "no evidence" conclusion are not published very often in the scientific journals. The researcher goes back to the drawing board and tries to figure out a way to do the research better the next time.

In environmental topics, research that finds no evidence of risk obviously does not necessarily mean that there is no risk. The conclusion could be a false negative error. If you are thinking that in environmental topics it would be good to make sure that the likelihood of a false negative error is as low as the chance of a false positive error, some scientists think so too.[42] A research study that is not very sensitive will usually result in the "no evidence" conclusion. A study could be insensitive because the sample size is too small or the measures used are not well refined, or it might just be sloppy. A study with low sensitivity has a high chance of a false negative error. An insensitive study will usually give us a "no evidence" conclusion, but it will not really tell us anything because the chance of a false negative error is quite high.

Here is the bottom line: good science tells us not only the conclusion but also the chance that the conclusion is wrong. For a positive conclusion, the chance of being wrong is preset (less than 5 chances in 100) and is also reported in the published paper. For a negative conclusion, the chance of being wrong depends on a lot of details pertaining to exactly how the study was conducted.

One upshot of the decision standards that scientists normally use is that the public should not hesitate to question scientific experts. When an expert at a public hearing on an environmental controversy says, "There is no evidence that substance X is harmful to people," someone should ask, "What's the chance that conclusion is wrong?" If the scientific expert cannot give at least an approximate answer, or claims to be absolutely certain the conclusion is correct, then he or she is not a very credible expert. A scientist's job is to provide the public and policy makers not just with a conclusion but also with information about the *uncertainty* of that conclusion.

Here is one more complexity about decisions in science. The preset cutoff for a false positive error is indirectly linked to the chance of a false negative error (see Appendix for more detail). If a scientist uses a very strict false positive cutoff (say, only 1 chance in 1,000), then the chances of making a false negative error are going to be very high. Setting such a strict cutoff means that the researcher would need a very sensitive piece of research in order to reach a positive conclusion. But if he or she did reach a positive conclusion, we could pretty well bank on it because it only has 1 chance in 1,000 of being wrong.

Kehoe's Smoking Gun Paradigm

The smoking gun standard, that leaded gas needs to be *proven* harmful before it is withdrawn from the market, should be thought of in light of the way scientists normally make decisions. The normal standard of "proof of harm" is that the conclusion "Substance X harms the public" should have 5 chances or less in 100 of being a false positive error. The problem is that where there are many studies conducted on a topic (as there are with lead and its effects on children) there will always be some studies that yield the conclusion "The data provide no convincing evidence that substance X is harmful." One very important reason this happens is that not every study is sensitive enough. The proponents of a new technology can point to the inconsistencies among research studies as lack of proof of harm. But remember that in most research the chance that a "no evidence of harm" conclusion is wrong is normally much higher than 5 in 100. Only if a study is exceedingly well carried out with a very large sample will the chance of a false negative conclusion be smaller than the chance of a false positive conclusion.

Ernhart and Needleman's conflict was based partly on what decision criterion should be used. Had Ernhart decided to take a 7 in 100 chance of concluding that lead is harmful when it may not be, then she and Needleman would have both concluded that subpoisoning lead exposure is harmful to children. But that is not what happened, and the conflict between them did not end.

Conflict Erupts Once More   After exchanging critical commentaries in the scientific journals, Ernhart and Needleman clashed again in 1990 as expert witnesses on opposite sides in a Superfund[43] case involving lead pollution. It seems Ernhart also took the opportunity to earn a fee in exchange for her testimony. The judge ordered Needleman to allow Ernhart and another scientist (Professor Sandra Scarr of the University of Virginia, who was also a consultant to a polluter in the Superfund case) access to his original data in his laboratory. Ernhart and Scarr apparently found unreported statistical analyses which they thought could change the interpretation of Needleman's results. They also questioned the criteria that Needleman used to include or exclude the data of particular children. Were the criteria decided before seeing the results, or afterward? They found a published graph that was slightly in error, and Needleman eventually published a correction.[44] Ernhart and Scarr reported their suspicions of scientific misconduct to the proper authorities, the National Institutes of Health. Any scientist who suspects another of "fudging" results has a responsibility to report it. In a public lecture after bringing the accusations, Scarr said, "We feel there are significant deviations from normal scientific practice here and we feel that the data has been massaged, to put it mildly."[45]

Needleman was investigated by the University of Pittsburgh and the National Institutes of Health's Office of Scientific Integrity. He felt the panel selected by the University of Pittsburgh to investigate him was biased. Two members of it were close associates of Scarr or Ernhart. On May 27, 1992, the *Wall Street Journal* reported that Needleman's data were "cleared by the panel."[46] Needleman was never accused of making up data, only of omitting certain cases and misrepresenting procedures for selecting cases. Remember that in her own reanalyses, Ernhart had also removed one case that was included in the original publication.

In an article on the debate, the editor of the journal *Pediatrics* wrote, "I am confused. Dr. Needleman believes he has been found not guilty. The government (Environmental Protection Agency) and other scientists also believe this, but others may not (see . . . the preliminary report of the Inquiry Panel). How long must this go on? Has Dr. Needleman been victimized over a difference of opinion about the quality of his science? . . . Conflicting opinions are common and very important in science. . . . Many studies are needed before one side convinces the other that they are right."[47]

How Long Must It Go On?   Ernhart's latest publications include commentaries criticizing not just the work of Herbert Needleman but of others who conclude that low amounts of lead exposure have an effect on children's intellectual development.[48] For his part, Needleman[49] continues to conduct research on lead[50] and to write commentaries emphasizing the importance of reducing children's lead exposure as well as exposure to other pollutants such as mercury.[51]

What the Research Shows

In the meantime, a lot of other research on how lead affects children's development has accumulated. By the 1980s there were many long-term studies under way in different parts of the United States and other countries. The studies involved children from a variety of nationalities, income, and racial/ethnic backgrounds. There are conflicting reviews of the effects of lead on cognitive functioning.[52] The most recent reviews conclude that there is at least a small effect of low-level lead exposure on children's IQ scores. Raising lead exposure from 0 to 10 µg is linked to a decline of about 7 points in IQ scores for 5-year-olds.[53] Increases in lead above 10 µg are linked to further decreases in IQ scores, but the damage levels off a bit.[54] In other words, the dose-response curve for lead is steepest for increases from very low levels of lead. And there seems to be no threshold for lead's damage to cognitive function.

My own assessment of the research on lead exposure at low levels is that it does show *at least* small effects of lead exposure on *both* cognitive functioning and children's behavior problems such as inattention, restlessness, and aggression. And in a later section of this book, I will explain why I believe status quo thinking has led many researchers to underestimate the effects of lead on children's development.

Before I explain my conclusion in more detail, I want to concentrate on the effects of lead on restlessness and inattention. I chose two studies, both from "down under"—New Zealand and Australia. I chose these examples because they meet the highest scientific standards in many ways. They used good, large sample sizes (approximately 800 and 500 children) and measured other important variables that are presumed to influence children's behavior and scholastic performance. I could have chosen other well-conducted studies as examples, but these studies have not been controversial, and they were conducted away from the heat of the regulatory controversy over lead that has occurred in America.

Christchurch Child Development Study, New Zealand

All New Zealanders in the Christchurch urban area who were born in 1977, more than 1,000 children, were enrolled in a longitudinal study. To measure lead exposure, the children's parents were asked to save one of their child's baby teeth. Because lead is stored in our bones and teeth, tooth lead provides a measure of lead exposure over time. The children were tested at 8, 9, 13, and 18 years of age with standardized tests of reading and IQ. Standardized questionnaires were also given to the teachers and to the mothers to measure the children's behavior. The research team also gathered extensive data on other variables that can influence children's school and IQ test performance, such as parent education, family socioeconomic status, ethnicity, and social quality of the home environment. Such family background variables are called either "confounders" or "covariates" because they vary along with the target variable the investigator is interested in: either lead exposure itself or the children's IQ scores, school performance, and behavior. The sample had relatively low tooth lead. Thus, the Christchurch study provides a more conservative test of the effects of low-level lead exposure than studies with higher lead levels, such as Needleman's and Ernhart's studies.[55]

Cognitive and School Performance   At ages 8 and 9 years, IQ scores were not related to lead exposure, but reading scores and teacher ratings of school performance were. Lead exposure accounted for between about 1% and 0.6% of the differences in the children in these variables. At age 12, lead exposure accounted for about 0.5% of standardized tests of academic performance and for about 2% of teacher ratings of academic performance. At age 18 the number of School Certificate passes, leaving school with or without "formal quali-fications," number of years of secondary education completed, standardized reading test scores, and whether the person scored below the 12-year-old level on reading were all associated with tooth lead at age 7. Higher lead and lower performance were associated, even after taking into account all the relevant family background variables, such as family income.

Restlessness and Inattention   In the Christchurch study at ages 8 and 9 years, tooth lead was related to both teacher and mother ratings of inat-tention and restlessness. Again, this relationship held even when background variables were included. Lead exposure accounted for between 3.5% and 0.6% of the differences among children in restlessness/inattention. The lower esti-mate, 0.6%, is exceedingly conservative because it includes pica (attempting to eat dirt and other inedible things, a source of lead exposure) as a back-ground variable. The results also held at age 12 to 13 years. Even after includ-ing a multitude of background variables, tooth lead accounts for about 1.5% of the differences among children in attention and restlessness.

Chicken or Egg?   Lead exposure and children's behavior is a classic "chicken or egg" problem. Which came first, lead exposure or inattentive and rest-less behavior? One idea is that children who are inattentive and restless are likely to do things to expose themselves to more lead than children who are not so restless. The Christchurch team tested this statistically. Between 1.5% and 0.6% of restless behavior is attributable to lead *after* taking into account the idea that behavior may have come first, and lead exposure may result from behavior. Hence, these estimates of the effects of lead are exceedingly conservative. They are exceedingly conservative in another way. The Christchurch study included not just a history of pica but also "residence in old weatherboard housing" as a background variable. Residence in old weatherboard housing is a source of lead exposure, due to deteriorating lead paint. But actual measured lead in the children's baby teeth was associated with behavior at age 12 to 13 over and above the kind of housing the child lived in.

Researchers' Conclusions   In 1997 the Christchurch team concluded, "The significance of the present study is that it shows that the harmful effects of early lead exposure are not short-lived but extend into young adulthood, having impacts on later educational and life opportunities."[56]

The Port Pirie Study, Australia

The Port Pirie study enrolled 90% of children born in the area between 1979 and 1982. There is a lead smelter there, so lead exposure is a concern to the local residents. Children's blood lead was measured at birth in the umbilical cord, and periodically until ages 11 to 13 years. The average was about 14 µg/dl, just barely below Australia's action level of 15 µg/dl.

Although the Port Pirie study has shown consistent relationships between lead exposure and cognitive functioning from age 2 to ages 11 to 13 years, here I will concentrate on the results for children's behavior problems. Children's behavior problems were measured by having the mothers fill out a standardized questionnaire called the Child Behavior Checklist.[57] The Child Behavior Checklist assessed a wide range of children's everyday functioning, including disobedience, destructiveness, sleeping habits, moodiness, toilet habits, temper tantrums, stealing, anxiety, bad dreams, immaturity, and so on.[58]

Total behavior problems reported by the mother were related to lifetime blood lead even after including family background variables such as father's and mother's education and employment, mother's IQ test score, and mother's psychological adjustment. The results are shown in Figure 1.2. The effects of lead on boys and girls differed in the ways you would probably expect. For boys, lead was associated with aggressive and delinquent behaviors, and for girls it was associated not just with aggression but also with attention, anxiety/depression, social withdrawal, and thought problems. "Internal" problems such as anxiety, depression, and social withdrawal are usually more characteristic of girls than of boys.

The Port Pirie research team drew this conclusion: "Overall, the results indicate that any deleterious effect of environmental lead is not likely to be large and that only a small fraction of the overall variation in childhood emotional

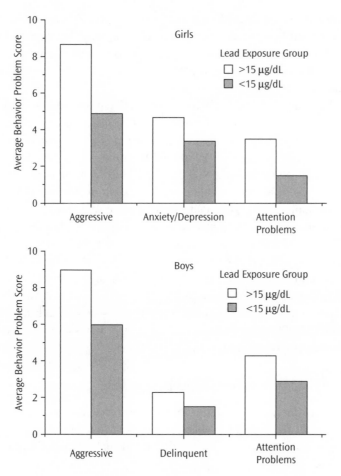

FIGURE 1–2. Mean behavior problem scores for boys and girls in the Port Pirie, Australia, lead exposure study. *Source:* Data from Burns et al., 1999.

and behavioral problems can be attributed to past exposure. Nevertheless, the social consequences of such an effect are not negligible....Until such time as compelling evidence to the contrary is available, policy makers should treat low level lead exposure as a potential source of harm to children."[59]

Other Behavior Problems: Juvenile Delinquency and ADHD

Other research supports the conclusion that childhood lead exposure is related to behavior problems. In a predominantly African American sample of 15- to 17-year-olds from the Cincinnati area, delinquency reported by the youths on a questionnaire was associated with lead exposure measured in umbilical cord blood at birth and with lead exposure measured at 6 years of age. Prenatal lead exposures accounted for approximately 5% of the differences among the youths in delinquency.[60] The first study to document the association of lead with juvenile delinquency was done by Herbert Needleman.[61]

If higher lead is associated with higher restlessness and inattention, does that imply that lead could be a cause of ADHD? The evidence suggests yes. The Centers for Disease Control and Prevention (CDC) conducts nationwide health screenings known as the National Health and Nutrition Examination Survey, or NHANES. The NHANES uses a careful sampling procedure so that the results are representative of the U.S. population. Using NHANES data, researchers found that higher lead was associated with a higher incidence of ADHD diagnosis.[62] A striking aspect of these results is the low levels at which these effects occur: 2 µg. When blood lead was between 2.0 and 5.0 µg the risk of ADHD was 4.5 times higher than for the lowest lead level in the study group. The authors concluded that 290,000 cases of ADHD in U.S. children 4 to 15 years old can be attributed to lead greater than 2 µg. If the child came from a home with a smoker, the chance of ADHD was also significantly higher. A total of 480,000 cases of ADHD in the United States can be attributed to either lead or environmental tobacco smoke.

Are the Effects of Lead Exposure Reversible?

Both the Port Pirie and Christchurch studies showed long-term effects of lead exposure. Does this mean the effects are irreversible? The research does not really tell us. But children with high lead at one point tend to have high lead later in life.

What happens if children's lead exposure is lowered? There are prescription drugs to lower blood lead called chelating agents. These drugs also remove some essential minerals from the bloodstream. A major clinical trial of one particular chelator, succimer, has been done. Over seven-hundred children 13 to 30 months old with blood lead between 20 and 44 µg were randomly assigned to receive either placebo or succimer. All children received a vitamin and mineral supplement. The results showed that chelation did not improve the children's cognitive function at 7 years of age.[63] Chelation also did not improve attention or behavior problems. None of the statistical tests came close to the $p = .05$ cutoff. Chelation did lower blood lead for approximately 6 months. Was lower blood lead related to better cognitive scores? In the placebo group only, test scores improved as lead declined. For the chelated children, change in test scores was not related to blood lead change.[64] These studies say that blood lead should be reduced by getting the lead out of the child's environment. The results also bolster the conclusion that enforcement of lead paint removal and abatement regulations is important.

How Much Are Children Exposed to Lead Now, and How Does Exposure Occur?

One piece of good news is that since lead was phased out of gasoline in the early 1980s, and from paint in 1978, the average amount of lead in people's blood in the United States has plummeted from approximately 16 µg/dl in 1976 to less than 3 µg/dl by 1990.[65] What a dramatic drop! But lead poisoning still occurs. And the research shows that there is not a specific threshold below which lead is known to be safe. Most children's lead exposure is highest (average of 1.9 µg/dl) when they are less than 5 years old, the age when they are walking and starting to talk and are putting almost everything in their mouths if you let them.[66] So it is important to know the routes by which children are exposed to lead. Sources of lead include house dust, old lead-based paint in the home, drinking water, lead finishes on older dishes and utensils, dirt outside (especially near the foundations of the houses built before 1978), use of lead in hobbies and sports, and tracking home of lead by family members who work in lead industries, construction, or locksmithing.

To see how lead gets into children, researchers at the University of Rochester tested children's blood lead and measured lead on the children's hands, lead dust in the home, lead in the soil near the home, amount of lead-based paint in the home, lead in home water, personal housekeeping habits of the family, and whether the child tried to eat dirt. The children were 1 to 3 years old and from Rochester families. About half of those families were single-parent households with income below about $16,000 (in 1991). Almost two-thirds were renters, and about 40% of the sample was African American.

The personal characteristic coinciding most with high blood lead levels was being an African American. Ethnic group accounted for 19% of the differences in children's blood lead.[67] In their sample the African Americans had lower average income than the whites and were more likely to be renters living in homes with lead paint. Low income and renting one's home were also associated with higher blood lead. Other important variables were the presence of lead paint (which contributes to lead in house dust and lead on children's hands), concentration of lead in house dust, and income. Children who play outside and eat dirt (or try to eat dirt) have higher lead levels, and lead in drinking water in the home also adds to children's blood lead.[68]

Other researchers have found that maternal involvement with the child is associated with lower lead exposure.[69] This is probably because parents who are able to be more involved with a child can prevent the child from eating dirt and mouthing dirty objects that may have lead dust on them. And perhaps more attentive parents wash their children's hands more often. But blaming children's lead exposure on parenting is not fair. Today's parents are not responsible for the presence of lead in their homes. If the home did not have lead in it, and in the soil around it, then the children would not be exposed. Public policy that allowed the unregulated use of lead in paint and gasoline for approximately 60 years is what put the lead in the environment of today's children. (See Chapter 8 for ways to help "get the lead out.")

The Social Injustice of Lead Exposure in Children

National surveys of children's blood lead show dramatic income and racial/ethnic differences. Figure 1.3 shows the percentages of African American and white children with different income levels who have blood lead exceeding 10 µg/dl (the current cutoff for "undue lead exposure"). These racial, ethnic, and income differences were also true in an earlier national survey in 1976. This is shocking and a shame to America. The graph shows that children

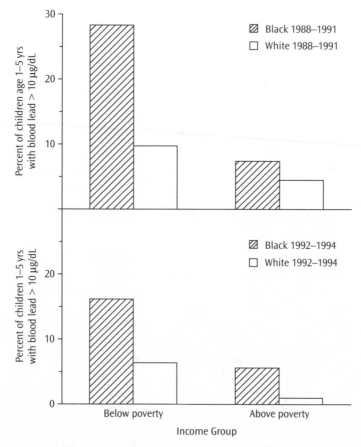

FIGURE 1–3. Racial/ethnic and income disparity in lead exposure of children 1–5 years old in the United States. *Source:* Data from EPA (2000).

living in poverty have much higher lead exposure than those who are more well-to-do, and that the most likely recipients of high lead exposure are poor African American children.[70] And in the 2001–2004 national sample, African American children in poor families still have the highest lead exposure of any group.[71] I want to make it perfectly clear that excessive lead exposure does occur in white children from middle class and well-to-do families, but it is less likely. My friend whose son was lead poisoned is middle class and white. And, of course, excessive lead exposure does not occur universally among African American children living on incomes below the poverty level, but in 1994 it was almost *14 times* more likely than for well-to-do white children.

What are the sources of the racial/ethnic and income disparities in lead exposure? Remember that the Rochester researchers[72] found that African Americans in their sample were more likely to have lower incomes and to be living in rental housing with lead paint than nonblacks in their sample. Renters with low income have little control over the condition of the paint in their homes. And I will pose another question: Is there still housing discrimination in America? Two other influential factors are poor nutrition and health care. Improving calcium and other mineral intake can reduce lead absorption.[73]

Lead Paint Abatement Controversy

Strict enforcement of regulations for lead paint in housing protects residents from lead exposure. Researchers at the Massachusetts Department of Public Health and Harvard University studied two adjoining states where the enforcement policies differed. In the state with strict enforcement, when a child was lead poisoned, the authorities automatically did three things: (1) notified the state lead poisoning prevention program, (2) notified the property owner that there were unsafe levels of lead in the building and that abatement was required or that penalties would be imposed, and (3) notified all tenants in the building that a child had been lead poisoned and informed them of the process for obtaining a lead inspection of their own units. In the area with limited enforcement, the only inspection was of the particular living unit in which the poisoned child lived, no penalties were assessed against property owners, and other tenants were not notified of the presence of lead hazards.

Over a 5-year period, the researchers studied the lead exposure of children 6 years or younger who were living at the same addresses at which a child had previously been lead poisoned (blood lead greater than 25 µg/dl). In the limited-enforcement state, addresses with a previously lead poisoned child were 4.6 times more likely to house a child with high lead exposure (>10 µg/dl) and 6.6 times more likely to have a child with blood lead 25 µg/dl or greater compared to the strict enforcement state.[74]

The authors concluded, "…residents are 'exposed' to the public policies in force in their communities. For lead poisoning, these policies include

abatement of lead hazards in individual units, property owner liability, notifi-
cation and referral for services of affected parties, screening, and public education.... This study...suggested that strict enforcement of lead poisoning
prevention statutes is an effective primary prevention strategy."[75]

The CDC stated that "deteriorated leaded paint and elevated levels of lead-
contaminated house dust...are found in an estimated 24 million U.S. dwellings, 4.4 million of which are home to one or more children aged <6 years.[76]
Lead hazards are especially common in homes built before 1960 (58%)."[77]

But controversy still surrounds how to deal with the lead-based paint hazard
in housing. In 1991 the U.S. Public Health Service proposed a strategy to end
lead poisoning that had four elements: (1) establishment of child lead poisoning
prevention programs in states across the country, (2) abatement of lead paint
and dust in older housing considered to be high risk, (3) reduction of lead exposure from sources other than paint, and (4) national surveillance and reporting
system for elevated blood lead in children. As part of the plan, the CDC called
for blood lead screening of all children 1 to 5 years of age. One reason for widespread screening is that the symptoms of lead poisoning in children even at
50–60 µg/dl—lethargy, headache, sporadic vomiting, and constipation—are
nonspecific and easy to mistake for other conditions. Current blood lead measurements are the only sure way to diagnose lead poisoning.[78] But this plan
was abandoned. Blood lead screening is required only for Medicaid-eligible
toddlers. Only 19% of Medicaid-eligible toddlers are screened.[79]

After the Public Health Service abandoned the plan to end lead poisoning,
controversy re-emerged. Herbert Needleman published a paper in 1998 that
seemed to blame racism (stereotyping lead poisoning as exclusively a problem
of poor African Americans), as well as political and industry forces, both within
and outside of government agencies, for the abandonment of the 1991 strategy. In addition, he chided the Alliance to End Childhood Lead Poisoning (of
which he was the founding chairman) for having "adopted a diluted position
toward the abatement of leaded properties. Like HUD [Department of Housing
and Urban Development], the alliance recoiled from the cost of true abatement.
It began to seek avenues of rapprochement with realtors and insurance agencies."[80] Needleman also pointed out that the National Center for Lead-Safe
Housing (NCLSH) was created by a grant to real estate businesses involved in
low-cost housing, along with the Alliance to End Childhood Lead Poisoning, to
offer "a real alternative to 'all-or-nothing' solutions [to the problem of lead paint
poisoning] that usually mean nothing gets done to help the millions of children
at risk."[81] Needleman pointed out that the NCLSH focused on the *costs* of abatement and ignored the societal monies that would be *saved* through abatement.
Indeed, another study[82] found that strict enforcement of lead abatement creates
a net savings to society of approximately $46,000 per child.

Needleman's 1998 paper was controversial. The editors of the journal that
published it received an "unusual number" of letters to the editor in response to

the article. The journal published six of them, along with Needleman's reply, and the editors' own comments on the controversies.[83] Some applauded Needleman's article, while others were exceedingly critical. Some researchers worried that universal screening would cause anxiety in parents. Another[84] commented that "lead hazard control actually costs more than the CDC estimated in 1991" but that "an updated analysis of the benefits and costs of modern lead hazard control still shows a large net benefit of investments to make homes lead safe."

The journal's editors wrote that "some writers presume that there is one straightforward and objective account of—in this case—the history of lead poisoning prevention policy, on which all right-thinking people would agree. We question this presumption.... By and large, professional historians do not seek to present a misleadingly 'objective' review of events because they are quite aware that other equally well-trained historians will soon challenge their construction of events and their claims to objectivity.... We believe that such debate can be useful not only for advancing historical discourse but also for clarifying options in policy discussions within public health."[85]

Underestimating the Lead Problem with Status Quo Science

Scientists still disagree about low-level lead exposure, in spite of all the evidence that it harms children's development. Scientists do agree that lead in high doses is toxic. Ernhart wrote in 1986, "Perhaps we will find reliable and consistent effects of low level lead exposure. If so, they will probably be small and of relatively little consequence in the overall complex of conditions that affect children, particularly the disadvantaged children who tend to evidence somewhat higher levels of exposure."[86] This quote by Ernhart succinctly expresses the widely held belief that family background factors like race and ethnicity, income, parental IQ and education, and social quality of the home environment are the most important influences on children's IQ test scores. Because these variables cannot be completely separated from lead exposure, some researchers feel that conclusions that low-level lead causes problems in children are unwarranted. Researchers regularly argue about whether a specific study included proper statistical control of such family background factors so that conclusions about the effects of lead exposure can be valid.

But entrenched ideas about IQ scores are involved too. It is striking that published critical commentaries on lead research center mostly on IQ scores, not the behavior problems of inattention and restlessness, and ADHD. Behavioral scientists seem hesitant to draw conclusions pointing to lead exposure as a key influence on IQ.

Here is the standard way many researchers think. First, they "know" that a major part of IQ is inherited from the parents.[87] Second, they "know" that the social quality of the home also predicts part of the differences in children's IQ scores.[88] Third, they "know" that socioeconomic status affects children's IQ scores.[89] And other features of life in poverty are related to children's IQ scores, such as changing residence frequently and growing up in a single-parent family or families that have a change of parent figures.[90] And, fourth, they "know" that on average African Americans score lower than Euro-Americans.[91] Because they "know" all this, researchers would judge the effect of lead to be important only if it *added* some information about children's IQ scores above and beyond those family background factors.

But wait. Are these researchers putting the cart before the horse? The income and racial/ethnic inequities in lead exposure in the United States are huge. So when the researchers consider the effects of family background factors *before* looking for the effect of lead, haven't they really removed most of the sources of lead exposure itself? Lead exposure itself is related to many family background variables, including socioeconomic status, parent IQ, parent education, and social quality of the home.

Here is some evidence. In one of Ernhart's studies with families in Cleveland (35% African American, and half of them selected for having a history of alcohol problems), she found that up to 13% of the differences in children's lead exposure were related to maternal IQ scores, 18% to the social quality of the home environment (including maternal involvement, home organization, and play materials), and 8% to parental education.[92] These results are startling, given that Ernhart and her colleagues found that maternal IQ scores only accounted for about 12% of the differences in the children's IQ scores (a more typical finding is about 25%). In the Christchurch study in New Zealand, the researchers found that maternal IQ scores, social quality of the home, and socioeconomic status were all significantly related to lead exposure.[93] In Cincinnati, researchers found that up to 21% of differences in the children's blood lead levels were related to their measure of the social quality of the home, and 14% of the differences in lead were related to socioeconomic status.[94]

I want to turn the standard reasoning around and put the horse in front of the cart. Suppose that a lot of what psychologists think they know about children's IQ scores is really analogous to what geologists thought they knew about the age of the earth before Clair Patterson figured out that lead pollution had been spoiling everyone's results. After all, almost all of what behavioral scientists know about the influences on children's IQ scores was found in a lead-polluted environment—between 1925 and now. Remember that when Patterson got the lead out of the air in his lab, his results revised scientists' estimates of the age of Earth enormously.

Here is my revisionist interpretation. First, from the research on lead, we "know" that income and race/ethnicity are related to lead exposure. Look at Figure 1.3 again, which shows that poor African American children have 14 times the chance of having high lead compared to whites who are well-to-do. This inequity has been true for as long as researchers have been measuring lead exposure in America. What accounts for the disparity in lead exposure? The primary source is poor-quality housing containing deteriorated lead paint. Lower-income children are also more likely to have poor nutrition, which will cause them to absorb more of the lead to which they are exposed, and African Americans are more likely than Americans of European descent to be lactose intolerant. Lactose intolerance means that cow's milk is eliminated as a major source of calcium. Lower calcium intake is tied to higher lead absorption.[95] Lower-income children are also more likely to be unsupervised or be placed in lower-quality day care, or perhaps day care in a building that contains lead paint. Many low-income parents work at two or three low-wage jobs in order to make ends meet. Working so many hours implies an inability to be home to supervise children closely. It is also difficult to breast-feed an infant when working extended hours at a low-wage job without a space available to pump and store one's breast milk.

Second, we also "know" that the social quality of the home is related to lead exposure.[96] Why might this occur? Attentive parents are more likely to stop a child from chewing and mouthing painted surfaces or dirty objects that contain lead dust. Measures of the social quality of the home also include whether there are appropriate toys and educational materials available. Availability of toys may also help deter children from handling and mouthing lead-contaminated objects and surfaces. Also, social quality of the home includes how organized the home appears to the researcher. Homes that appear more organized may also be cleaned more regularly, and regular cleaning reduces lead dust in the home.[97]

Third, we "know" that parental IQ scores and/or education are related to children's lead exposure.[98] One cause of this is that smarter, better-educated parents may simply know more about the hazards of lead exposure and how to prevent it. Maybe their doctors mentioned it; maybe they learned it from other parents or read about it in a parenting magazine or in a child development course in college. In addition, parent IQ scores and education are related to socioeconomic status, which in turn is related to residence in housing with lead paint.

My argument is that it is valid to think of family background variables as *causes* of lead exposure. In fact, when researchers examine the sources of lead exposure itself (rather than IQ scores), this is exactly how they do their research. Figure 1.4 shows a composite conceptual model of lead exposure based on the research conducted in both Rochester and Cincinnati.[99] In this diagram, I have drawn all the arrows pointing toward blood lead because lead exposure is the variable being predicted. When researchers are trying to predict lead exposure, the family background variables of race, income, and social quality of the home are conceptualized as potential indicators of or causes of lead exposure.

Now compare Figure 1.4 with Figure 1.5. Figure 1.5 represents the status quo science approach to finding out whether lead exposure has an effect on

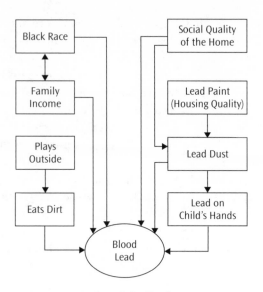

FIGURE 1–4. Composite conceptual model of lead exposure. *Sources:* Based in part on Lanphear & Roghmann, 1997, and Bornschein et al., 1985.

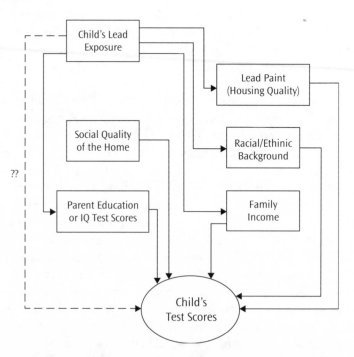

FIGURE 1–5. Conceptual diagram of the "status quo science" estimate of the effects of lead on children's test scores. Lead can influence test scores only after the association of lead with other family background variables is removed.

children's test scores. In Figure 1.5 lead exposure can have an effect on test scores only if it contributes something *over and above* the effects of all the other family background factors. Housing quality and social quality of the home is usually included as a family background factor, even though it is an obvious source of lead exposure.

My proposal is that lead exposure should be conceptualized as a *cause* of lowered IQ scores and altered restlessness, inattention, and aggression, and that family background factors should be conceptualized as potential causes of lead exposure. Animal research supports the causal link between lead and altered development. In monkeys lead causes learning problems and alterations in other behavior down to about 10–15 µg/dl (comparable to the average levels of exposure in Americans two decades ago).[100] And in another project monkeys with higher lead showed "sensory defensiveness."[101] Sensory defensiveness may be a key part of ADHD in children.[102]

If examined in the way I suggest, virtually every study of children's lead exposure and IQ scores or behavior would show much stronger effects of lead. For example, in a 1974 study conducted by Perino and Ernhart with black preschoolers in New York, preschool lead exposure would account for almost 18% of the differences among the children's IQ scores at age 9, and in Ernhart's 1989 long-term study in Cleveland, blood lead measured at age 2 (when lead exposure for most children peaks) would account for about 14% of the children's IQ scores at age 4. Many other studies would yield results in the same range. The estimates that lead is responsible for about 10%–18% of differences among children in IQ scores are quite different from the conclusions drawn from the status quo approach, 1% to about 5% (see the appendix for a brief explanation of how overlapping variables alter correlations).

I do not expect my revisionist interpretation to be accepted by most social scientists. Ideas about heritability and racial differences in IQ scores are much too entrenched to be shaken. I am not saying that family background factors are unimportant in child development. Poverty and racial discrimination are unfair experiences that saddle children with many disadvantages. My main point is that we do not know how much of the disadvantage of poverty and race is linked to lead exposure, because lead exposure, poverty, and race are inextricably intertwined.[103] Status quo science about IQ is a major element of the conclusion that lead has only a small effect.[104]

Arguments that Lead Exposure Is Trivial, and Counterarguments

Q1: If lead is so bad, how did people survive such high exposures in the past seemingly unharmed?

A: Many people *didn't* survive, literally. Death rates for lead poisoned children and lead workers were not trivial in the past. And in New Hampshire in March 2000 a 2-year-old girl died from lead poisoning due to peeling lead paint in the home.[105] In Minneapolis in 2006 a preschooler died after swallowing a heart-shaped charm that turned out to be 99% lead.[106]

Q2: But death from lead poisoning is less likely now, so isn't the problem pretty much solved? Isn't fussing over small changes like 1% in IQ scores and children's behavior silly?

A: Lead exposure is a "silent" malady, like high blood pressure and coronary artery disease.[107] And because human behavior is so complex and determined by multiple causes working together, doctors or psychologists will never be able to point to a specific behavior in a specific child and say, "See, that behavior shows that child got too much lead." Restlessness, inattention, ADHD, aggression, lowered IQ scores, and poor school performance come from many sources combined. Lead is one of those sources. Any contaminant or ill effect that impacts a large population is a public health problem. As Perino and Ernhart wrote in their 1974 paper, "While the effects of subclinical lead intoxication may not be noted in the individual cases seen in a pediatric clinic, analyses of group data indicate quite clearly that performance on an intelligence test is impaired."[108]

Q3: Don't the costs of cleaning up lead outweigh the benefits?

A: Economic estimates are that there will be a net gain from reducing lead exposure in homes.[109] Even in 1991, the CDC's "Strategic Plan for the Elimination of Childhood Lead Poisoning" estimated that each 1 µg/dl reduction in lead exposure would save approximately $2,000 per child in avoided special education and medical costs.[110] And these savings are *just dollars*—there would also be a lot of heartbreak saved for parents who avoid the agony of having a child who just cannot do well in school, has behavior problems, or has developed a seizure disorder after lead poisoning. Heartbreak does not have a dollar value and is not easily accommodated in cost-benefit analyses.

Lead and Adults

Other health effects of lead exposure are documented in adults. At 10–20 µg/dl lead interferes with the body's ability to synthesize heme, a component of red blood cells. The risk of a pregnant woman having a premature baby increases at 10–20 µg/dl. Nerve conduction slows when blood lead is 30–40 µg/dl, and kidney damage occurs at 40 µg/dl.[111] Higher systolic and diastolic blood pressure is also associated with lead exposure.[112] The effects of lead on blood pressure seem to plateau—that is, increases from low lead levels have a larger effect on blood pressure than increases from high lead levels.

In former lead industry workers, the higher the bone lead, the worse the workers performed on tests of manual dexterity, eye-hand coordination, and IQ.[113] In that study, the workers had not been exposed to lead for more than 17 years. Other studies also show more rapid cognitive decline or worse cognitive function for those with higher lead.[114] Finally, the children of *men* who work in lead industries have been found to be at risk for low birth weight.[115] Low birth weight is associated with a variety of negative developmental outcomes.

Lead and Wildlife

Lead is toxic to wildlife, too. Yellowstone National Park banned the use of lead for fishing sinkers because of the effects on wildlife. In 1991, the U.S. Fish and Wildlife Service banned lead shot for hunting ducks and geese. Ducks and geese were being lead poisoned by accidentally eating lead shot off the bottom of wetlands. Researchers tested the lead levels in ducks in 1996 and 1997 and found that lead levels had declined 64% since the lead ban went into effect. The team of scientists estimated that the ban prevented more than 1 million ducks per year from being poisoned by lead. The lead shot ban also indirectly lowers lead exposure in other animals, birds, and people that prey on ducks.[116] A recent study also suggests that lead shot in mourning doves can be a hazard to hunters who eat them (hunting mourning doves is not permitted in all states). A study of loons in New England found they are very susceptible to lead poisoning from fishing tackle.[117] As reported in 2008, twenty-five percent of bald eagles treated at the Raptor Rescue Center in Minnesota were found to be lead poisoned.[118]

Decision Criteria Again

Earlier I explained that false positive and false negative errors have different consequences. What if the "best estimate" of how much lead affects children's IQ scores and behavior is a false positive error? Then as a nation we would be spending money preventing further lead pollution and cleaning up lead (and perhaps forcing some lead-related businesses into bankruptcy while creating jobs for workers to do lead removal) and we would not get any benefits from the cleanup, financial or otherwise. We would spend health care money screening children's blood lead, when that health care money could be spent on something else.

Whenever an agency like the EPA decides to regulate a form of pollution or technology, it is taking the risk of a false positive error. But when regulation is based on sound research, we know the chance of that error, and usually it is much smaller than 5 in 100, often more like 1 in 1,000 or less. Good science tells us not only the conclusion but also the chance that it is wrong.

On the other side, if the currently allowable levels of lead exposure are a false negative error, the government is allowing lead exposure that is harmful. False negative errors have consequences for public health. When a government agency decides to allow a pollutant (either a new one or an old one) because it has not been "proven" hazardous, the risk is a false negative error. False negative errors subject people and the environment to hazards while the businesses benefit financially from the product. Robert F. Kennedy Jr. calls this "Getting rich by making other people poor."[119] For leaded gasoline, the companies profited and the public drove faster, more fuel-efficient cars. But the public also received involuntary exposure to high lead without information about the risks. History shows that when the 1925 Surgeon General's committee used Kehoe's smoking gun criterion to decide there was not enough evidence to ban lead in gasoline, it took too high a risk of a false negative error. The only research was hastily done with small samples. It was insensitive research. In 1925, the Surgeon General's committee handled this by asking that follow-up studies be done. Such studies were not done. If the United States had signed an international agreement restricting lead paint from indoor use in 1925, many childhood deaths due to lead poisoning would have been avoided. The current epidemic of ADHD would be less severe. There would not be an ongoing debate over what to do about old lead paint inside homes built after that date.

Because businesses often bear a major part of the costs of cleanups, businesses want to avoid false positive errors. The public usually wants to avoid exposure to involuntarily imposed hazards (like lead in the air), and it is the government's job to protect the public. But the public seemingly also wants fast cars, cheap gas, paint that covers in one coat, cheap food, and new technologies like wireless Internet and self-cleaning windows with nano-coatings. In order to innovate in business, risks must be taken—both financial and environmental. The issues involved in regulation are inherently value-based. Scientists cannot be expected to tell us as citizens what regulatory decisions to make, because regulatory decisions involve personal ethics and values. In Chapter 7 of this book, I tackle some more complexities of societal risks and examine how values and ethics of both scientists and the public enter into that process. Ethics enter the process even when they are not discussed openly. It is important that the values and ethics be brought out into the open, where all of us can discuss them. Your values are as important as mine in a democratic society.

CHAPTER 2

Mercury: Not Just a Fish Story

M ercury was the Roman god of commerce as well as the fleet-footed, wily, deceptive, changeable, and quick-witted messenger of the gods. One of Mercury's jobs was to conduct the dead to Hades. But the mercury that concerns us here is the metallic element—changeable like the Roman god and the only metal that is liquid at room temperature. Many of us remember playing with liquid mercury as children, pouring it from one palm into another, and perhaps wiping the residue on our pants. Also known as "quicksilver," mercury has seen a variety of industrial uses over the years.

Like lead, mercury has been known to be toxic in high doses since Roman times. In this chapter I tell you a little of the history of mercury pollution. Then I review the research on how mercury in small amounts affects people, especially children who are exposed prenatally. There is also scientific controversy over mercury, but it is more routine than what we saw for lead. Research studies show slightly different results. The different results have all been used in a risk assessment of prenatal exposure to mercury. When I walk through it, you will see that risk assessment is not cut and dried. Judgment calls must be made at virtually every step. But what are the judgment calls, and how do they affect what our government tells us about the hazards of mercury? Ultimately, the judgment calls embedded in the risk assessment influence how mercury is regulated.

Mercury Poisoning

Mercury poisoning has been called "Mad Hatter's syndrome" or "Hatter's shakes." The substance was used in the process of felting wool for hats, and workers exposed to enough of it developed serious symptoms. "Mad Hatter's syndrome" in particular emphasizes that some of the toxic effects are psychological.

The symptoms of mercury poisoning in adults are shown in Table 2.1. The first symptoms to appear are usually sensory. But, mercury is a "sneaky" poison because its effects are delayed. You could be exposed to a toxic dose today, but it would take about 30–60 days for the neurological effects to show up. A silent pollutant with latent effects, mercury lends itself to the "I always did this before and was fine" fallacy. This is why it may be difficult for people to connect their mercury exposure to their symptoms.

Mercury Is "Natural"

Like lead, mercury is an element. Its symbol in the periodic table in your chemistry textbook is Hg. Because it is an element, it exists naturally in rock and in water. So how does mercury become a pollution problem?

Mercury is deposited in the environment in unnaturally large quantities in several ways: industrial uses of mercury in gold mining, chemical and paper

TABLE 2–1. Symptoms of mercury poisoning in adults

Sensory effects	Loss of sensation in extremities
	Loss of sensation in areas around mouth
	"Tunnel" vision
	Loss of hearing
	Altered taste and smell
Motor effects	Abnormal reflexes
	Loss of coordination in walking
	Slurred speech
	Tremor
	Loss of fine motor coordination
Cognitive	Lowered intelligence
	Forgetfulness
Psychological	Irritable temperament
	Social withdrawal
	Anxiety
	Depression
Other	Insomnia
	Fatigue
	Joint pain
	Increased reflexes
	Blushing easily
	Headache
	Dizziness
	Weight loss
	Convulsions and seizures, coma
	Death

Sources: Compiled from Weiss, 1983, 2000 (Bakir et at., 1973: Harada, 1995)

manufacturing, and medicine; accelerated erosion out of rock when water is highly acidic (as can be caused by acid rain); releases into the atmosphere from burning coal; and improper disposal of mercury-containing consumer products (button batteries, latex paint produced before 1991, silent light switches, older thermometers and thermostats, and broken fluorescent light bulbs).

Once mercury is released into the environment in abnormal quantities, it can damage living things. But in order to damage human beings mercury must be inhaled, absorbed through the skin, or eaten. Most mercury exposure to both people and wildlife is through the food chain (see Figure 2.1). Microorganisms in aquatic sediments convert mercury into an organic form of mercury called methylmercury. Methylmercury then concentrates as it moves up the aquatic food chain. Ocean mammals, large predator fish, and fish-eating birds are likely to have the highest levels of methylmercury. People are exposed to methylmercury by eating fish and other aquatic foods that carry methylmercury. Methylmercury in food is almost entirely absorbed by the digestive systems of mammals, including people.

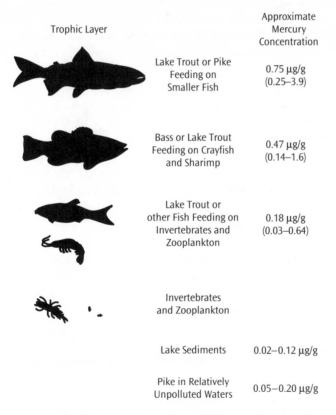

Trophic Layer		Approximate Mercury Concentration
	Lake Trout or Pike Feeding on Smaller Fish	0.75 µg/g (0.25–3.9)
	Bass or Lake Trout Feeding on Crayfish and Sharimp	0.47 µg/g (0.14–1.6)
	Lake Trout or other Fish Feeding on Invertebrates and Zooplankton	0.18 µg/g (0.03–0.64)
	Invertebrates and Zooplankton	
	Lake Sediments	0.02–0.12 µg/g
	Pike in Relatively Unpolluted Waters	0.05–0.20 µg/g

FIGURE 2–1. Food chain diagram illustration that biomagnification of methylmercury in fish depends on trophic level. *Sources:* Values compiled from Bjorklund et al., 1984, and Cabana et al., 1994.

Home accidents that release mercury can also expose people to mercury. One family vacuumed up the mercury from a broken thermostat. The mercury kept evaporating out of the carpet and the vacuum cleaner. Since metallic mercury is readily absorbed by inhaling it, the teenager in the family, who did most of the vacuuming, developed mercury poisoning symptoms.[1] (See Chapter 8 for information on how to clean up a mercury spill at home.)

In some Caribbean religions such as Santeria and espiritismo, small amounts of mercury are spread around the house as a protective ritual.[2] Then the mercury vapors can be inhaled, absorbed through the skin and ingested by children who play on the floor and put their hands in their mouths. If you have a lot of metal fillings in your teeth, you are inhaling small amounts of mercury from them.[3] So far no studies have found that mercury-based dental amalgams harm children's intelligence or behavior function.[4]

What happens when you are exposed to methylmercury in food? Biological processes gradually convert methylmercury to inorganic mercury, which is eliminated in urine and feces. The half-life of mercury in the human body is estimated to be 70–80 days. The half-life is the time it takes to eliminate one-half of the body burden of mercury. While the mercury is in your body, it is not only in your blood but crosses the blood-brain barrier, where it can influence the nervous system. So it is important to keep mercury exposure low. The time it takes the human body to eliminate mercury is one factor in how much mercury exposure the government agencies say is safe. A major difference between mercury and lead is that mercury is eliminated from the body much more rapidly.

Because mercury is natural, there will always be some background level of methylmercury in aquatic ecosystems. But the concentrations of mercury in the environment have risen dramatically over the last 175 years or so.[5] The question now is not whether mercury is toxic but how much exposure is too much for whom. This is where the controversy occurs both among researchers and in government regulations.

The Road from Poisoning to Regulation   Too often it takes a pollution disaster to show us the dangers of toxic substances in the environment. Current regulations on mercury releases from factories only came after dramatic poisoning incidents in the twentieth century.

A mysterious condition called "pink disease" sometimes afflicted infants in the United States and elsewhere (Table 2.2 lists the symptoms) in the first half of the 20th century. But it was not until 1947 that the cause was identified as mercurous chloride (calomel) in teething powders.[6] And not all infants exposed to mercury in teething powders got pink disease. In the 1950s reports of other mercury poisoning incidents started to appear in medical journals. Mercury ointments for fungus infections on the skin also caused poisoning incidents. When methylmercury was used as a fungicide on wheat

TABLE 2–2. Symptoms of "pink disease" in infants

Apathy and irritability
Rashes and sloughing of skin
Red cheeks and nose
Cold, blue fingertips and toes
Profuse sweating and high blood pressure
Itching and burning sensations
Loss of appetite
Tremor and stiff muscle tone
Muscle twitches
Light sensitivity
Pain and loss of sense of touch in extremities

Source: Condensed from Weiss, 2000

seeds, outbreaks of mercury poisoning occurred in Iraq in 1960, Guatemala in 1963–65, and Pakistan in 1969.

But two mercury poisoning disasters were studied by researchers in order to pin down the cause and relate the dose of mercury to the symptoms in the poisoning victims: Minamata, Japan, in the late 1950s, and Iraq in 1971–72.[7]

Methylmercury Poisoning Disaster in Minamata, Japan   A tourist Web site declares, "Minamata is a city with bright sunshine, clear blue skies and picturesque jagged coastline enveloped in greenery and with flowers bursting into bloom. Minamata's magnificent scenery unfolds before your very eyes with the Shiranui Sea in the foreground and the islands of Amakusa dotting the sea in the distance."[8] Tragically, Minamata in the 1950s was the site of a major poisoning that occurred from consumption of seafood heavily contaminated with mercury. The mercury was used in a chemical plant. Waste containing mercury was discharged into the Shiranui Sea, an inland sea (see Figure 2.2). Effects on fish were noticed as early as 1925, when the chemical company paid compensation to the Fishermen's Union. Pollution from the plant was obviously the problem, but no one really knew that it was the mercury in particular.[9]

The first cases of "Minamata disease" were reported as early as 1953. But it was not until 1959 that the cause was found to be mercury poisoning due to seafood from the bay. Not only were the people of Minamata sick; fish were described as spinning out of control and floating belly up. Fish-eating birds reportedly fell from the sky. The cats drooled, walked in zigzags, had convulsions and violent, disordered movements that sometimes made them to fall into the sea and drown. The cat population in the area was decimated during the 1950s. The cats helped pin down the cause. Public health researchers brought healthy cats to Minamata and fed them locally caught fish as a test to see if they would get sick and act like the native cats in the area. About

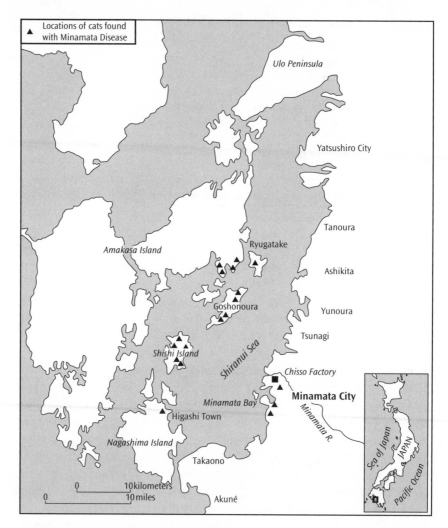

Locations of cats found
with Minamata Disease

Ulo Peninsula

Yatsushiro City

Tanoura

Ryugatake

Amakasa Island

Ashikita

Goshonoura

Yunoura

Tsunagi

Shishi Island

Shiranui Sea

Chisso Factory

Minamata City

Minamata Bay

Higashi Town

Minamata R.

Nagashima Island

Sea of Japan

JAPAN

Takaono

Pacific Ocean

0   10 kilometers
0   10 miles

Akuné

FIGURE 2–2. Map of the Minamata, Japan, area. *Source:* Information on locations of cats with Minamata disease reprinted with permission from M. Harada, 1994, *Critical Reviews in Toxicology, 25,* 1–24, copyright CRC Press, Boca Raton, Florida.

30–60 days after starting on a diet of Minamata seafood, the new cats developed symptoms. Autopsies revealed high concentrations of mercury in the cats' brains, livers, kidneys, and hair.[10] Today there is a monument to the cats in Minamata in honor of their role in revealing the extent of the poisoning, and in solving it by serving in experiments.

The mercury at Minamata also affected children who were exposed during gestation. This is called "congenital Minamata disease." An unusually high number of children had cerebral palsy: 9% of births, versus a national rate of about 1%–2% in Japan at that time. Mental retardation affected almost 30%

of children born in the most heavily contaminated areas between 1955 and 1958. These same children showed clumsiness, indicating that their motor systems were affected as well. And congenital Minamata disease occurred in children whose mothers showed no noticeable symptoms.[11] Imagine almost 30% of a neighborhood's children showing mental retardation. Translate this to your own local school—3 out of 10 children mentally retarded—and the tragic nature of the Minamata disaster is clear.

The Minamata pollution disaster taught the world two things. First, industrial pollution can be a hazard even to those who are not workers. Second, prenatal effects can occur even if the pregnant mother does not become ill. Unfortunately, the source of the poisoning at Minamata was not discovered early enough to prevent similar poisoning from heavily contaminated fish and seafood in Niigata, Japan, in the early 1960s.[12]

Seed Grain Disaster in Iraq   In 1971–72 another mercury poisoning disaster occurred, this time in Iraq. People ate seed grain that had been treated with the fungicide methylmercury. Methylmercury had been used since the 1940s to fumigate seed grain. Most of the chemical disperses into the ground after the grain is planted. It is no longer used in the United States or in most other parts of the world.

The Iraqi government bought wheat seed from the United States and Mexico because of a severe shortage.[13] Iraq had specifically asked that the seed grain be fungicide treated. The toxic grain was distributed to farmers, who were told that it had been treated with a poison. But the bags were labeled only in Spanish or English, and some of the grain was not in bags when it was distributed. Many people ended up eating the wheat, even though it was intended for seed. Before eating it, some farmers tested the grain by feeding it to livestock to see if it would make the animals sick. However that test did not work because it takes mercury about a month to build up enough to show in an animal's behavior. The government verified that at least 6,500 people were hospitalized, and more than 500 poisoning victims died.[14] At that time many areas of Iraq did not have good access to hospitals or medical care. Other estimates are that 5,000 people eventually died, and as many as 50,000 may have been severely poisoned.[15]

What did the world learn from the Iraqi seed grain disaster? Because the source of the poisoning was identified relatively quickly, scientists began measuring both exposure and effects in the poisoning victims. The amount of mercury a person has absorbed can be measured in hair. So researchers gathered blood samples but also analyzed hair samples. On average, hair grows about 1 centimeter (slightly over a third of an inch) per month. Chopping a strand of hair into segments will reveal the approximate time line of a person's mercury exposure. Because of this, a hair sample taken from a pregnant mother when she gives birth can provide a history of prenatal mercury exposure for the child. Researchers located mothers who said they had eaten the grain during pregnancy, took a hair sample, and quizzed the mothers about details of their children's development. Did they remember when their children reached certain developmental milestones, such as crawling, walking, talking, dressing themselves, toileting independently, and so on? Then the researchers constructed dose-response curves for prenatal exposure. The dose of mercury in the mother's hair was graphed against the age at which the child reached each developmental milestone. The dose-response graphs showed that children's developmental delay increased with mercury exposure.[16] Until recently, the data collected in Iraq provided the best dose-response estimates of the effects of exposure to mercury.[17]

The effects of mercury poisoning in Iraq were similar to those found in Minamata. Adults with severe poisoning died. Those with mild poisoning showed sensory, motor, and psychological disturbance. Children who were exposed prenatally showed developmental retardation of varying degrees. But after the Iraqi victims were studied, it became clear that the severity of the prenatal effects of methylmercury depend on dose.

What the Research Shows: Recent Studies of Methylmercury Exposure from Fish

Most of us will not be exposed to the high levels of mercury seen in Iraq and Minamata—either prenatally, as children, or as adults. For most people living in industrialized countries such as America, unless we eat relatively large quantities of mercury-tainted fish, mercury exposure is in small amounts in the fish we eat occasionally. But as with lead, the controversy is not over whether mercury is toxic in high doses but over whether eating fish with low levels of mercury is harmful.

After the poisonings in Minamata and Niigata, researchers around the world began testing ocean and inland fish for mercury. In the 1970s a graduate student at the University of Western Ontario in Canada was surprised when he found that fish in the Great Lakes area had relatively high levels of methylmercury, even without a specific industrial plant discharging mercury waste. Worse than that, in the northern part of the United States, most inland lakes are now polluted enough with mercury that consumption warnings were instituted for eating fish caught in those lakes. There are now statewide warnings due to mercury for certain kinds of sport-caught fish in 43 of the 50 United States. In Wisconsin, where I live, the warnings now apply to all freshwater lakes.[18]

The United States has had standards for mercury levels in commercially sold fish for several decades. The FDA (Food and Drug Administration of the

U.S. Department of Health and Human Services) and EPA issued warnings in 2003—women who are or plan to become pregnant should not eat swordfish, shark, king mackerel, or tilefish. The FDA recommends that pregnant women eat certain other kinds of fish, such as tuna, only in moderation to avoid excess mercury buildup.[19] Fish consumption warnings should not be taken lightly, even by men. The State of Wisconsin has found mercury poisoning in four adult men from eating either commercial or sportfish.[20] But the FDA and EPA warnings specifically point to women who may become pregnant and says that eating too much fish that contains mercury may be harmful to the fetus.

We know there were effects of the high amounts of mercury in both Iraq and Minamata. But what about lower amounts of mercury? To study this, researchers began long-term studies with people who eat fish frequently in the Faroe Islands and the Republic of Seychelles. I describe the Faroe Islands and Seychelles studies for you here. The Faroe and Seychelles studies were both conducted very well, with large samples (more than 700 children) and good follow-up study of the samples over time. The results of these studies are not completely consistent with each other.

Faroe Island Studies

The Faroes, part of the kingdom of Denmark, are in the northern Atlantic Ocean. Home to about 48,000 people, most Faroese are descendants of Nordic settlers from the ninth century. The per capita gross domestic product is about $31,000 per year, and the major industry is fishing.[21] The Faroese eat a lot of fish, but that is not the source of most of their mercury exposure. The types of fish they eat are rather low in mercury. Pilot whales, however, are higher in mercury, and there is a Faroese tradition of hunting pilot whales and sharing the meat.

Phillippe Grandjean, from Odense University in Denmark, led an international research team in a long-term study of fish consumption and children's development in the Faroe Islands. In 1986–87, more than 1,000 children were enrolled in the study at birth. To measure prenatal exposure to mercury, the researchers took blood from the umbilical cord and also sampled hair from the mother. Hair from the children was also assayed for mercury when the children were 1 year old and 7 years old. When the children were 7 years old, they participated in a battery of psychological tests: motor performance (finger-tapping test and hand-eye coordination), memory and cognitive functioning (including some parts of an intelligence test, the Wechsler Intelligence Scale for Children, and attention tests that measure the time it takes children to react to a signal on a computer screen). And, of course, the researchers assessed a variety of family background factors. Also, because whale blubber contains PCBs (polychlorinated biphenyls) and certain pesticide residues, and some Faroese do eat whale blubber when it is available, the research team measured PCBs as well (see Chapter 3 on PCBs).

The results showed that the higher the mercury in umbilical cord blood at birth, the worse the performance. This was true for finger tapping, reaction time, digit span memory, and two other tests of verbal learning. These results held up even after the family background factors and PCB exposure were taken into account in the data analyses. The researchers concluded that a doubling of mercury leads to a developmental delay of about 1½ to 2 months. As mercury increases 10 times, the developmental delay of children is about 4–7 months.[22] At age 14 years there was still a negative effect of prenatal mercury on finger tapping, reaction time, and vocabulary.[23] The Faroe team also found slower brain wave responses to sounds, called auditory evoked potentials.[24]

The average mercury exposure in the Faroe Islands study was much less than at Minamata, yet an association was found between higher mercury and lower performance on the tests. For children classified as having congenital Minamata disease in Japan, the mothers' hair mercury was between 2 and 191 ppm (parts per million). But their mercury was not measured until 5 to 8 years after the birth of the affected child. In the Faroes, hair mercury ranged from .04 to 26 ppm. The Faroe Islands study provides evidence that even at relatively low levels, mercury can have negative effects. In fact, Grandjean and his colleagues also analyzed the results by excluding those children whose mothers had hair mercury above 20 ppm. The results were similar, though a bit weaker.

The researchers concluded, "A discernible effect seems to be present at exposures currently considered to be safe, viz., up to 10–20 µg/g in maternal hair.... This insidious effect does not resemble the severe clinical manifestations of Minamata disease...but it bears considerable resemblance to effects of early exposure to lead."[25]

Republic of Seychelles Child Development Study

The Republic of Seychelles is an archipelago in the Indian Ocean, north of Madagascar, off the coast of Africa at about the equator. Because of its coral reefs and tropical climate, tourism is a large part of its economy. In 1770 the French established a settlement there. Prior to that the islands were uninhabited, though visited occasionally by ships. The Seychelles were used as a holding area in the slave trade, and both the French and British sent political outcasts there. Now populated by about 82,000 people mainly of African backgrounds, the Republic of Seychelles is an industrially developing country. Small commercial fishing operations are a major part of the local food supply. Large commercial fishing for tuna is linked to a new processing plant owned by Heinz Corporation. Health care and education are free, but the infant mortality rate is about 14 per 1,000 live births (compared to about 6.5 per 1,000 in the Faroes). Per capita gross domestic product is about $16,600 per year. Over 90% speak Creole, and the adult literacy rate is 92%.[26]

Fish is a major part of the Seychellois diet. Professors David Marsh, Philip Davidson, and Gary Myers from the University of Rochester, along with others, formed a research team in collaboration with the Ministry of Health of Seychelles to study the effects of mercury on child development. Marsh, a medical doctor, had also researched the Iraqi methylmercury poisoning. Approximately 700 children and their families in the Seychelles were involved in the study when it started, prior to the births of the children. People in the research sample reported eating fish an average of 12 times each week. The researchers measured mercury in the mothers' hair at prenatal visits, when the children were born, and in the children's hair when they were 5 years old.

The results in the Seychelles showed that mercury levels in the mothers' hair at birth and in the child's hair at age 5 were *not* related to cognitive functioning when the children were 5 years old. None of the results met the criterion for statistical significance of less than 5 chances in 100 of a false positive error. Most of the results weren't even close. The kinds of tests the children took were similar to those in the Faroe Islands study. The maternal hair mercury levels were very similar to the Faroes, from 0.5 to 27 ppm, and the children's hair mercury at age 5 ranged from 0.9 to 26 ppm. However, the researchers did not test motor function, such as finger tapping.

The Seychelles research team concluded that mercury exposure from eating fish in large quantities is not necessarily harmful to child development. They said that the nutritional benefits of eating fish should not be foregone to avoid the "small risk of adverse effect at the levels of MeHg [methylmercury] found in ocean fish on the U.S. market."[27] Similar results have been found in the follow-up when the Seychellois children were 9 years old.[28]

Resolving the Conflict

We have two well-conducted long-term studies of children from cultures that eat a lot of fish. The two studies showed different results for mercury exposure. The controversy was not rooted in the use of different decision criteria for drawing conclusions, as was the case with lead exposure. Instead, this time the results simply differ, so the researchers drew different conclusions. But this time the researchers worked it through. Let's see what happened.

"The Devil Is in the Details" Scientists' first strategy for resolving differences in results is to examine the details with a fine-tooth comb. The two research teams have published some of their arguments about which details really matter and therefore who is right. These kinds of arguments lead to new hypotheses and new research. And in 1998 the two teams met for a workshop sponsored by

agencies of the U.S. government. At the workshop the teams bantered about the differences and similarities between their studies, and their alternative hypotheses. The final panel report had many suggestions for both research teams.[29]

The Seychelles team thought that the different sources of mercury and peak doses of mercury exposure might make the difference. People in the Faroes eat whale meat periodically, but whale is 10 times higher in mercury on average than are the fish in the Seychelles. So mercury exposure in the Faroese is likely to have a higher peak than in the Seychellois. For example, in the Faroe Islands a woman might have the same average mercury exposure as a woman in the Seychelles based on a 9-month segment of hair. Because whale meat is relatively high in mercury, if the Faroese woman ate whale, then a small segment of hair that grew after the woman ate whale would reflect the high concentration of mercury in the whale meat, while the rest of the 9-month segment of hair would be rather low in mercury content. At present, it is not really known whether a single large dose of mercury (as from whale meat) has more serious effects on development than chronic low-level exposure. Another issue is that whale blubber contains other pollutants (PCBs and organochlorine pesticide residues).[30]

On their side, the Faroe team called attention to three details of their own study that they thought made it better than the Seychelles study. First, they believe umbilical cord blood is a better measure of prenatal mercury exposure than maternal hair because it is most closely related to the mercury exposure of the developing fetal brain.[31] Also, hair mercury at birth and umbilical cord blood are indices of exposure at slightly different times during fetal development. Maternal hair closest to the scalp at birth reflects mercury exposure during approximately the second trimester. Umbilical cord blood represents third-trimester exposure. In reanalyses of their own results, the Faroe research team found that mercury in maternal hair at birth was slightly more strongly predictive of poor performance on the finger-tapping and eye-hand coordination tests, whereas cord blood was a slightly stronger predictor of performance on the verbal tests.[32] This could reflect different damage at different points in prenatal development. Second, the Faroe team says the Seychelles team did not use the best measures of children's development. The Faroe team said, "The sensitivity to neurotoxicants is likely to be lost when using American omnibus tests, translated into Creole, administered in another culture by nurses, and then adjusted according to American norms."[33] Third, the Faroe team thought that because several languages are official in Seychelles (Creole, French, and English), it is possible that bilingualism could affect the test scores.[34] The Seychelles researchers replied that these latter criticisms were based on opinion and that scientifically based facts, not opinion, should be the center of scientific disputes.[35]

An important idea from the 1998 Workshop was that the omega-3 fatty acids found in fish may protect against the negative effects of methylmercury.

Researchers started testing this idea. A study of women in Massachusetts found that eating more fish during pregnancy is associated with better infant functioning, but that higher mercury is associated with lower infant functioning.[36] The project was part of a study of a thousand women and how nutrition during pregnancy affects child outcomes. More fish meals during pregnancy predicted better infant memory. But higher hair mercury also predicted lower infant memory.[37] In 2001 the Seychelles researchers started measuring omega-3 fatty acids and other important nutrients in pregnant women. The Seychelles team found better psychomotor functioning at 9 months of age for infants whose mothers had higher levels of omega-3 fatty acids even when mercury was included in the analyses. At 30 months of age, the toddlers whose mothers had lower mercury during pregnancy showed better psychomotor functioning. The authors concluded that "... because of... the adverse effect of MeHg [methylmercury], the beneficial effect of omega-3 LCPUFA [long-chain polyunsaturated fatty acids] and other nutritional factors from fish is likely to be underestimated."[38]

The scientific jury has returned its verdict. Fish is good, but mercury is bad. What we need is low mercury fish. How low should the mercury be? In the "Risk Assessment" section later, I describe how the National Academy of Sciences (NAS) committee worked to answer that question.

Mercury Pollution in the Amazon

It is not just sport fish in North America and ocean fish that have become polluted with mercury. Mercury is still used in gold mining in the Amazon, in French Guiana (on the Caribbean coast of South America), and in other areas around the globe. In this type of gold mining, sediments from a river are run through sluice boxes with mercury in them. The gold ore sticks to the mercury, and then the mercury-gold goop is cooked over a fire to drive off the mercury, leaving the gold. But the process puts mercury into the water as well as the air. Several tributaries to the Amazon are gold rush areas.

Children along the Tapajos River

The Faroe Islands research team also studied mercury exposure in four villages on the Rio Tapajos in Brazil. The people rely on subsistence farming and fishing. Most of the people eat fish from Rio Tapajos once or twice a day. The area is relatively poor, access to health care and electricity are minimal, and almost 50% of the mothers in the study only went to primary school.

The four villages are at different distances from a gold-mining area. Mercury in hair was measured for approximately 350 children 7 to 12 years old and in their mothers. But the study didn't measure mercury while the mothers were pregnant, only later. The villagers farthest from the mining area had an average of 3.8 ppm in their hair, and the upstream villagers had averages of 12, 18, and 25 ppm, respectively. The closer to the source of the mercury pollution, the higher the level of mercury exposure.

What kind of assessments would be best to use in a nonindustrialized part of the Amazon, where not many people spend much time in school? The researchers chose tests that do not rely very much on school learning: two tests of fine motor dexterity (the finger-tapping test and the Santa Ana form board test), and three measures of short-term memory taken from different intelligence tests (one was the digit span test, a test of memory for strings of numbers, and the other two were memory for designs).

The results showed that higher mercury exposure was related to worse performance on both the motor and cognitive tests. The researchers interpreted their results as showing that there would be approximately a 6-month to 2-year developmental delay (depending on which test was considered) due to a 10-fold increase in mercury concentration. Let's translate this into an example. A child with hair mercury of 10 ppm would be likely to lag at least 6 months behind a child with hair mercury of only 1 ppm, other things being equal.[39]

Sensory Functioning in Adults along the Tapajos River

Another international research team from Quebec and a university in the state of Para, Brazil, also studied the effects of mercury exposure in the Tapajos River. Table 2.1 at the beginning of the chapter shows that one symptom of mercury poisoning is altered vision. Based on the poisoning disasters in Japan and Iraq, doctors believed that mercury poisoning symptoms require hair mercury above 50 ppm. But this conclusion is based on clinical assessments of mercury poisoning by physicians. A person can be affected by mercury without being ill enough for a doctor to diagnose mercury poisoning. Laboratory assessments are more sensitive than a doctor's clinical assessments, especially for sensory and cognitive functioning. To test vision, this research team in Tapajos set up a portable lab that did not require electricity. They assessed color vision, visual contrast sensitivity (ability to detect small areas of change from light to dark), and sensitivity of peripheral vision.

The results showed relatively large effects of mercury exposure on color vision (14% of the differences among people in their color vision were associated with mercury exposure). Those with the highest mercury exposure also had worse visual contrast sensitivity and peripheral vision than people with lower mercury exposure. These results were the first to establish that mercury can have toxic effects on vision when hair mercury is considerably below 50 ppm. The mercury levels in the Tapajos villagers ranged from about 6 to 38 ppm, and most of the values were below 20 ppm.[40]

In studies of monkeys, the same type of measures of vision were more sensitive to mercury exposure than learning tests.[41] And researchers in Germany found that visual contrast sensitivity in children 5 to 8 years old was related to mercury levels even though the children's exposure was low.[42] These results are very important because this type of sensory testing was not done in the children in either the Faroe Islands or Seychelles studies when the children were tested at 5 and 7 years of age, respectively. In recent studies in the Faroe Islands, measures of visual contrast sensitivity have been included.[43]

Why didn't the Faroe Island and Seychelles researchers measure vision? I think one reason is that our society and the agencies that give out grant money tend to emphasize IQ and other cognitive tests. If a pollutant affects children's IQ scores negatively, it is implicitly regarded as the psychological equivalent of cancer. IQ scores are viewed as important because they are related to school performance and are also correlated with what kinds of jobs people end up in.

Do sensory effects such as visual contrast sensitivity, peripheral vision, and color discrimination have implications for everyday functioning? They certainly do if you are reading fine print, driving your car, or doing art work.

Just a Few Biological Details

In experiments with animals, as well as with animal tissues, researchers have been trying to find out exactly why mercury is toxic to the nervous system and how it damages the nervous system prenatally. During prenatal development, methylmercury changes the processes by which individual nerve cells in the brain migrate to different locations and make their connections. Neuron migration to and connection at the right locations are key to normal brain development. Methylmercury also seems to disrupt nerve cell growth and division processes that are critical for prenatal development. It may do this by disrupting the timing of signals that turn genes on and off. It is not known whether methylmercury itself is toxic to nerve cells or whether the toxicity occurs when methylmercury is metabolized into inorganic mercury.[44]

Does Mercury in Vaccines Cause Autism?

One of the hottest controversies is whether mercury in vaccines could be one cause of autism. This issue could fill a book by itself. Remember that mercury is a powerful anti-fungal and anti-bacterial agent. That's why methylmercury used to be added to seed grain, and that is also why Thimerosol (a mercury-containing preservative) used to be added to vaccines. Thimerosol contains more than 40% ethylmercury.[45] Ethylmercury is metabolized differently than is methylmercury. Ethylmercury has a much shorter half-life than methylmercury: about 7 days in blood and 17–30 days in the brain. However, more of the mercury in ethylmercury than in methylmercury enters the brain before it is processed out of the body.[46] As a safety measure Thimerosol was phased out of vaccines for children in the U.S. starting in 1999.

Researchers have looked for a link between the rate of autism diagnosis in the population as a whole and the phase-out of Thimerosol from vaccines. If Thimerosol caused an increase in autism, then we should have seen a drop in autism when it was removed from vaccines. That hasn't happened.[47] But let's look at the complexities of the issue. First, the autism studies have examined the rate of autism diagnosis in the population. Some scholars think that the autism rate has gone up because of changes in awareness of autism or the criteria doctors use to diagnose autism.[48] Second, although the rate of autism diagnosis has increased, it remains a rare condition, affecting between 1 in 1,000 to 5 in 10,000 children.[49] It is difficult for research to detect changes in rare disorders. Even for studies using a national database, "...if only one child who has autism [and] did not receive Thimerosol - containing [vaccine] were misclassified...the risk ratio would be reduced by half and the p value would be $>.05$ [$p = .05$]."[50]

At present the best research evidence is that autism is not increased by mercury in vaccines.[51] There are contrary views.[52] A hypothesis that has not really been tested is that a particular sensitivity to mercury in certain infants would make them susceptible to autism from a large dose of mercury in vaccinations. None of the studies of large groups of children have actually measured mercury exposure. Instead, they have estimated mercury exposure from the number of mercury-containing vaccinations the child received. This means that the studies are likely to have a high chance of missing an effect of mercury. And if data are misclassified, there is also a higher chance than $p = .05$ of finding an effect of mercury when there really isn't one.

People differ in their ability to metabolize mercury and eliminate it from the body.[53] For instance, less than 1 in 500 infants given mercury in teething powder got "pink disease." Whether differences in mercury sensitivity relate to certain genetic characteristics is not known. But the same mercury exposure will likely lead to a more toxic buildup for someone whose body is slow to eliminate mercury than for someone whose body more efficiently eliminates mercury. There is a hint of evidence that children with autism may be less efficient at eliminating mercury. A sample of 82 Chinese children with autism showed significantly higher blood mercury compared to non-autistic children.[54] But the children with autism also showed *lower* hair mercury than expected from their blood mercury. Another piece of evidence relevant to autism and mercury is genetic. Research with mice shows that genetic characteristics influence the degree to which ethylmercury can damage the developing brain and later behavior.[55] For all these reasons, I think the jury is still out on whether mercury might turn out to be *one* potential cause of autism. And like most developmental disorders, it is likely there will ultimately be more than one cause of autism.[56]

With respect to Thimerosol in vaccines, my own values are that we should take a precautionary stance. Mercury-free vaccines have reduced one route of mercury exposure. Mercury is a known neurotoxin, so less exposure is better. The downside is that vaccines without Thimerosol must be refrigerated and distributed in single-dose vials instead of cheaper multi-dose vials.

Chelation, Mercury and Autism   The first line of defense with any toxic substance is to stop the source of exposure. Remember that mercury is processed out of the body relatively quickly. When a patient has dangerous poisoning symptoms, a physician can prescribe chelating drugs to remove heavy metals from the body. As I mentioned in Chapter 1, these drugs can also remove other essential minerals. Chelators have unpleasant and potentially dangerous side effects such as digestive problems, kidney damage, heart rhythm problems, rashes, and peripheral nerve damage.[57] Because of the side effects, unless there are clear symptoms of mercury poisoning, chelation should not be used. A diagnosis of autism is not a symptom of mercury poisoning. Studies of chelation have not found reduction in poisoning symptoms.[58] The chelators do take some of the metal out of the body, but unless the source of exposure is stopped, the toxic load will likely rebound. Research with rats showed that giving chelating drugs without exposure to a heavy metal caused problems in attention and learning.[59] Finally, even if mercury were a cause of autism, it would not necessarily mean that chelating mercury after the onset of the condition would be a cure. The U.S. National Institute of Mental Health (NIMH) began a randomized trial of chelation treatment versus placebo for children with autism. Only children who had detectible but not high mercury were eligible to participate. The study was suspended after the 2007 research showed that rats were harmed by the chelator if they had not been exposed to lead.[60]

What We Don't Know
1. Is it the peak mercury level that is damaging during prenatal development, or does low-level chronic exposure also damage a child's developing nervous system?
2. Are there certain time periods during fetal development when mercury can do the most damage to the developing brain? Different recommendations for pregnant women would follow from different kinds of results.
3. What happens with aging? When animals age they may show toxic effects of mercury exposure even though no toxic effects showed earlier in life.[61] In Minamata, the elderly showed more severe effects.[62] Did they have higher exposure or are the elderly more sensitive to mercury?
4. What are the sensory effects in children of low-level mercury exposure, either prenatally or postnatally? Are the sensory effects long-lasting, or do they dissipate when exposure is lowered?

Risk Assessment by the National Academy of Sciences

In 1999 the U.S. Congress directed the EPA to have the National Academy of Sciences (NAS) prepare a risk assessment that would help determine how much mercury exposure should be allowed for the public. The NAS is an association of scientists with the purpose of advancing knowledge and advising the federal government. It operates under a charter from the U.S. Congress that was issued early in the twentieth century, and it is one of the most highly respected scientific institutions. Congress requested that the NAS get involved because of disagreements about the effects of mercury exposure and about the costs and technical difficulty of reducing mercury emissions from power plants that burn coal.

The NAS convened a committee of top scientists to look at the latest research on the effects of mercury exposure. The report was published in 2000. The NAS committee had to cope with the different effects of mercury that were found in the studies in the Faroe Islands and Seychelles. The next section reviews the steps in the risk assessment process for methylmercury. This will teach you about the procedures that U.S. agencies use to set cutoffs for exposure to toxicants. Also, by reviewing the steps in the risk assessment for mercury, we will see where judgment calls have to be made in the risk assessment process.[63] Any judgment call involves someone's values. In a democratic society, an important issue is whose values are represented in those judgment calls.

The risk assessment process has an "alphabet soup" of acronyms for technical terms: RfD, BMD, BMR, BMDL, LOAEL, NOAEL, and so on. Table 2.3 summarizes the acronyms, but each one is described below. This book deals with very few "yes or no" outcomes such as having or not having cancer. Measurements of visual perception, motor function, IQ scores, school performance, and developmental behavior problems are all a matter of degree. The exposures are also a matter of degree. Risk assessment for these types of outcomes is a little different than risk assessment for discrete events such as a diagnosis of cancer. My goal is to help my readers see where assumptions enter. What assumptions are based on values, what assumptions are based on pretty good evidence, and what assumptions have to be made simply because there is not enough evidence? A risk assessment is not just number crunching. It involves a lot of value judgments.

Steps in the Risk Assessment Process

1. Define an Adverse Event   This step essentially determines how much damage from pollution is considered to be too much. The risk assessors decide two things: (1) the cutoff for what counts as an "adverse event" in an unexposed or "normal" population, and (2) what percentage increase in adverse events to

TABLE 2–3. Acronyms used in risk assessment

BMD	Benchmark dose. The dose of a toxic substance that is estimated to produce a specified increase in adverse events that is equal to the benchmark risk (BMR).
BMR	Benchmark risk, or benchmark response. The rate of increase in adverse events that is chosen in the risk assessment as indicating a negative response to the toxic substance. It functions as the level of "acceptable risk" from exposure (see NAS, 2000, p. 298).
BMDL	The lower 95% confidence limit of the BMD calculated using inferential statistics. The BMDL was proposed partly to replace the NOAEL.
LOAEL	Lowest observed adverse effect level. The lowest dose that yielded a negative effect in the experiment chosen for the risk assessment.
NOAEL	No observed adverse effect level. The highest dose in an experiment that yielded no negative effect.
RfD	Reference dose. The dose that can be "safely" consumed, or consumed without an appreciable increase in the risk of a negative response, if the assumptions in the risk assessment are accepted as correct or as reflecting one's own value system and ethics.

Sources: Compiled from NAS, 2000a; Crump et al., 1995; Crump et al., 2000

allow due to exposure (this is called the "benchmark rate," "benchmark risk," or BMR). The cutoff for an adverse event is interpreted as the "background" rate of adverse events in an unexposed population. For methylmercury, the NAS report chose the bottom 5% on any cognitive test as the background rate of adverse events. The NAS committee also chose 5% of the population as the BMR, or the increase in low scores to be considered to be an adverse event *due to* prenatal exposure to methylmercury.

Think about these assumptions with the Minamata pollution disaster in mind. Assume that we could turn back the clock so that before the pollution at Minamata became severe we could use the NAS report to decide to regulate mercury pollution. Assume that those scoring in the bottom 5% on the cognitive tests would need special education services. The risk assessment by the NAS would allow the rate of special education students due to prenatal methylmercury to double. So if the other assumptions in the risk assessment are in the ballpark, the most highly exposed communities near Minamata would have been spared the heartbreak of having 30% mentally retarded children. Instead, that number would be somewhere between 5 and 10%. Most risk assessors tend to think that other assumptions and built-in safety margins are good enough that the number would be closer to 5% than to 10%.

I explain below how risk assessors build in safety margins, but here are a couple of other things to think about. Suppose that the cutoff for an adverse event was the bottom 10% rather than the bottom 5%. Keep the BMR or benchmark increase at 5%. Now the risk assessment based on the same research will yield results that call for *stricter* regulation of the toxic substance. This is because most psychological test scores approximately follow a bell-shaped curve.

Here are two questions that members of the public should ask about risk assessments. How was the background rate of adverse events defined, and what increase in adverse events was allowed in the risk assessment? Because about 15% of children in the United States receive special educational services of all kinds, an argument could be made for using the bottom 15% as an adverse event for anything that impacts school performance. In the city of Madison, WI, where I live, approximately 15% of white and Hispanic students receive special education services, and almost one-third of African American students are classified as needing special education.[64] Is adding another 5% to the number of children who need special education an acceptable risk?

Whose Values? What constitutes an adverse event is decided by the risk assessment team, not the public. This decision is called a "science policy" decision—a decision about an aspect of the scientific procedures involved in the risk assessment. But it is a value decision. Using 5% as a criterion is relatively standard in science. The NAS methylmercury committee rejected a 10% increase in the BMR because a potential tripling (5% background + 10%) of adverse events seemed too high to them. Does an additional 5% of the population scoring low on cognitive tests seem too much to allow due to mercury exposure? I am not arguing for a particular value of the BMR, and I am not criticizing the NAS methylmercury report—it is excellent. My intent is to help people understand where value judgments are involved in risk assessment.

It is very important to realize that the definition of an adverse event and the selection of the BMR are value decisions that dramatically influence the results of a risk assessment. One of the originators of these numerical risk assessment techniques, Kenny Crump,[65] thinks EPA should set guidelines for the BMR. Crump has given examples that show how the outcome of the risk assessment will depend on the BMR and the definition of the abnormal event.[66] And a risk assessment only protects the public from an increase of adverse events at least as large as that selected by the risk assessment team. Is doubling adverse events an acceptable risk to you, or only to the risk assessment team?[67]

2. Choose the "Best Science"   The second step in the risk assessment is to choose the best research. Risk assessment should use a study that tested a "sensitive" subpopulation and used the most sensitive outcome measure. The NAS chose the Faroe Islands project as the best study of children who were prenatally exposed to methylmercury. But because of the conflicting results, the committee also did their risk analysis for all three major studies (Faroe Islands, Republic of Seychelles, and another study done in New Zealand).

3. Estimate the "Benchmark Dose" (BMD)   The BMD is the dose that yields the BMR increase in adverse events (5% more low scores in this case). To find it, the risk assessor first estimates a dose-response curve. Then the risk assessor looks at the curve to see how much toxic exposure it takes to increase the percent of adverse events. That is the benchmark dose.[68]

NAS estimated dose-response curves for all outcome measures (finger-tapping, memory, etc.) in the studies in the Faroes, Seychelles, and New Zealand. The different outcome variables in the three studies all yielded slightly different BMD values. The NAS report concluded that if the results of the three studies are combined, the average BMD is approximately 21 ppm in maternal hair.[69] That implies that when maternal hair mercury reaches 21 ppm, the percentage of children scoring very low on the tests is expected to be 10% instead of 5%.

All estimates in science are uncertain. Therefore, the next step is to find a boundary below the BMD that will establish a low chance of making a false negative error in setting the BMD. This is called the BMDL ("L" for lower boundary). A false negative error would occur if the BMD were higher than what is needed to protect the public. This step takes into account the uncertainty of the dose-response curves. It also uses a 5 in 100 cutoff. Based on all three studies together, NAS concluded that the BMDL was approximately 7 ppm in hair.[70] But using just the single best study (Faroe Islands), and the single most reliable and sensitive outcome from that study (NAS chose the Boston Naming Test), NAS calculated that the BMDL was 12 ppm in hair for the Boston Naming Test. Based on the NAS report, women who are pregnant (or intend to be in the near future) should keep their mercury exposure low enough that mercury in their hair is below 7–12 ppm.

Notice that the judgment call of what study and what outcome to use can have a rather large effect on the end result: 12 ppm is 71% more than 7 ppm! Yet another step would be to do a risk assessment for the results from the visual sensitivity testing, not just the language, cognitive, and motor tests. The lowest BMD actually comes from a perceptual task in the New Zealand study.[71] Also, notice that the BMDL could have been determined by a criterion of *fewer* than 5 chances in 100 of a false negative error. This would make the risk assessment more protective of the public.

4. Use the BMDL to Predict Safe Consumption   The safe exposure level is called the "reference dose" or RfD. It is the dose that will keep a person's exposure below the BMDL. The assumptions that go into this calculation include the rate at which methylmercury is eliminated from the body, the percentage of methylmercury absorbed from the food, the average body weight of the person eating it, and a factor for working backward from mercury in hair to mercury in blood. These assumptions are well grounded in research. But these assumptions are based on *averages*. Some people excrete mercury more or less quickly than others. In 1997 the EPA did these calculations using 11 ppm in hair as the BMDL. The calculations yielded the recommendation that the average woman could consume 1.1 micrograms of methylmercury per kilogram of body weight per day. But if 7 ppm is the BMDL, the RfD would be about 0.7 µg/kg per day. So, based on results in the NAS report, the reference dose for mercury should be between 0.7 and 1.1 µg/kg per day. A bit later I turn all this into pounds of fish.

5. Include "Uncertainty Factors"   Uncertainty factors give an additional margin of safety to protect the public. The RfD can be divided by an uncertainty factor to take care of what science doesn't know yet. The EPA used an uncertainty factor of 10. This gives a reference dose for methylmercury of 0.1 µg/kg of body weight per day, or a total of about 6 micrograms of methylmercury per day for an average woman who weighs 132 pounds (or 60 kg).

The EPA chose an uncertainty factor of 10 to account for differences in how people metabolize methylmercury, lack of scientific knowledge of what happens during long periods of low-level methylmercury exposure (e.g., lifetime), and the fact that studies have not tested what happens with the next generation after the children who are exposed prenatally. Starting with the BMDL of 0.7 µg/kg would give 0.07 µg/kg per day after dividing by the uncertainty factor. That is a daily dose of 4.3 µg of methylmercury for an average woman.

6. Make Recommendations to the Public   Now we need to convert the reference dose to the average serving of specific kinds of fish. The FDA kept it simple and issued a warning that said pregnant women should not eat certain kinds of fish at all but could eat 12 ounces of other fish per week.[72] Most state sport-fish consumption advisories list different kinds of fish and how often you can safely eat them.

For those of us who object to simply being told what to do, understanding the why behind such recommendations is essential. I've calculated some examples here. I used mercury values from the FDA Website. Canned light tuna averages about 0.12 ppm of methylmercury, or 0.12 µg/g (see Table 2.4). Assume an average tuna salad sandwich you make at home is about 4 ounces. In that tuna sandwich there are about 14 micrograms of methylmercury (114 g serving multiplied by 0.12). That one tuna sandwich is about 3 times higher than the more conservative daily RfD of 4.3 micrograms per day, and more than 2 times the EPA's RfD of 6 micrograms per day.

Tuna steak averages about 0.38 ppm. An 8-ounce steak has about 86 micrograms of mercury. That's 14 times the EPA's daily RfD, and 20 times the more conservative RfD. Four ounces of fresh tuna in sushi would be 7 times the EPA RfD. That's about a week's worth of mercury in one meal.

Now let's see why the FDA decided that swordfish is a definite no-no for pregnant women. Swordfish averages 0.88 ppm, or 0.88 µg/g. The average 8-ounce swordfish steak (a serving of 225 g) has about 200 µg of methylmercury in it. That is about 35 times the EPA's *daily* reference dose, and almost 50 times the RfD of 4.3µg. More than a month's worth of mercury in one meal.

TABLE 2.4. Mercury concentrations in commercially sold fish

SPECIES	RANGE (ppm)	AVERAGE (ppm)	AVERAGE μg MERCURY/SERVING	
			4 OZ SERVING 113.5 g	8 OZ SERVING 227 g
Most frequently purchased fish in U.S.				
Shrimp	ND–0.05	ND	—	—
Tuna (canned light)	ND–0.85	0.12	14	27
Salmon (fresh/frozen)	ND–0.19	0.01	1	2
Pollock	ND–0.78	0.04	5	9
Catfish	ND–0.31	0.05	6	11
Tuna (canned albacore)	ND–0.85	0.35	40	79
Tilapia	ND–0.07	0.01	1	2
Pregnant Woman and Children Should Not Eat				
Shark	ND–4.54	0.99	112	225
Swordfish	ND–3.22	0.98	111	222
King Mackerel	0.23–1.67	0.73	83	166
Tilefish (Gulf of Mexico)	0.65–3.73	1.45	165	329
Fish that exceed EPA mercury weekly reference dose for an 8 oz serving for a 132 lb woman				
Bass (sea, rockfish, striper)	ND–0.96	0.22	25	50
Chilean Sea Bass	0.09–2.18	0.39	44	89
Bluefish	ND–0.96	0.22	25	50
Croaker (Pacific)	0.18–0.41	0.29	33	66
Grouper	ND–1.20	0.47	53	107
Halibut	ND–1.52	0.25	28	57
Lobster (North America)	0.05–1.31	0.31	35	70
Mackerel (Gulf of Mexico)	0.07–1.56	0.45	51	102
Marlin	0.10–0.92	0.49	56	111
Orange Roughy	0.30–0.86	0.55	62	125
Snapper	ND–1.37	0.19	22	43
Sea Trout (Weakfish)	ND–0.74	0.41	47	93
Tuna (canned albacore)	ND–0.85	0.35	40	79
Tuna (all fresh or frozen)	ND–1.30	0.38	43	86
Tuna (Yellowfin)	ND–1.08	0.33	37	75
Tuna (Skipjack)	0.21–0.26	0.21	24	48
Tuna (Bigeye)	0.41–1.04	0.64	73	145
Tuna (Albacore)	ND–0.82	0.36	41	82
Other fish				
Clam	ND	ND	—	—
Anchories	ND–0.34	0.04	59	

(Continued)

Table 2.4. (Continued)

Species	Range (ppm)	Average (ppm)	Average μg mercury/serving	
			4 oz serving 113.5 g	8 oz serving 227 g
Cod	ND–0.42	0.09	10	20
Crab	ND–0.61	0.06	7	14
Crawfish	ND–0.05	0.03	3	7
Croaker (Atlantic)	0.01–0.15	0.07	8	16
Flatfish	ND–0.18	0.05	6	11
Hake	ND–0.48	0.01	—	—
Mackerel Chub (Pacific)	0.30–0.19	0.09	10	20
Mackerel (Atlantic)	0.02–0.16	0.05	6	11
Oysters	ND–0.25	0.01	1	2
Perch (freshwater)	ND–0.31	0.14	16	32
Perch (ocean)	ND–0.03	ND	—	—
Salmon (canned)	ND	ND	—	—
Salmon (fresh or frozen)	ND–0.19	0.01	1	2
Sardines	ND–0.04	0.02	2	4
Scallops	ND–0.22	0.02	2	4
Squid	ND–0.40	0.07	8	16
Trout (freshwater)	ND–0.68	0.07	8	16
Whitefish	ND–0.31	0.07	8	16

(from U.S. FDA, 2006, http://www.cfsan.fda.gov/~frf/sea-mehg.html)

The U.S. limit for commercial fish is 1 ppm for methylmercury alone. Canada's action level for commercially sold fish is 0.5 ppm total mercury, with exceptions for swordfish and fresh and frozen tuna (as "gourmet" foods), which generally exceed the limit. WHO guidelines. Limit of allowable contamination: 0.5 μg/g =.5 ppm

Your allowable daily intake = (your weight in pounds)/(132 lb average woman) × (6μg/day). *Your weekly allowable intake* = 7 × daily allowable intake for your weight.

Examples:
You weigh 150 pounds (before pregnancy).
  Your allowable daily intake = (150)/(132) × (6 μg/day) = 6.8 μg
  Your allowable weekly intake = 7 × 6.8 μg = 47.6 μg
You weigh 110 pounds (before pregnancy).
  Your allowable daily intake = (110/132) × 6 μg = 5.0 μg
  Your allowable weekly intake = 7 × 5.0 μg = 35 μg

Omega-3 fatty acids in fish

Species	Average (mg/g)	Average mg omega-3/serving	
		4 oz serving 113.5 g	8 oz serving 227 g
Most frequently purchased fish in U.S.			
Shrimp	3.15	357	714
Tuna (light)	2.70	306	612

(*Continued*)

Species	Average (mg/g)	Average mg omega-3/serving	
		4 oz serving 113.5 g	8 oz serving 227 g
Pollock	1.39	158	316
Catfish (farmed)	1.77	201	402
Tuna (albacore)	8.62	978	1956
Tilapia	1.20	136	272
Fish that exceed EPA mercury weekly reference dose for an 8 oz serving for a 132 lb woman			
Halibut	4.65	528	1050
Lobster (North America)	0.84	95	190
Snapper	3.21	364	728
Tuna (Skipjack)	2.70	306	612
Tuna (Albacore)	8.62	978	1956
Other fish			
Anchovies	20.55	2332	4664
Clam	2.84	265	530
Cod	1.58	179	358
Crab	4.13	469	938
Mackerel (Atlantic)	10.59	1202	2404
Mahimahi	1.39	158	316
Oysters	6.88	781	1562
Salmon (wild)	10.43	1184	2368
Sardines	9.62	1092	2184
Scallops	3.65	414	829
Trout (freshwater)	9.35	1061	2122

For cardiovascular health one recommendation is at least 250 mg/day of omega-3 (Mozaffarian & Rimm, 2006). For pregnant and nursing mothers, consensus recommendation is at least an average of 200 mg/day of DHA, or 1400 mg/week (Koletzko et al., 2008). The table gives values of DHA+EPA from Mozaffarian & Rimm (2006).

The FDA's warning said pregnant women can eat up to 12 ounces per week of most kinds of fish but that a variety should be eaten. Because the human body does eliminate mercury over time, health professionals think that a weekly average intake is fine. A weekly total of 12 ounces of canned tuna would give you a weekly average intake of 68 micrograms of methylmercury. The EPA reference dose for a week is 7 days × 6 µg/day = 42 micrograms. The more conservative weekly reference dose that I calculated from the NAS report is 30 micrograms. But if you ate 6 ounces of salmon and 6 ounces of canned tuna in a week, your total intake would be between 34 and 51 micrograms of methylmercury. The mix of tuna and salmon is still above the conservative reference dose I derived from the NAS report. Therefore, the more conservative reference dose derived from the NAS report implies that even canned tuna should be limited during pregnancy, perhaps by replacing it with other low-mercury species of fish.

Omega-3 Fatty Acids in Fish   Remember that the research shows that fish can be beneficial to prenatal brain development. Just as different kinds of fish are polluted with different amounts of mercury, beneficial omega-3 fatty acids also vary across fish species. Omega-3 fatty acids are beneficial for prenatal development.[73] In addition, omega-3s are associated with better cardiovascular health in adults.[74] Information on omega-3s in fish is not readily available. I have put some omega-3 information in Table 2.4.

You can find a handy Weekly Methylmercury Counter in Figure 2.3. By using this with the table of fish mercury and omega-3 in table 2.4, a pregnant or nursing woman can keep track of both. Choose fish low in mercury and high in omega-3. It is tragic that we now need to count pollutants in our food. Unfortunately, the mercury content is not labeled even on canned fish, and the EPA and FDA do not test very many samples of fish annually. In a later

Weekly Methylmercury Counter

Day	Item	Today's Merc
M	Tuna sandwich	14
T	---	---
W	---	---
T	Fried catfish	11
F	---	---
S	Small crab appetizer	Approx. 3
S	---	---
Weekly Total		28

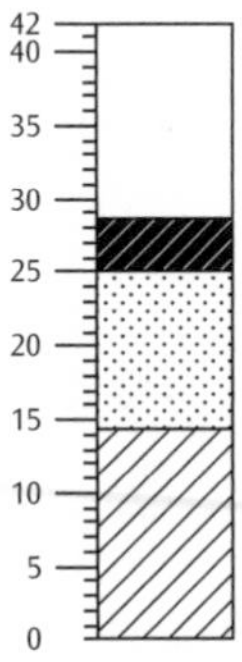

It's like counting calories.

Step 1: Each time in a week you eat fish, write it in the chart. Using Table 2.4 (or more recent information), estimate the amount of mercury in your meal (e.g., if tuna salad sandwich is approximately 4 oz. of tuna, you ate 14 μg of methylmercury), and enter the amount as Today's Merc. If a fish you ate isn't on the list, fill in 15 μg for a 4 oz. serving and 30 μg for an 8 oz. serving. An exception is that for sport fish *not* on your state's fish advisory you can put 0.

Step 2: Each day you eat fish, color in the bar at the right up to the total you've eaten so far this week. Try to keep your weekly total below the top of the bar. Adjust the bar to your body weight—for every 10 pounds more than 132 that you weighed before becoming pregnant, you can increase by 3 μg per week. If you weighed less than 132 pounds before becoming pregnant, decrease the allowable dose by 3 μg per week for each 10 pounds.

Step 3: Keep track across weeks. If you went over one week, try to compensate by keeping your mercury consumption low for the next week or two.

Remember: Fish is good for you, but mercury isn't. Choose low mercury high omego-3 fish when you can. More omega-3's are good.

FIGURE 2–3. Chart to use to count weekly methylmercury exposure from fish.

section I touch on another very important risk issue: what are the relative risks of eating fish versus other foods?

Uncertainty Factors: The Margin of Safety   As you saw above, the risk assessment by the EPA errs on the conservative side by including an uncertainty factor. Uncertainty factors are "fudge factors" that are included to account for things scientists do not yet know.

When a risk assessment is based on animal research, the uncertainty factors are very important. First, the risk assessor finds the lowest dose that affected the animals. This is called the LOAEL, or lowest observed adverse effect level. Notice that the scientists' $p = .05$ decision criterion enters here. A researcher who uses a very strict criterion (1 chance in 100 or fewer) will start with a higher LOAEL than a researcher who uses 5 in 100 as the cutoff, other things being equal. Also, the chance of a false negative error is important. An insensitive experiment with a high chance of a false negative error will give a higher LOAEL than a better-conducted experiment (e.g., one with a larger sample size or more reliable measurements).

The EPA risk assessment procedures say to divide the LOAEL dose by 10 if there was no dose in the research that failed to produce a damaging effect. Then the dose is divided by 10 again to account for the fact that results are being extrapolated from animals to humans. Finally, the dose can be divided by 10 once more to account for the possibility of individual sensitivity and other unknowns. So together the uncertainty factors can divide the dose used in an animal study by 1,000! This is why some industry researchers and representatives think that risk assessments are too conservative.

But there are arguments on the other side too. One reason I present some history of each type of pollution is so that we see the exposures that industrialized societies around the world used to consider safe. The idea of mercury compounds in baby teething powders is appalling to us now, but there were no formal numerical risk assessments in the 1940s. This type of ignorance, which is revealed only by hindsight, is one argument against believing that the risk assessments are protective enough: "The experts said higher levels of exposure were safe, and they were wrong. Why should we believe the experts this time?"

Another argument is that risk assessments do not consider the multiple sources of exposure that people experience in real life. The risk assessment for fish consumption assumes that fish is the only source of mercury exposure. But there are other potential sources of mercury exposure that are thought to be more minor, including preservatives that are still used in multi-dose vaccines,[75] the fillings in our teeth, and household sources of exposure. So although the fish consumption advisory might be fine for protecting us from methylmercury in fish, the next step for the risk assessors would be to figure out how to protect us from the combination of all sources of mercury.

A slight twist on this argument is that the risk assessments for single toxic substances do not protect us from the multiple exposures to toxins we experience. We know very little about how different toxic exposures combine—lead plus mercury plus PCBs, plus pesticides, and so on. Perhaps some exposures have dramatically worse effects when combined.

Judgment Calls along the Way   I pointed out places where the risk assessment requires the scientists to use their own judgment. Here's a list: how to define an adverse event; what increase in adverse events to allow (BMR); choice of the best science; which population is the most sensitive; which outcome measure is the most sensitive; which mathematical shape to fit to the dose-response function to find the BMD; what false negative error criterion to use in setting the BMDL (the lower limit of the benchmark dose); what uncertainty factors to apply; and, finally, what kinds of recommendations to give the public.

I have just listed all the steps in the risk assessment process as having some kind of judgment call or value judgment embedded in them.[76] Does that mean that risk assessment is just someone's opinion veiled in fancy science and a lot of equations? No. Risk assessments do involve the results from research studies. The calculations synthesize research findings in ways that could not be done otherwise. But more important, when the steps of a risk assessment are laid out, the judgment calls are brought out from under the rug and put on the table for discussion.[77] Without the formal risk assessment, only vague arguments can be made about acceptable risks or how bad the effects of mercury are. Judgment calls are hidden when the public is told by an expert, "This is too complicated for you to understand, so leave it to us." To that comment the public should respond, "Try me."

Once the basic steps in risk assessment are known, questions can be asked about any risk assessment. How was the adverse event defined? What rate of increase in adverse events was deemed acceptable in the risk assessment? Are there other research studies that the risk assessor might have used instead of the one that was chosen? How much does a different fitted dose-response curve alter the results? Should another uncertainty factor be used to account for exposures from other toxins that may affect similar aspects of psychological functioning? Because risk assessment is not a cut-and-dried process, the experts need to be open in saying how much the results depend on the specific choices made along the way (this is sometimes called "sensitivity analysis"). As I pointed out in Step 3 above, the data in the NAS report itself lead to two different BMDL values that differ by 71%.

Weighing Risks, Costs, and Benefits

Researchers and the U.S. government focus on what to tell the public about the safety of eating fish, especially the safety for pregnant women. Mercury in fish is not like lead in paint and gasoline. We are not manufacturing fish, so we cannot choose whether or not to put mercury in them. The mercury that the past few generations have discharged into waterways and the atmosphere is now in the food chain for our generation, and for several generations after us. And our own generation is still discharging a lot of mercury into the atmosphere, primarily by burning coal.

But we all have to eat something. In some places, such as villages along the Rio Tapajos in Brazil, or in the Seychelles and the Faroes, people do not have many choices about what to eat. In Rio Tapajos, fish is the staple of a subsistence economy. In the Seychelles, drastically reducing the amount of fish eaten is not an option. The Seychelles is an industrially developing country without the cash to import other kinds of foods and without the natural resources to grow much of its own food. The Faroe Islands are much better off economically than the Seychelles or Rio Tapajos area, but fish is still a mainstay, culturally, nutritionally, and economically. In Seychelles and along the Rio Tapajos, the relative risk of eating fish during pregnancy should probably be compared to a diet inadequate in both protein and calories. Perhaps moving downstream is an option for some residents of villages on the Rio Tapajos. But leaving home because of pollution is a tragedy, especially for indigenous peoples. I will address this topic in Chapter 6.

The real issue is not how much fish to eat each week but how to avoid poisoning ourselves and ecosystems that are sensitive to mercury. Since the disaster in Minamata, the hazards of mercury in the food chain have been known. The human costs of the Minamata pollution disaster were high. The costs of not being able to go fishing and freely eat the fish you catch are also high. The costs of worldwide pollution of the food we get from our planet's oceans are very high. My opinion is that as a global community the human race needs to reduce the amount of mercury released into the environment. Less mercury from industrial emissions is better for the future.

Controls on mercury emissions from coal-burning power plants will cost money. We will all pay for it through higher prices for electricity and products that require electricity in the manufacturing process. That includes almost everything. How much more are you willing to pay for electricity in order to have the coal-burning plant clean the mercury out of its smokestack emissions? Or to have electricity generated in other ways, maybe purchasing solar panels for your house, or installing a hydrogen fuel cell in your basement, or purchasing more energy-efficient appliances? What benefits will come from cleaning up mercury emissions? Will we see the benefits in our lives or will it take several generations? Your children did not put the mercury in the atmosphere. Is it fair for your children to have to live with the consequences of our generation's industrial choices? In Chapter 7 I will discuss some of these ethical issues. I also touch on the problem of global warming in the Afterword.

Environmental Injustices

Mercury contamination is global in many species of fishes. Even people living with only minimal benefits of technology are being subjected to mercury waste products created by others.

In Wisconsin and Minnesota, the Ojibwe people retained fishing and hunting rights when they ceded land to the United States in the nineteenth century.[78] The fish to which they retained their rights are now listed in Wisconsin's fish consumption advisory. No more than one meal per week of walleye. In the western United States, research on mercury contamination of fish on tribal lands is just beginning. On the Wind River Reservation (Shoshone and Arapahoe tribes) in Wyoming, surveys of the mercury content of fish began in 2001.[79]

Mercury in Wildlife

Because fish are a main source of mercury exposure for people, anything that eats fish is also susceptible to mercury toxicity. Walleye are a very popular sport and food fish in the Great Lakes area. When young walleye eat mercury-tainted prey, their growth is slower and their stress hormones change, compared to walleye fed untainted prey. In addition, testicles atrophied in male walleyes that were fed mercury-tainted prey.[80]

Mercury pollution is widespread. We visit our national parks to get away from pollution and enjoy pristine nature. Some tourists go to the Republic of Seychelles for the same reasons. Fish in many U.S. National Parks are now polluted enough with mercury that they show liver and spleen damage. The fish also contain enough mercury to harm otters, mink, and fish-eating birds.[81]

Is it news that mercury can damage wildlife? Effects of mercury on wildlife have been recognized since the 1960s. In Minamata Bay, the fishery and bird populations were harmed as well as the people. In Sweden in the mid-1960s this chain of evidence pointed to mercury as the cause of a decline in bird populations: (1) the dead birds, including species that do not feed on fish, had high mercury concentrations, (2) Swedish grain and agricultural products had higher mercury content than those same products in other European countries, and (3) in Sweden, methylmercury was used extensively as a grain fungicide. Use of methylmercury as a fungicide was banned in Sweden in February 1966, and other mercury pollution was also strictly regulated. The mercury content in bird eggs then started dropping.[82] Regulations on mercury pollution were enacted in Japan at about the same time.

In 1968 the researchers who discovered how bacteria convert mercury to methylmercury in aquatic ecosystems made this prescient comment: "We feel that the example set by these two countries [Sweden and Japan] should be followed elsewhere before concentrations of mercury reach a point where methyl-mercury is being titrated in humans as well as fish."[83] Methylmercury is now being titrated in humans.

CHAPTER 3
PCBs: Another Global Pollutant

Just as with mercury, two pollution disasters called attention to the effects of PCBs (polychlorinated biphenyls) and related chemicals on children's development. Following those disasters, researchers initiated long-term studies of the prenatal effects of low-level exposure to PCBs. PCBs are extremely stable chemicals. Like mercury, they are distributed around the globe and biomagnify up the food chain.

Most PCB exposure now occurs through food. But older electrical equipment can have PCBs inside. When I was a graduate student at the University of Illinois, sometimes I studied in a building where the ballasts on the fluorescent lights were gradually going bad. When one went out we would first smell a slight burning odor. Then a gooey black spot would gradually appear on the translucent light cover overhead. Some kind of oily gunk had dripped onto the light cover from above. The burning and dripping goo very likely contained high concentrations of PCBs. But in 1975 we had no idea the oozing black stuff might be toxic. Now the EPA recommends immediately evacuating a room with a broken fluorescent ballast. The room should be ventilated as well as possible. Then the mess should be cleaned up while wearing gloves and a special respirator. The cleanup rags and gloves should be double-bagged and taken to your community's toxic disposal program.[1] But in 1975 that didn't happen. We just kept studying. Usually it was a couple of weeks before the light was repaired.

What is the chain of evidence leading the EPA to recommend evacuation, ventilation, and special disposal of PCB contaminated items?

Three conditions are normally necessary for scientists to conclude that a specific type of pollution is harmful to people. *First*, the results of a "landmark" human epidemiological study or poisoning incident need to be supported by other studies that find similar effects. *Second*, researchers need to find effects on laboratory animals that are analogous to the effects on humans. This piece of the puzzle is needed because environmental pollution in humans and wildlife does not occur just one chemical at a time. Epidemiological studies of humans cannot draw strong causal conclusions because there are always confounding variables. People are exposed to multiple pollutants, smoking, alcohol use, family background, and so on. Of course these other variables can be included in the analyses. But in principle there is always the possibility that another unmeasured confounding variable could account for the results. Finding a correlation does not imply causation. Laboratory animals can be randomly assigned to different levels of exposure to a pollutant. Random assignment to conditions in an experiment is one key to drawing a cause-effect conclusion. *Third*, laboratory research needs to show that the pollutant affects biological processes in a way that could plausibly account for the effects in people. When committees of scientists review pollutants for the EPA, they look at all three kinds of evidence. Then they decide whether the weight of the evidence supports regulating a pollutant.

In evaluating evidence against a pollutant, the decision criteria discussed in the preceding chapters are important. There are no hard-and-fast rules for deciding that a pollutant has effects on children's development that are serious enough to require regulation. Is it enough to have one long-term study of humans that shows negative effects, even if the results of other studies do not completely agree? Is it enough to have laboratory evidence that a pollutant disrupts brain chemicals and hormones in developing animals? Is it enough to show that a pollutant disrupts a biological process such as the gene signaling involved in normal brain growth?

The first study of the long-term effects of prenatal exposure to low concentrations of PCBs was conducted in Michigan starting in 1980. This study became the centerpiece of debates over how much PCBs should be regulated, and how thoroughly they should be cleaned up. Partly because of the Michigan study, there are warnings about consuming certain Great Lakes fish. I will describe this study in some detail, along with the scientific differences of opinion about the interpretation of the findings. But first, background and a pollution disaster.

Some Background on PCBs

The family of chemicals known as polychlorinated biphenyls (PCBs) were first synthesized just before the start of the twentieth century. PCBs have desirable electrical properties, are chemically stable, and resistant to exploding when heated. Because of their desirable properties, by the 1930s PCBs came into wide use in electrical transformers, for lubrication, and in many other products. For a short time PCBs were even used in paints in farm silos.[2] A partial list of uses of PCBs is in Table 3.1. Prior to the advent of PCBs, explosion

TABLE 3–1. Some former uses of PCBs

Sealant paints (used in farm silos, as well as other applications)
Electrical capacitors
Electrical transformers
Coating on electrical wire (sometimes mixed with asbestos)
Hydraulic fluids in aircraft, naval systems, and other heavy industrial equipment
Natural gas pipeline turbine lubricants
Vacuum pump oil
Microscope immersion oil
Plasticizers (to increase flexibility of plastic)
Sealants (caulks)
Safety glass
Paints and varnishes (improves weatherability, adhesion, and fire-retardant properties)
Textile coatings on ironing board covers
De-lustering agent in rayon
Textile flameproof treatments
Paper coatings (carbonless copy paper, heat-transfer copy papers)
Combined with chlorinated and nonchlorinated insecticides (e.g., aldrin, dieldrin, chlordane, malathion)

Source: Compiled from Meigs et al., 1954; Broadhurst, 1972

was a problem in large electrical transformers. The use of PCBs in electrical transformers was required by building codes in certain situations. PCBs probably saved lives by reducing fires,[3] and were regarded as a very safe family of chemicals because their *acute* toxicity is low.[4]

In 1966 PCBs attracted the attention of scientists as a global environmental concern. Soren Jensen, a chemist at the University of Stockholm, was studying residues of the pesticide DDT in wildlife. Researchers around the world knew that when they tested for DDT and its metabolic products (called DDE), there were always other similar chemicals present that sometimes caused problems in their methods. But no one knew exactly what the other chemicals were. Jensen figured out that they were PCBs. In his analyses he was able to separate PCBs from DDT. He found that PCBs were in fish throughout Sweden, both inland and in the Baltic Sea. He found PCBs in an eagle that was found dead in Stockholm. He found PCBs in the hair of all his family, including his 5-month-old daughter. He speculated that his daughter must have gotten the PCBs from breast-feeding. Jensen then studied bird feathers from a Swedish museum and found that PCBs first started showing up in eagles in detectable amounts in 1944.[5]

Jensen's breakthrough allowed other researchers to separate PCBs from DDT and DDE residues. A burst of research in the late 1960s found PCBs in wildlife around our planet.[6] The manufacture and new use of PCBs in the United States was phased out starting in 1976. In the United States, PCBs may stay in closed systems where they are already present, such as older electrical transformers. But PCBs continue to be released into the environment from old electrical equipment, waste disposal sites, and evaporation from PCB hot spots around the globe. PCBs are transported around the earth in the air.[7]

PCBs are a large family of chemicals, with 209 possible congeners, or chemical variants. The congeners vary slightly in molecular shape, how many chlorine atoms are attached to the molecule, and where the chlorine atoms are attached. The 209 PCB congeners differ in their toxicity to people and other animals, but the details of how and why are not fully understood.[8]

PCBs are in the general category of organochlorine chemicals (OCs). This category includes dioxins, and also many pesticides that are banned in many parts of the world: DDT, chlordane, and dieldrin. PCBs and other OCs are lipophilic. They bind to fat molecules in people and animals. PCBs concentrate as they move up the food chain. Top-of-the-food-chain predators such as eagles, polar bears, and whales that eat fish and other animals are likely to be high in PCBs. Because of PCB contamination, there are fish consumption warnings for many types of fish in the Great Lakes, the Hudson River, and many other areas (see Chapter 8).

Some History of Human PCB Exposure

The first hint that PCBs could be toxic came not long after they were invented. As early as 1899, workers who had contact with chlorinated chemicals similar to PCBs got an acnelike skin malady that is now called "chloracne."[9] Other organochlorines can cause chloracne. Soldiers and citizens in Vietnam got chloracne from dioxins in the hebicides sprayed by the U.S. military. Then more cases of chloracne occurred in 1936 in factories using PCBs,[10] during World War II,[11] and in the 1950s.[12] Many more cases of chloracne reports appeared in the 1970s.[13] At least one worker death was reported from liver failure attributed to PCBs.[14] As with mercury, the onset of symptoms after exposure to acutely toxic levels of PCBs is delayed by anywhere from 6 weeks to several months. With PCB exposure, chloracne and changes in liver enzymes are quite variable across different people.[15] By the 1960s PCBs were generally acknowledged to be an industrial hazard to workers, transmitted via skin contact or inhalation.[16] In the United States in 1957, chickens were poisoned by feed with PCB oil in it.[17] And a similar chicken poisoning incident occurred in Japan in 1968.[18] But it took two pollution disasters involving people to focus the world's attention on PCBs. The poisonings also showed that PCBs have effects on children exposed prenatally.

Polluted Cooking Oil in Japan and Taiwan

Yusho* or "*Oil Disease*" *in Japan Doctors in 1968 in the Nagasaki and Fukuoka areas of Japan knew something was amiss when an unusual number of patients with acnelike skin eruptions began requesting treatment. A research team of doctors and scientists from Kyushu University began tracing the cause of the illness. They studied and interviewed the poisoning victims. The source of the poisoning turned out to be contaminated rice oil. Then they studied the chemical constituents of the contaminated oil and did animal experiments using the contaminated oil itself. And the scientists also did experiments with PCB mixtures like those used in the rice oil factory.[19] During the refining process, the rice oil had become accidentally contaminated with PCBs that were inside part of the machinery. The PCBs were supposed to stay separate. But pin-hole sized leaks in the heat-transfer equipment allowed small amounts of PCBs into the rice oil. The contaminated rice oil was sold, and people used it for cooking. Adults developed darkened patches of skin, chloracne, weak vision, and numbness in the extremities. Some people died of acute poisoning. Adult poisoning victims who survived had persistent fatigue, weight loss, headache, numbness, and pain in the limbs, or neuralgia.[20]

The final count of victims is uncertain. In the Kyushu area in 1968, approximately 1,800 adults and their children were treated for exposure to PCBs and related chemicals.[21] The symptoms of oil disease are summarized in Table 3.2.

TABLE 3–2. Symptoms of "oil disease"

Acute symptoms of "oil disease" (Yusho and Yu-Cheng)
Poor appetite, nausea, vomiting
General fatigue or weakness
Blackheads and acnelike rash on the skin (called "chloracne" because it is
    caused by contact with organic chlorine chemicals)
Pigmentation of the lips, gums, eyelids, and nails
Swollen eyelids
"Cheeselike" secretions from the glands in the upper eyelid
Disturbed vision

Long-term symptoms of "oil disease" in adults
Blackheads and acnelike rash on the skin
Abnormal secretions from the glands in the upper eyelid (stones and
    cheeselike)
General fatigue
Poor appetite and weight loss
Abdominal pain
Headache
Numbness and pain in the limbs
Cough

Sources: Acute symptoms compiled from Hsu et al., 1984; Chia & Chu, 1984; Masuda, 1985; Rogan, 1989; long-term symptoms adapted from Masuda, 1985

But the worst was yet to come. PCBs and related chemicals readily cross from a pregnant mother's bloodstream through the placenta to the developing child. Some prenatally exposed infants died shortly after birth. Exposed infants had eye discharges and swollen eyelids, deformed nails, and abnormally dark pigment in the skin. An unusually high percentage of infants were born with teeth. The prenatally exposed infants who survived were described by researchers as listless and dull. They had low IQ scores (around 70) when tested 7 years later.[22]

Yu-Cheng or Oil Disease in Taiwan　A similar oil poisoning incident occurred in Taiwan in 1978–79. More than 10 years after the poisonings in Japan, in Taiwan, more than 2,000 people were affected. Again, contaminated rice oil was the culprit, and 24 patients died.[23] The epidemic was first noticed in a school for blind children in May 1979. Poisonous cooking oil was finally identified as the cause. But because PCB poisoning does not show up immediately, cases had occurred as early as December 1978. The latency period between the beginning of consumption of the oil and the onset of symptoms was between 1.5 and 6 months.[24]

In the Taiwan incident, the prenatally exposed children were studied extensively. The children tended to be smaller at birth and were developmentally delayed in achieving milestones such as turning over, walking, and so on. The children had lower IQ scores than comparison children matched on family background variables.[25] The presence of symptoms of PCB-induced birth defects (especially nail deformity) was related to the degree of developmental delay.[26] The children also showed higher activity levels and behavior problems at 11 years of age.[27]

When mass poisoning incidents occur it can be hard to pin down the cause. For the Yusho and Yu-Cheng victims the big question has been whether PCBs are really the culprit, or whether other contaminants could be to blame. First, the mix of chemicals in these poisonings is not completely certain. When PCBs are heated above a certain temperature, some are transformed into polychlorinated dibenzofurans (PCDFs or furans) and dibenzodioxins (PCDDs or dioxins). Most commercial mixtures of PCBs (called "Arochlors" in the United States and "Kanechlors" in Japan) contained trace amounts of furans and dioxins. Both dioxins and furans have higher overall toxicity than do PCBs.[28] Second, the PCBs were used as the heat-transfer fluid inside the rice oil refining equipment. This could have created more furans and dioxins. Third, the contaminated oil was used in cooking. Cooking could make more dioxins and furans out of the PCBs. Adult Yusho and Yu-Cheng victims had elevated levels of PCDFs, though the prenatal victims showed only PCBs in their blood. Years later, PCDFs were not detectable in the adult victims, only PCBs and polychlorinated quarterphenyls, or PCQs.[29]

But the overall pattern of developmental delay was clear, and the results persisted with age. The poisonings in Japan and Taiwan showed that prenatal exposure to the PCB family could be teratogenic (cause birth defects). The exact cause was not clear: was it the PCBs themselves, or related chemicals such as dioxins and furans? What about lower levels of exposure? Two major long-term studies of the effects of lower levels of PCB exposure were launched in the United States between 1978 and 1980.

What the Research Shows

Exposure from Lake Michigan Sport Fish

In 1980 Sandra and Joseph Jacobson, Greta Fein, and other colleagues from Wayne State University in Detroit began a long-term study of prenatal exposure to PCBs from Lake Michigan sport fish. The researchers interviewed 8,000 pregnant women in the maternity wards of hospitals, mainly in the Grand Rapids area. They asked the women whether they ate Lake Michigan sport fish and, if so, how much they ate. From the 8,000, the research team found about 240 women who had eaten at least 26 pounds of fish in the preceding 6 years. That means they had eaten a minimum of 8 or 9 half-pound fish meals per year. On average, the fish-eaters ate about 15 pounds of Lake Michigan fish per year (about 1.25 pounds per month). They had been eating Lake Michigan sport fish for about 16 years, or since the mid-1960s. To assure that the sample would include some individuals with lower exposure to PCBs, a group of 71 women was randomly selected from those who said they never ate Great Lakes sport fish. Then at birth PCBs were measured in the umbilical cord blood and in the mothers' breast milk. Family background variables were collected. The women were predominately white and middle class (only two women were not categorized as white). The children were tested while they were infants and several other times before the end of elementary school.

The Michigan Research Results   Higher maternal fish consumption was associated with worse newborn functioning and reflexes. Maternal fish consumption accounted for about 8% of the differences among the babies in their functioning as newborns.[30] The more highly exposed newborns showed weak muscle tone and poor reflexes. Newborns have certain reflexes at birth (see Table 3.3 for a list of some newborn reflexes). Testing for normal newborn reflexes is one way of looking at brain and nervous system development. About one-third of the most highly exposed infants showed weak reflexes, classified as a "worrisome" indicator, compared to about 15% of the infants whose mothers ate no Great Lakes sport fish. The more highly exposed infants also showed slower motor development, on average, and were less active during testing.[31]

TABLE 3–3. Some newborn reflexes

REFLEX	DESCRIPTION
Sucking reflex	When an object touches the lips or mouth, the infant sucks.
Rooting reflex	When an object touches the infant's cheek, the infant turns in that direction and begins sucking.
Stepping reflex	When the infant is held upright and the feet are allowed to touch the floor, the infant lifts the legs alternately in a stepping motion.
Grasping reflex	When an object touches the palm, the infant grasps it firmly.
Swimming reflex	When held prone (stomach down) the infant makes arm and leg movements that resemble swimming.
Moro reflex	When startled (by noise, or vestibular change) the infant spreads arms, arches the back, and then brings the arms in to the chest in a grabbing motion, and cries.

When the infants were 7 months old, the researchers found that PCBs in the umbilical cord blood were related to performance on an infant test that measures memory. In the Fagan Test of Infant Intelligence, the infant is first shown two pictures that are identical, and then one of the pictures is changed. Because at that age infants have a preference for novelty, the amount of time the infant spends looking at the new picture indicates whether the infant noticed that it was different from the original one. Even after adjusting for family background variables such as socioeconomic status, maternal education, maternal vocabulary, and social quality of the home, PCBs in the umbilical cord accounted for about 10% of the differences in infant memory.[32] The infants in the very highest exposure group showed no evidence of noticing the switch between the original and the novel pictures at 7 months of age. Because infant performance on the Fagan test has been shown in other research to have a positive relationship to IQ test scores in childhood, these results are a cause for concern.

Do the differences due to prenatal PCB exposure last into childhood? Yes. When the children were 4 years old, the researchers assessed them very thoroughly. They administered standardized intelligence tests, plus laboratory tests of sustained attention and memory, and response speed in a visual discrimination task. They measured the children's physical growth and collected behavior ratings from the mother and from the research staff member who assessed the child. Higher prenatal PCB exposure was related to worse performance on memory tests. The more highly exposed children were less efficient cognitive processors overall. They responded more slowly on the visual discrimination task. Growth and behavior were also affected. The most highly exposed children weighed slightly less and were less active.[33]

When the children were 11 years old, higher prenatal PCB exposure was still related to lower IQ test scores, especially on the verbal parts of the test. Reading comprehension on a standardized reading test was also lower for the more highly exposed children. These results are shown in Figure 3.1. The findings held up when the family background variables were included. Spelling and arithmetic scores on the standardized test were not associated with PCB exposure. And total postnatal PCB exposure from breast milk did not relate to the children's scores after prenatal exposure was considered. The prenatal period seems to be a key time when PCB exposure can harm development.[34]

In humans, pollutants occur in combinations. The Michigan researchers tested whether exposure to DDE (the metabolite of DDT found in virtually all

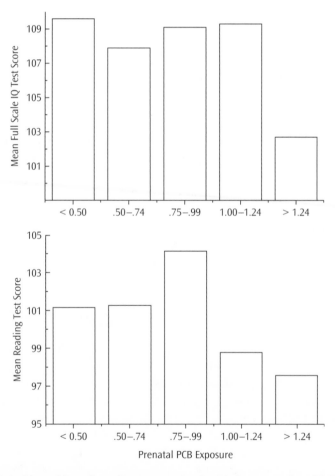

Figure 3–1. Mean IQ and reading scores in the Michigan study of PCB exposure. *Source:* Adapted from Jacobson & Jacobson, 1996, *New England Journal of Medicine, 335,* 783–789, copyright 1996, Massachusetts Medical Society. All rights reserved.

of us), polybrominated biphenyls (PBBs, a family of chemicals similar to PCBs that were used in fire-retardant coatings), mercury, and lead could account for their results. DDE did show a relationship to performance, but it was not related to test performance over and above the effects of PCB exposure. As expected, lead measured at age 4 was related to lower scores at age 11 on the verbal part of the IQ test and on the reading comprehension test, even after including family background variables. Mercury at age 11 was related to worse performance on the spelling tests.[35] (The researchers had not measured prenatal exposure to mercury at the outset of the study.)

Researchers' Conclusions   After assessing the children when they were 4 years old, the researchers concluded, "Although there was no evidence of gross functional impairment at the levels of exposure in our subjects, the poorer memory performance seen in the study indicates diminished poten-tial....These deficits, although subtle, could have a significant impact on school performance in later childhood."[36] Based on the 11-year testing, the research team said, "There was no evidence of gross intellectual impairment among the children we studied....Nevertheless, there was a substantial increase in the proportion of children at the lower end of the normal range...who would be expected to function more poorly in school. This intellectual deficit seemed to interfere particularly with reading mastery."[37]

North Carolina Study of
Background Exposure

In 1978 another long-term study of the effect of PCBs on children's develop-ment was started in North Carolina with more than 800 babies born between 1978 and 1982. The children were followed until they were in approximately 3rd grade. These researchers measured PCBs and DDE in umbilical cord blood, in a sample of the mother's fat tissue, and in breast milk soon after birth. Their sample was also predominantly of European American background (94%) and middle class.

When tested as newborns during the first 2 weeks of life, those with the highest prenatal exposure to PCBs scored lower on measures of muscle tone, activity, and reflexes. Similar results were found for DDE, except that activity was not related to DDE. These results were very similar to those reported in the Michigan study.[38]

At 3 to 5 years of age, the researchers gave standardized tests to the children, and in 1988, when the children were in 3rd grade or older, the researchers wrote to the parents asking for school report cards. There was no statistically discernible relationship between either prenatal PCB or DDE exposure and the assessments at age 3 years or older. These results differ from those found in the Michigan study. The tests used at 3 to 5 years of age were the same ones used by the Michigan researchers, but some of the criteria for giving and scoring the tests may have differed across the two studies.[39] Report card grades are probably not a very sensitive indicator of school performance for many reasons. Schools, and even teachers in the same schools, differ in their grading policies, and schools differ in their quality. It would have been easier to compare the Michigan and North Carolina studies if the North Carolina researchers had the funds to use some standardized tests or lab tests of cognition when the children were in 3rd grade.

Other PCB Evidence

Several other studies of exposure to PCBs are being conducted around the world. In Europe, investigators in three countries—the Netherlands, Germany, and Denmark—are cooperating to study the effects of PCBs.[40] In Oswego, New York, pregnant women who ate sport fish from Lake Ontario were sampled.[41] These studies all show an association between prenatal exposure to PCBs and newborn functioning (reflexes, muscle tone, and adaptation to new stimuli) and at later ages. The details of the studies differ in many ways: the specific measures of exposure (umbilical cord blood, maternal blood, or breast milk during the first weeks of life, and what chemical variants or congeners of PCBs were measured), what other chemical exposures were measured (dioxins, furans, DDT metabolites, mercury, and lead), the specific tests used to assess the infants, whether the mothers were sampled from the general population or from those who were expected to be more highly exposed (Lake Ontario sport-fish eaters), and exactly what family background and other confounding variables were included in the data analyses.

In the Dutch study results are available up to when the children were 9 years old. At 18 months of age, the effects of early PCB exposure were still evident in neurological exams that measured mainly motor functioning.[42] When the children were 3.5 years old, effects of PCB exposure were no longer evident in the neurological tests.[43] But cognitive tests at age 3.5 years did show deleterious effects of prenatal PCB exposure, even when social quality of the home, parent education, and parent IQ scores were included.[44] This result replicates the findings of the Michigan study, but for a sample from the population at large rather than just those who eat contaminated sport fish. At age 6, prenatal PCB exposure was not related to the children's scores on standardized tests of cognitive functioning (the McCarthy Scales of Children's Abilities). However, higher prenatal PCB exposure was associated with worse performance for children who came from families in which the social quality of the home was judged to be worse, parental IQ was lower, or the mother was younger. The Dutch research team concluded that "the adverse effects of prenatal PCB exposure were more pronounced when parental and home characteristics were less optimal, whereas these effects were not evident when parental and home characteristics were more advantageous."[45] Finally, at age 9 children with higher prenatal PCBs had slower and more variable reaction times in an attention task, performed worse on a problem solving task, but not on a verbal learning task.[46] The highly exposed 9-year-olds also showed slower brain wave responses called the "P300 event related potentials" (or ERP).[47] The ERP result reflects a particular pattern of brain activity when a person perceives or evaluates an event. This replicates a difference found for children prenatally poisoned with the rice oil in Taiwan.[48]

PCBs in Breast Milk: To Breast-Feed or Not?

PCBs and other contaminants are present in breast milk. Is exposure to PCBs in breast milk harmful to infants? Before addressing that question, we must ask another important question. How does breast-feeding by itself affect children's cognitive development or other aspects of development?

Breast-Feeding and Cognitive Development A review of more than 20 research studies concluded that *breast-feeding is beneficial to children's cognitive development*, especially for low-birth-weight babies.[49] Also, the results showed that more months of breast-feeding is associated with greater cognitive benefits. The benefits start to show up in tests after the children are 2 years of age.[50] The positive effects last into the teenage years.[51] More months of breast-feeding provided more cognitive benefits to the child. How large are the benefits? IQ test scores averaged about 3 to 5 points higher for normal-birth-weight babies who were breast-fed, and about 5 to 8 points higher for low-birth-weight babies. The lower numbers are the results *after* adjusting the scores for family background factors. Including family background variables in studies of breast-feeding is important because women from less advantaged economic backgrounds are less likely to breast-feed their babies. And on average lower income women breast-feed for fewer weeks than women who are more economically privileged.

The differences in cognitive test performance associated with breast-feeding are what researchers call a small to medium effect size. A 3- to 8-point difference in IQ test scores is enough to reduce the chances of mental retardation (IQ score lower than 70) by about half. The same difference is enough to raise the percentage of gifted children (IQ score above about 130) by about half. So for the whole of society, breast-feeding could reduce the number of children with very poor cognitive abilities and raise the number of outstanding achievers. A difference in IQ score between 75 and 80 might make the difference between holding a job and living independently as opposed to needing special social services for the rest of one's life.

Why is breast-feeding beneficial to children's cognition? Family backgrounds might differ for those who breast feed or not. But researchers have included differences due to family background variables. Beyond that, there are two ideas: nutritional effects or social effects. Human milk has substances in it that are *not* in infant formulas. Some hormones from the mother are present in her milk, such as thyroid hormone. Mother's milk also contains polyunsaturated fatty acids (PUFAs), including omega-3s. The PUFAs are now added to infant formula, although they used to be absent.[52] These hormones and fatty acids are all known to be important for normal brain growth and development. One study with premature babies who were sick enough that they had to be tube-fed showed that those tube-fed breast milk progressed better than those who received formula.[53] That eliminates the social interaction as the cause of benefits. The World Association of Perinatal Medicine and the Early Nutrition Academy both recommend breast-feeding. If breast milk is not available the formula should have two specific PUFAs added, DHA and AA.[54] But the social interaction of breast-feeding may also be beneficial. In one long-term study, teenagers who were breast-fed as infants felt closer to their mothers than those who were not breast-fed.[55]

One more thing. Most studies of how breast-feeding affects children's cognitive development were carried out in industrialized countries. Virtually all of us in industrialized countries have been exposed at least slightly to PCBs and other organochlorine chemicals (see my section below on PCB exposure). Most of the long-term studies of the effects of breast-feeding were begun in the 1970s, when the levels of these pollutants in people in industrialized countries were higher. In other words, if the *average* levels of all kinds of pollution that children get from breast-feeding were worse for cognitive development than not breast-feeding, the results would have shown negative effects. Instead, breast-feeding shows positive effects. But this in no way implies that the pollutants themselves in breast milk are good for people.

PCBs and Breast-Feeding: The Michigan Study   Can the PCBs in breast milk be harmful? Is there a concentration of PCBs in breast milk above which nursing would be deleterious to the infant? This is a complex question for research. First, there are socioeconomic differences associated with which women breast-feed their infants. Second, mothers whose breast milk is high in PCBs have babies who were exposed to more PCBs prenatally. Therefore, prenatal and postnatal exposure to PCBs will usually be confounded. But not every mother breast-feeds, and mothers breast-feed for different amounts of time—some for only 6 weeks or so, and others for 2 years or longer. So the Michigan researchers measured both the length of time an infant was breast-fed and the concentration of the pollutants in the breast milk. This estimates the quantity of PCBs received postnatally.

The results of the Michigan study give some good news and some bad news about breast-feeding and PCBs. The *bad news* is that the number of weeks of breast-feeding is a major determinant of the concentration of PCBs in the child's blood at age 4 years. Children who were breast-fed longer also had higher levels of DDT metabolites at age 4. PCBs and organochlorine chemicals in the mother are transferred to the infant during breast-feeding. These chemicals are lipophilic, and mother's milk is high in fats. Other studies show this too.[56] The *good news* is that 4-year-olds who were breast-fed longer had lower levels of lead.[57] The relationship between lower lead levels and breast-feeding, though, is partly due to socioeconomic differences (see Chapter 1).

What about the effects of breast-feeding on the children's behavioral and cognitive development? Again, there is good news and just a little bad news. The only *bad news* is that higher PCBs in breast milk were related to lowered activity levels at 4 years of age.[58] This may be because PCBs in breast milk were measured very soon after birth and therefore reflect prenatal exposure. But the number of weeks that the child was nursed did *not* relate to the children's activity levels or to any other outcomes. The *good news* is that when the children were 4 years old, longer breast-feeding was *positively* related to performance on the laboratory tests of sustained attention[59] and to memory on the standardized test.[60] But there is just a hint that the very highest concentrations of PCBs in breast milk may have slightly negative effects. Infants breast-fed by those women with more than 1,250 ng/milliliter (1.25 ppm or 1,250 ppb) of PCBs in their milk performed very slightly worse on memory at age 4.[61]

PCBs and Breast-Feeding: The Dutch Study The Dutch project selected women so that there would be a 50–50 split of those who intended to breast-feed their infants and those who did not. In the first two weeks after birth, infants with higher PCBs, dioxins and furans in breast milk showed weaker muscle tone. At 7 months of age, the infants with the highest PCB and dioxin exposure from breast milk had motor development that was equivalent to the formula-fed infants. But breast-fed infants with lower PCBs in their milk showed positive effects of breast-feeding.[62] By 18 months of age the results showed positive effects of length of breast-feeding on the toddlers' motor development.[63] When the children were 3.5 years old, the formula-fed children showed more pronounced negative effects of prenatal PCB exposure than the breast-fed children. For the breast-fed children there were no negative effects of PCB concentration in breast milk on the children's test scores.[64] It looks like the benefits of breast-feeding are overriding the negative effects of the PCBs in the breast milk. But the results could also be due to the fact that the parents of the breast-fed children had slightly higher education levels and IQ scores than the parents of the formula-fed infants, even though these family background variables were included in the analyses of the data.

The researchers in the Dutch study drew these conclusions from the results on their 3-year-olds: "Children who were breast-fed performed better on cognitive tests and were from a more advantaged socioeconomic environment than their formula-fed counterparts. Fetal exposure to PCBs and related compounds should be lowered by reducing maternal body burden rather than discouraging breast-feeding. Our data demonstrate the continuation of a toxic impact received in utero on cognitive functioning at toddler age."[65] At age 6, the breast-fed children still performed better than formula-fed children. But remember that the breast-fed children were from slightly more advantaged backgrounds. Within the breast-fed group, exposure to PCBs and dioxins in breast milk was not associated with test performance.[66] At 9 years of age, the children who were breast-fed the longest and whose mothers had the highest PCBs scored lower on the problem solving task. This is the only task that was affected negatively by post-natal PCBs in breast milk.[67]

What about other health effects of PCB exposure through breast-feeding? The Dutch research also examined the children's immune system functioning, both by physiological measures (such as lymphocytes, T-cells, and other indices of immune function) and by allergies and frequencies of illness. When the children were 3.5 years old, higher total PCB body burden was related to greater likelihood of chicken pox and recurring middle ear infections but less likelihood of shortness of breath and wheezing (symptoms linked to asthma). The researchers concluded that "the effects of perinatal background exposure to PCBs and dioxins persist into childhood and might be associated with a greater susceptibility to infectious diseases. Common infections acquired early in life may prevent the development of allergy, and therefore PCB exposure might be associated with a lower prevalence of allergic diseases.... Perinatal exposure to PCBs, dioxins, and related compounds should therefore be lowered by reducing the intake through the food chain at all ages, rather than by discouraging breast-feeding."[68]

So far, based on the studies of children's cognitive and immune functioning, it looks like breast-feeding is the way to go, unless a woman has been exposed to exceedingly high levels of PCBs (perhaps from contact with a spill or on the job). Can the animal research clarify the issue further?

Animal Research on Postnatal Exposure to PCBs   Animal experiments can separate prenatal and postnatal exposure by cross-fostering. For example, rat dams that were exposed to PCBs while they were pregnant are given litters from unexposed dams to nurse, and vice versa. Experiments like this have shown that prenatal exposure has negative effects on rats, but low-level postnatal exposure to PCBs does not seem to have detectable damaging effects.[69]

In one experiment with monkeys, macaques were separated from their mothers at birth and given approximately 50 ppb (parts per billion) of PCBs in their monkey formula until they were 20 weeks old. The control monkeys were also separated from their mothers but were given formula with no PCBs. The 50 ppb exposure was chosen because the Canadian government said that breast-feeding is fine as long as the PCBs in the mother's milk are below 50 ppb.[70] The monkeys fed PCBs in their formula learned more slowly on some types of tasks. The blood levels of PCBs in these monkeys were in approximately the same range as the children of the fish-eaters in the Michigan study; exact comparisons cannot be made because of different lab techniques for measuring PCBs.[71]

Can we apply the findings obtained from research on monkeys to people? Yes and no. The results do show that primate learning can be harmed by low-level postnatal exposure to PCBs. People are also primates. But the monkeys were not breast-fed at all. They were formula-fed. And although the monkeys were socialized as infants through play sessions with their peers, monkeys that are raised without a mother develop differently than monkeys raised with their mothers. Separation from the mother at birth was the easiest way to treat the animals with PCBs in formula. We also don't know whether beneficial substances in breast milk would balance the negative effects of the PCBs.

The monkey research underscores the fact that it would be ideal for women planning to have children to reduce their body burden of PCBs prior to pregnancy. In principle, a diet low in PCBs and other OCs for several years will reduce the woman's body burden of these chemicals prior to pregnancy. This should result in lower exposure of the infant to these chemicals, both prenatally and postnatally (in breast milk). PCBs and other OCs are carried in other animal fats, not just fish.[72]

Controversies

Criticisms of the Michigan Study

The Michigan study is the landmark study of the effects of PCB exposure on children's development. It was the first study of subpoisoning exposure with a good sample size that followed the children all the way to the end of elementary school. Because the Michigan study found lasting effects and because the North Carolina study did not find lasting effects, the details of the Michigan study have been closely scrutinized. Results from the Dutch study now corroborate the Michigan evidence. But before the Dutch results were published, the Michigan study had a prominent role in public policy. The scrutiny of the Michigan Study has come partly from industry and partly from other scientists. Careful review of research is part of environmental policy. But also keep in mind that without the Yusho and Yu-Cheng disasters in Japan and Taiwan, PCBs might still be widely used today.

The General Electric Corporation (GE) posted a harsh critique of the Michigan study on the Web.[73] GE is responsible for cleaning PCBs out of the Hudson River. After reciting a variety of criticisms of the Michigan study, GE summarized its view this way: "These problems have led various reviewers to question whether any valid conclusions can be drawn about developmental risks related to consumption of contaminated fish, or, more specifically, whether any conclusions can be reached about the potential role of PCBs in determining the results."[74] Where did GE get the criticisms it posted on its Web site? They came primarily from a review published by Schantz[75] and from commentaries on the review by at least 10 other scientists. The "open peer" commentaries on Schantz's review indicate the variety of viewpoints in the scientific community on how best to conduct any specific research project. The criticisms are part of the everyday scientific process of *striving* for the unreachable goal of the unimpeachable research study.

What are the main issues that GE had in its Web site? *First* is that the Michigan researchers did not use a "case-control" design. Instead, they randomly selected pregnant women in the same hospitals who said they had never eaten Lake Michigan sport fish. Because of the random selection from non-fish-eaters, those who ate no sport fish differed in some important ways from the fish-eating sample. They were about 9 pounds heavier on average before pregnancy, 1 year younger, and less likely to report that they drank alcohol, caffeine, or took cold medicines during pregnancy. *Second*, in their analyses of the data, the Michigan team did not always use *all* the family background and other confounders like smoking and alcohol. *Third*, some other potentially toxic contaminants in fish were not measured. Perhaps the other contaminants could account for the effects rather than PCBs.

Let's look at these criticisms by using Schantz's original review and the commentaries on it. Remember, Schantz's paper was the main source of most of GE's points. I also invite my readers to seek original sources themselves by using Internet resources such as PubMed (www.pubmed.com). PubMed contains original scientific research on health topics.

Issue #1: Are the Results Negated by Differences between the Comparison Sample and the Fish-Eaters? One way to design epidemiological studies is by the case-control method. That method requires the research team first to find a fish-eating pregnant woman, measure many important characteristics of that woman and her family (IQ score, social quality of the home, smoking and drinking during pregnancy, marital status, husband/partner's education and employment, and so on). Then the researchers select a control woman who is similar to the fish-eater on all these key variables except fish consumption.

But there are disadvantages of the case-control method for studying PCB exposure.[76] It is impractical to match on more than a few variables. The Jacobsons measured 73 background variables in the Michigan study. Matching on certain variables assumes we know which confounding variables are the most important, a presumptuous assumption. Even if a case-control design could be constructed for exposure to pollutants in sport fish, after childbirth things can change. How many fish-eaters and non-fish-eaters will get divorced, quit smoking, start smoking, lose their jobs, or remodel a home with lead-based paint? It is impossible to match samples in advance variables that change later. Many scientists think that random selection of a comparison group is most appropriate for a study of PCB exposure such as the Michigan project.[77] This is exactly how the Michigan study was designed: "The controls were selected at random from among the women who did not eat fish, and were distributed among the four participating hospitals in proportion to the number of exposed participants from each hospital."[78] The GE Web site did not cover these counterarguments.

If the non-fish-eaters were randomly selected, why did they differ in key variables from the fish-eaters? One answer is that fishing and eating sport fish are related to certain lifestyle variables that created these other differences. The Oswego study also found some socioeconomic and other differences between the fish-eaters and non-fish-eaters. Can the differences between fish-eaters and non-fish-eaters in these other variables (prepregnancy weight, weight gain during pregnancy, smoking, and alcohol) account for the differences in the children's performance on all the exams later? The Jacobsons say "No." Alcohol use during pregnancy was not related to any outcome variable in the study. Smoking during pregnancy showed a weak relationship to some outcome variables. In the study of the 11-year-olds, smoking was included in the analyses. This brings us to the next issue.

Issue #2: Did the Michigan Researchers Include All the Appropriate Confounding Variables? The Jacobsons and their collaborators used confounding variables in the statistical analyses if they were at least weakly associated with PCB exposure or with the children's test performance. The cutoff they used was a probability level of 0.10, or 1 chance in 10 of a false positive association, but sometimes they used $p = .20$ for including confounders. Critics say that variables like alcohol and tobacco during pregnancy ought to be included no matter what.[79] Other researchers argue that including variables that do not actually affect the results will reduce the sensitivity of the research. In extreme cases using too many meaningless variables can produce misleading results.

In a paper published almost contemporaneously with the criticisms, the Jacobsons did include all background variables that were even very weakly associated (a probability of 0.20, or a chance of 1 in 5 or less) with the outcome variables. This means that smoking and drinking were included in the analyses for the reading and spelling tests, and smoking was included for the verbal parts of the IQ test. Background variables such as socioeconomic status, maternal education and vocabulary, and social quality of the home were also used. In these analyses, higher PCB exposure was significantly related to lower verbal comprehension, lower overall verbal performance on the IQ test, and lower word comprehension on the reading test.[80]

This should answer the critics. But an excessively rigid critic might not believe the results until smoking and alcohol use are included in all analyses, whether or not they are related to the outcome variable being analyzed. Such differences of opinion are part of science. Here is what the Jacobsons concluded: "These deficits are not attributable to maternal drinking or smoking during pregnancy, the quality of intellectual stimulation by parents, postnatal exposure to lead, or numerous other control variables."[81]

Notice that the original commentaries on the Michigan study bring to the forefront the fact that there are important decision criteria embedded in the *entire* research process. There are no strictly right or wrong answers to questions involving the theoretical and statistical criteria for deciding how to analyze a set of data. These are the kinds of issues scientists enjoy puzzling over in their quests for further knowledge.

Issue #3: Are the Results Due to PCBs or to Some Other Contaminant in Lake Michigan Sport Fish? Epidemiological research with humans is incapable of answering this question definitively. Environmental exposure to pollutants occurs in complex combinations. In Canada researchers found that PCBs co-occur with dioxins and furans in people.[82] This finding has two important implications. First, laboratory measures of these closely related chemicals can easily be distorted by each other's presence unless the laboratory explicitly tests for the separate congeners of these chemicals. Remember that before Jensen's discovery in 1966, PCBs used to distort the readings for DDT. A second implication of the findings from Canada is that epidemiological studies need to consider the presence of the other organochlorine chemicals in drawing their conclusions.

For the Michigan study, the Jacobsons and their colleagues have always considered the possibility that other substances could be causing the negative effects on children. In 1996, they stated it this way:

> However, environmental exposure to polychlorinated biphenyls typically also entails exposure to polychlorinated dibenzofurans and dibenzo-p-dioxins, highly toxic byproducts of the manufacture and combustion of polychlorinated biphenyls that accumulate with them in fish and human tissue but are present only in trace concentrations and could not be measured. Moreover, polychlorinated biphenyls are complex mixtures of various congeners, each with its own unique molecular structure and potentially different toxic effects, which could not be identified by the analytic methods available for this study.[83]

But some of the epidemiological research has partially addressed the issue of what specific pollutants might be responsible. In the Michigan study, polybrominated biphenyls (PBBs) were also measured. PBBs were included in the data analyses even when the children were infants.[84] So PBBs are unlikely to be the culprit.

In the North Carolina study, DDE (the metabolite of DDT) was measured and was not related to infant functioning at 6 and 12 months of age, whereas PCB exposure was.[85] Newborn reflex functioning was worse when both PCB and DDE prenatal exposure were higher. But some aspects of newborn functioning were related only to PCB exposure.[86] The Dutch study found that 5 specific dioxin congeners, 2 furans, and 10 individual PCB congeners were all related to lower newborn functioning.[87]

In the Oswego fish-eater study, higher PCBs were associated with lower newborn functioning. But DDE, mirex (an OC pesticide), HCB (hexachlorobenzene, a fungicide, also used as a plasticizer), lead, and maternal hair mercury were not.[88] (Mercury was very low in the women in this study.) The Oswego study also tested separate PCB congeners.The most heavily chlorinated PCB congeners (those with more chlorine atoms) were related to the behavioral decrements found in the newborns. Thus, the Oswego study comes a step closer to finding what specific congeners of PCBs affect children negatively. However, the Oswego study did not report on dioxins and furans.

Summary

Each of the three main criticisms of the Michigan study could, in principle, be a reason for the results to lack credibility. Further research is always necessary. At the same time, each of the criticisms has been answered at least partially. These nuances in the arguments are important in science. In the arguments the value differences among scientists become evident. As I said at the start of this chapter, a single epidemiological study cannot seal the case against a pollutant. The latest results from the Dutch and Oswego studies have so far confirmed the Michigan findings. Each of those studies also has flaws that can be criticized. But the trend that began with the poisoning incidents in Japan and Taiwan is being confirmed. Prenatal exposure to PCBs and the related pollutants normally found in conjunction with PCBs can have deleterious effects on children's mental development.

I now turn to the other two types of evidence scientists require in order to conclude there is a causal connection between a pollutant and altered development. Does animal research show similar developmental effects of PCBs, and what biological processes might cause these developmental effects?

Is Behavior Altered by PCBs Alone in Animal Research?

As humans we are exposed to complex mixtures of pollutants. Animal research in the laboratory allows the study of specific pollutants. Differences in lifestyle and socioeconomic status are also not an issue in lab animals. The animal research that has used PCBs has produced results analogous to the human studies. Animals with early developmental exposure showed deficits in learning and related behavior, at least for some types of tasks. Tilson, Jacobson, and Rogan (1990) summarized the animal research this way: "Developmental exposure to PCB results in persistent neurobehavioral alterations in monkeys and nonprimates, similar neurological or behavioral effects are observed across species, and neurological and behavioral effects can be observed in the absence of reduced body weights or gross signs of PCB intoxication during development."[89] And "the most common finding in the animal studies was that developmental exposure to PCBs results in behavioral hyperactivity; this was particularly true if the testing period allowed for habituation to occur....That developmental exposure to PCB can influence higher cognitive processes or learning was observed in rats, mice, and Rhesus monkeys."[90]

Is There a Plausible Biological Process?

At least three biological processes have been proposed as causes of the neurodevelopmental effects of PCBs: thyroid functioning, retinoids (chemicals related to vitamin A), and the concentrations of certain neurotransmitters (chemicals that carry signals between neurons).

The concentrations of thyroid hormones, both prenatally and during early development, are very important in brain development. Lack of thyroid hormone due to iodine deficiency was one of the first causes of mental retardation ever discovered. The chemical structure of PCBs is similar to the two active thyroid hormones known as T3 and T4. PCBs bind to some of the same proteins that thyroid hormones do. Therefore, PCBs can interfere with the normal effects of thyroid hormones. Animal experiments have shown that PCBs can alter the size of the thyroid, the levels of TSH (thyroid-secreting hormone) produced by the pituitary gland, and can decrease the levels of T4 in serum, and increase excretion of T4 in bile. In addition, the kinds of changes in animal neurological function that result from prenatal PCBs are similar to those found with prenatal hypothyroid (underactive thyroid) function. One reviewer concluded, "Consequently the possibility exists that these toxins [PCBs and dioxins] could be altering neurological development via their action on thyroid hormone availability during critical brain developmental periods. This hypothesis remains to be tested."[91]

PCBs also affect concentrations of retinoids.[92] Retinoids are very important in early brain development. If concentrations of retinoids are either too low or too high it can dramatically alter development. Retinoids influence the timing of gene expression. Gene expression during development influences virtually all aspects of development, including the brain and nervous system.[93]

Another biological process by which PCBs may affect the nervous system was proposed by Richard Seegal, a scientist at the New York State Department of Health.[94] Seegal and his colleagues have done research in which rodents were given a diet of regular chow mixed with freeze-dried fish from Lake Huron or Lake Ontario. The offspring showed changes in the levels of dopamine in certain areas of the brain. Dopamine is a neurotransmitter that is important in many aspects of behavior, including motor functioning, depression, and learning. Seegal concluded, "The magnitude of the decreases in both prefrontal cortical and striatal DA [dopamine] concentrations following developmental exposure to these diets provides a possible mechanistic basis for changes in motoric (e.g., reflexive) and cognitive deficits observed in both humans and experimental animals following exposure to the contaminants present in Great Lakes fish."[95] The prefrontal cortex and striatum are specific regions of the brain.

But are PCBs the key neurotoxicant in Great Lakes fish? Seegal has also done research in which rats were prenatally exposed to only the single most toxic PCB congener, 2,4,4',4'-TCB (called "TCB"). The single most toxic PCB congener was determined by testing the effects of congeners on nerve tissue in vitro (nerve tissue that had been extracted from animals). Studies of prenatal exposure to TCB showed that it requires much more TCB to create the same effects on dopamine that were found with the diets containing Great Lakes fish. From this Seegal[96] concluded that PCBs are not the only neurotoxicant in Great Lakes fish. Something else more potent than PCBs must also be responsible for the developmental effects. There are a number of key assumptions behind Seegal's conclusion. Interested readers should examine the original paper.

The Role of Decision Criteria

I have already described the role of various decisions and decision criteria in the Michigan study. In this section I discuss similar controversies among scientists not just on the developmental effects of PCBs but also some other health effects.

When reviewers draw different conclusions from looking at the same scientific studies, it is usually because they are applying different decision criteria. I will use a review of PCB effects by Marie Swanson and her colleagues[97] of Michigan State University to illustrate the important role of decision criteria in conclusions from a review. This is not merely an academic exercise. The EPA and other government agencies convene committees of scientists to review research on pollutants. These reviews form the basis for pollution policy. The implicit and explicit criteria that the scientists use in their review have a dramatic influence on their conclusions. We saw that pointedly in the arguments between Needleman and Ernhart over lead.

Swanson and her colleagues[98] stated that their goal was "to ensure that comparable criteria were utilized in evaluating each report."[99] They classified each study according to 9 criteria: study design; subject selection; participation rate of those eligible; exposure measures; appropriateness of statistical analyses; inclusion of confounding risk factors; the reviewers' own evaluation of whether the results show a negative effect of PCBs; whether the authors' conclusions were compatible with the reviewers' evaluations of study results; and, finally, the overall assessment of whether the study provided convincing evidence of health risk from PCBs.

Using their selected criteria, Swanson and her colleagues concluded that only 5 of 73 studies showed "suggestive" evidence of harmful human health effects. Those 5 all involved occupational exposure. Swanson and her colleagues classified 70% of the studies as "inconclusive" based on their methodological criteria and their own evaluation of the results. The authors concluded that "when an effect occurred from workplace exposure, it was often in the form of chloracne...and, for one study, malignant melanoma."[100]

How did Swanson and her colleagues evaluate the Michigan study of prenatal PCB exposure? Because the participation rate was less than 75%, Swanson categorized the Michigan study as having a biased subject selection process. Epidemiology textbooks recommend achieving at least 75% participation, but it is not a hard-and-fast rule. In the Michigan study about two-thirds of women approached in the hospital on delivery day agreed to participate. The women who agreed to be in the study were slightly higher in socioeconomic status and slightly better educated. This is true in most epidemiological research. Swanson and her colleagues also classified most of the results of the Michigan studies as "inconclusive," with one exception. Mysteriously, they classified the results at age 4 as showing "no association" of PCB exposure with IQ test scores. This was certainly not the conclusion reached by the Jacobsons and their colleagues.

Other Health Effects of PCBs

Swanson and her colleagues also stated that they disagreed with the assessment of PCB toxicity by the National Research Council and the International Joint Commission on the Great Lakes in 1991 and 1994. Swanson and her colleagues disagreed because these groups used evidence from the Michigan and North Carolina studies.

Swanson and her coauthors concluded:

The environmental and occupational studies of human exposure to PCBs performed to date appear to yield results that are not consistent with those obtained from laboratory animal research or with effects observed in wildlife. A large number of adverse biological effects including tumor formation, reproductive failure, developmental effects, immune system deficiencies, and neurological deficits have been observed in lower animals....The scarcity of reliable reports of human toxicity from PCBs is likely due to the fact that past environmental exposures to these chemicals were not high enough to produce harmful effects. More information on the relative sensitivities of humans and laboratory animals and the biochemical mechanisms of PCB action is needed to clarify what can be considered a human threshold for observation of PCB effects.[101]

Notice that Swanson's group explicitly assumes that there is a threshold for toxic effects. Although there may be a threshold, epidemiological studies usually cannot examine the shape of the dose-response curve in enough detail to discern thresholds. The conclusions of another review paper published in 1993 were very similar to those of Swanson, but the 1993 paper dealt primarily with occupational exposure.[102]

The conclusions of the Swanson and James reviews differ markedly from the reviews conducted by scientists for the Agency for Toxic Substances and Disease Registry (ATSDR) and the EPA in 1996. Of course, two years passed for more evidence to accumulate, but the reviews relied on many of the same studies. The ATSDR and EPA concluded that exposure to PCBs in humans has been found to be associated with the negative health effects listed in Table 3.4.

TABLE 3–4. Alterations in human health associated with PCB exposure

Shorter gestational time resulting in lowered birth weight
Developmental disorders and cognitive deficits in offspring of mothers exposed prenatally to Lake Michigan sport fish
Increased systolic and diastolic blood pressure
Increased triglyceride and cholesterol levels
Altered liver function
Altered immune system function
Joint pain and joint problems

Cancer findings:
- The ATSDR and EPA comment: "While the inconsistency of sites, except for liver and biliary tract cancer, across these studies prevents their providing conclusive evidence, concern remains owing to methodologic limitations (insufficient latency, inadequate power) in the current observations. The liver and biliary tract site appears similar to results seen in animals."
- Increases in cancer mortality in workers exposed to PCBs (including malignant melanoma, gastrointestinal tract cancer, liver cancer, gall bladder and biliary tract cancer, and cancer of hematopoietic tissue)
- Significant excess cancer of the liver in Yusho victims of rice oil poisoning
- Breast cancer related positively to serum PCB concentrations
- Leukemia in children related to concentration of PCBs in bone marrow

Source: ATSDR and EPA, 1996

Three Decision Criteria That Control the Conclusions of a Review   Three different kinds of decision criteria that drive the conclusions of a review are: (1) evaluation of the plausibility and importance of alternative interpretations of the results, (2) the consistency of the results across studies and the consistency of results across similar measures within a single study, and (3) whether the *differences* observed in a study are large enough to be interpreted as "harm."

First, the Swanson and James reviews placed great emphasis on the possibility that other toxic substances, such as dioxins and furans, might account for the results. But as James and his colleagues say, "Since the potently chloracnegenic PCDFs were generally present at part per million levels in commercial PCB fluids, low-grade PCDF exposure is common to all PCB-exposed cohorts."[103] From a basic science perspective, it is important to separate the effects of PCBs from PCDFs (furans). But because the pollutants are found together in both industry and the environment, epidemiology does not need to separate them. We could still conclude that commercial PCB fluids are hazardous.

Second, the James and Swanson reviews emphasized the inconsistencies among studies. For example, not all studies have found significantly altered liver function in PCB-exposed workers. But liver toxicity of chlorinated chemicals was discovered quite early, based on worker deaths as well as animal experiments in 1937.[104] The amount of exposure is key in determining the severity of effects and what portion of a sample will show effects. A specific study may fail to show significant effects for a variety of reasons. Poor measurement of exposure or outcome variables, and inadequate sample size to detect a small effect are the two most likely causes of an insensitive research project. Also, when an individual scientific study draws the conclusion that there is no evidence of a negative effect, the probability of that conclusion being wrong is usually much higher than $p = .05$. In contrast, when a study draws a positive conclusion, the probability of being wrong is .05 or less. Because studies differ in sensitivity, a review should consider more than whether all studies were statistically significant.

The criterion of consistency has sometimes been used to criticize the Michigan study. For example, the amount of fish consumed was significantly related to newborn functioning. But there was not a significant effect of umbilical cord PCBs or maternal blood PCBs on newborn functioning. However, the measures of PCBs in cord and maternal blood are not as reliable as the measures of PCBs in breast milk. This is because blood is much lower in fat than milk, and PCBs are lipophilic. Also this particular puzzling result has paled in importance as evidence has accumulated that prenatal PCBs are associated with newborn functioning. Further, when one subtest of an IQ test shows significant results and another subtest does not, that is not really an inconsistency. Instead, it is evidence that the toxic exposure affects one area of functioning more than another.[105] Again, decision criteria are important in interpreting results.

Third, some reviewers emphasize finding "clinically significant" alterations, not just evidence of negatively altered functioning. For example, some think that higher blood pressure or cholesterol is not important unless they would be classified as hypertension or hypercholesteremia. But are higher blood pressure and higher cholesterol bad for people? Current medical practice says yes. Were the PCBs in the Michigan study linked to diagnosed developmental disorders? No. But the Michigan study showed evidence of lower IQ test scores. Here is an analogy to lead exposure. If lead were viewed as toxic only if it caused clinically significant mental retardation, then rather high levels of lead exposure would be required to meet that decision criterion. Insisting that a study find an increased rate of diagnosed developmental disorder is asking for a body count.

Decision Criteria, Public Health, and Industrial Hygiene: 1936 and Now In 1936, when negative health effects of chlorinated chemicals on workers first became noticeable, a conference was held on PCBs. The conference was similar to the 1925 Surgeon General's conference on tetraethyl lead. Some physicians at the 1936 conference seemed to adopt the Precautionary Principle. They wanted to set exposure limits to prevent any negative health effects at all: "If there are any cases of acne or of this dermatitis occurring in a plant where halowax or the chlorinated naphthalene or chlorinated diphenyls[106] are used, then that shows that there is sufficient concentration of these substances in the air to cause plugging of the follicles and to cause a skin condition. If there is sufficient concentration to do that there may be sufficient concentration to cause systemic poisoning in the few people who are hypersensitive."[107] But at this meeting at least one industry representative seemed to adopt Kehoe's smoking gun approach. Dr. R. E. Kelly (of Monsanto Corporation) remarked, "Although on one occasion we did have a more or less extensive series of skin eruptions which we were never able to attribute to a cause...we have never had any systemic reactions at all in our men."[108] Kelly wanted to think of chloracne as entirely separate from the systemic liver effects. Liver lesions had apparently caused the deaths of some workers and had been found in animal experiments. Other industry representatives expressed more precautionary viewpoints. F. R. Kaimer, assistant manager at a GE plant, said, "We had 50 other men in very bad condition as far as the acne was concerned. *The first reaction that several of our executives had was to throw it out—get it out of our plant....* But that was easily said but not so easily done. We might just as well have thrown our business to the four winds."[109]

No one in the published transcript of the 1936 discussion advocated banning chlorinated biphenyls. Ironically, had PCBs been banned at that point, the playing field would have been leveled for the GE executives who did not want to risk toxic effects in their plant. All of the coated electrical wire manufacturers would have had to seek a substitute technology.

PCBs were banned from manufacture in Japan in 1972. Writing in 2001, Yasunobu Aoki, a scientist with the National Institute for Environmental Studies of Japan, advocated a version of the Precautionary Principle with respect to PCBs, dioxins, and furans (see Chapter 7 for more discussion of the Precautionary Principle). "To reduce the release of endocrine disrupters to the environment and their uptake by humans, if necessary, some regulation of them should be required *before* their endocrine disrupting activities are verified scientifically. Endocrine disrupting compounds have been created in the twentieth century. Their impact on health might not be revealed with the concepts and methods of traditional medical science and biology, but it is definitely necessary that research on the toxic mechanisms of endocrine disrupters should proceed in consideration of the future of humans."[110] This attitude takes suggestive evidence of harm as its decision criterion for regulation, even before the details are worked out. One goal of the Precautionary Principle is to prevent the occurrence of negative health effects even if the specific toxic substances or their mechanisms have been determined precisely. Certainly the poisonings in Japan in 1968 called for serious action, even though it was not known what particular aspects of the PCB-contaminated rice oil were responsible: PCBs, dioxins, or furans. Some aspects of what is now called the Precautionary Principle were expressed by the GE executives in the 1930s: if it's harming our workers, get the offending substance out of here!

2008: Controversy at EPA over PBDEs   A chemical family used as flame retardants, PBDEs (polybrominated diphenyl ethers), are now biomagnifying up the food chain. PBDEs are similar in some ways to PCBs. Americans are exposed to 10 times more PBDEs than people in other industrialized countries.[111] PBDEs were phased out in the European Union in 2006. In the United States one kind of PBDE, deca-BDE, is still used in electronic equipment, bedding, and children's pajamas. Most exposure comes from house dust as the substance is released from consumer products.

A controversy erupted over EPA's toxicology assessment for deca-BDE.[112] The EPA removed Dr. Deborah Rice from her position as chair of the expert panel to review the toxicology assessment. Rice is a scientist with a long track record of excellent publications on the toxic effects of mercury, lead, and PCBs. The expert panel she chaired was charged with giving the EPA feedback on the correctness of the EPA report on deca-BDE. An industry representative accused Rice of being biased against deca-BDE. Industry based the claim of bias on the fact that Rice had delivered a review of PBDEs to the Maine legislature as part of her job.[113] Rice is employed by the state of Maine. Her scientific judgment was that deca-BDE exposure in North America is dramatically increasing in both people and wildlife, it alters thyroid function and reproduction, and changes behavior in mice exposed early in life. The EPA not only removed Rice from the expert panel, but also expunged all of Dr. Rice's comments from the record.[114] If a scientist who has presented a review of research is regarded as biased, it undercuts the expertise of EPA panels.

Social Justice Issues for PCBs and Related Chemicals

Recall that in the United States, lead exposure is related to a number of family background variables: socioeconomic status, racial and ethnic background, and parent education and IQ test scores. The research on PCB exposure in the United States has been done with samples that are predominantly middle class and of European backgrounds. The Michigan study had only two infants who were black. In the North Carolina study 94% were categorized as white. There were no race differences in the North Carolina study in PCBs in the women's breast milk, but there were race differences in DDE (the metabolite of DDT). Only 5% of the white women while 45% of the black women had 6 ppm or more DDE in their milk.[115] The National Human Adipose Tissue Survey (NHATS), which I describe a bit more below, showed no differences in concentrations of PCBs, dioxins, and furans related to white or nonwhite race.[116]

PCBs, Traditional Foods, the Inuit People, and the Arctic

PCBs are chemically stable and lipophilic, and they bioconcentrate up the food chain. Arctic animals and fish are all especially high in PCBs. The concentrations of PCBs increase up the food chain, with the polar bear at the top. Polar bear fat had about 135 times the PCBs of arctic char (a fish related to brook trout), 13 times the PCBs of seal blubber, and about 7 times the PCBs of beluga whale blubber.[117] This puts arctic peoples at risk. Canadian researchers studied Inuit people living in arctic parts of Quebec along the Hudson Bay and the Hudson Strait.[118] The researchers measured PCBs and other chlorinated chemicals, including DDE, in the breast milk of Inuit women right after the birth of a child. The results were startling. The Inuit women had approximately 7 times the total PCBs and more than 3 times the DDEs of the women from southern Quebec. The Inuits from the Hudson Strait area are more traditional in their lifestyle than Inuits from some of the other areas of Quebec, and eat more traditional foods such as *misirak* (fermented blubber). The more traditional Inuit people from Hudson Strait had higher concentrations of chlorinated pollutants in their breast milk than the Inuit women living in villages with a less traditional lifestyle. It looks like the traditional diet is the major source of PCBs for Inuit women.[119]

Some of the same scientists from Canada examined the adipose tissue of Inuit Greenlanders and concluded that their environmental exposure to organochlorines was among the highest known.[120] Another sample of Greenlanders had PCB concentrations about twice as high as the Quebec Inuits, 25 times as high as the control sample from southern Quebec, and several times higher than the levels of Europeans.[121] The very high levels of certain PCB congeners in Inuits in Greenland were significantly related to eating marine mammals. The authors also found evidence that PCBs first entered the arctic ecosystems in the early 1950s.

The irony is that even though there has been little industrial use of PCBs and related chemicals in the arctic, the substances have bioconcentrated there. People who live in arctic areas and eat arctic foods are subject to the heaviest pollution from PCBs and other persistent organochlorine chemicals.

On May 23, 2001, an international treaty on "persistent organic pollutants" was signed by more than 100 countries. The treaty covers PCBs and related organochlorines as well as other chemicals. Canada's signature of the treaty carried the force of immediate ratification; it was the only country to take such quick action.[122]

Is Environmental Exposure to PCBs and Related Chemicals Declining in America?

Lead exposure has dropped in the last 30 years for all ages among Americans. What about PCBs and similar chemicals? The EPA's NHATS program collected nationwide samples of adipose tissue between 1972 and 1983 from hospitals in urban areas so that every census district was represented. The study took fat samples only from cadavers and surgery patients because fat sampling is painful. The manufacture and new use of PCBs was phased out in the United States starting in 1976, in the middle of the time span of the adipose tissue study.

In 1983 virtually *all* Americans had detectable levels of PCBs in their adipose tissue. However, the percentage of the population having greater than 3 ppm in adipose tissue peaked around 1977 at near 10%, and by 1983 it had declined to close to zero. Also, the percentage of the population with PCBs of 1 ppm or greater was about 60% in 1972, and by 1983 it had declined to less than 10%. So Americans' exposure to PCBs has declined, and relatively high exposures to PCBs were less frequent in 1983 than at the peak in 1977.[123] This can be interpreted as evidence of the effectiveness of the regulation of PCBs. Whether PCB exposure has continued to decline is not clear.[124]

What about age differences? The youngest age group (0–14 years; most of those samples were actually from infants) had the lowest levels of PCBs.

People older than 45 years had the highest levels of PCBs. The same trend was found for women in the North Carolina study. One interpretation is that these toxins build up in our tissues over life. The older you are, the more of them you have. Another interpretation is that older people accumulated their PCBs before they were regulated. A similar finding for age and exposure to dioxins and furans has been found for NHATS data from 1982 to 1987.[125]

There were not significant differences in PCB exposure between white and nonwhite tissue donors in the NHATS data up to 1983. There were also no race differences in the exposure to the other two chemicals, beta-BHC (Lindane) and HCB.[126] The NHATS program was discontinued.

What Are the Sources of PCB Exposure?

The Bad Old Days PCBs were used in a wide array of products. People were exposed to them routinely in many aspects of life. Japanese researchers did an interesting experiment on PCB-containing carbonless copy paper forms. Carbonless copy papers had different amounts of PCBs in them, from none to around 6% (by weight). Kuratsune and Masuda[127] had volunteers count 100 sheets of PCB-containing carbonless copy forms by hand. The researchers then cleaned the volunteers' hands with tissue containing a solvent to pick up PCBs. The people's hands contained an average of 30 micrograms of PCBs. Even if the people washed their hands with regular soap and water after handling the forms, their hands still retained an average of 20 micrograms of PCBs each. Workers in the early 1970s were routinely exposed to PCBs by handling these forms.

How would an office worker's exposure in 1972 from handling 100 forms compare to eating Lake Michigan salmon in the year 2001? A half-pound meal of Lake Michigan salmon from the Wisconsin shore would contain, on average, about 330 micrograms of PCBs, or 10 times the amount found on the people's hands in 1972 after counting 100 forms.[128]

The poisonings in Japan and Taiwan were from contaminated food. It is known that some paints or sealers used in farm silos contained PCBs. The PCBs were absorbed by the silage. When the cows were fed silage, their milk was high in PCBs. One researcher commented, "The number of farms with PCB-contaminated silos is not known. However, PCB contamination of milk is probably more prevalent than the regulatory record would suggest."[129] The FDA guideline at that time was 200 ppb in whole milk, or 5 ppm in milk fat (it is now 3 ppm). The European Union guideline is 200 ppb in fat (or 0.2 ppm). However, contaminants are not labeled on food containers and are not regularly tested.

Current Sources of PCB Exposure The primary source of exposure is eating food that contain animal fats.[130] This is mainly due to concentration up the food chain. But accidents still occur in which PCB-contaminated oils are mixed with animal feeds. In 1999 in Belgium animals were given feed that had been accidentally contaminated with PCBs and furans (less than 0.5% of the farms in the country were affected).[131] After this incident Belgium instituted a food monitoring program to test for PCBs and related chemicals. Samples of chicken and pork for export are tested, and so are the animal feeds. A group of Belgian researchers reported that 1.8% of the chicken and pork exceeded the limit and that 9% of the chicken and about 10% of the pork scored in the elevated range of 50 to 200 ppb in fat.

How do the PCBs get into the chickens and pigs, and what feeds are highest in PCBs? The main source is animal fat from slaughterhouses. The fat is collected by feed producers, melted, and added to feed to beef it up. Society is continuously recycling PCB and OC pollutants by using the slaughterhouse fats in animal feed. The Belgian researchers suggested that "the use of at least some types of recycled fat should be banned. Rigorous control of organochlorines in the animal feed production is a cost-efficient way to avoid large-scale feed contamination."[132]

Exposure to dioxins and related chemicals in Japan is estimated to be about the same as in Europe and the United States. Is exposure to these chemicals in food too much? In Japan, the government set a guideline of 4 picograms toxic equivalent (TEQ) per kilogram of body weight per day for tolerable daily intake (TDI) of dioxins, furans, and coplanar PCBs in foods. The average exposure from food is about 2.4 picograms TEQ/kg/day. The WHO estimates consumption is between 2 and 6 picograms TEQ/kg/day.[133]

In some parts of Europe there are strict standards for exposure to PCBs in air. In Germany some schools were closed because PCBs in the air exceeded the government standards. The PCBs were in the elastic sealants (caulks) and in the fire-retardant paint on the heating radiators. Smaller amounts of PCBs were found in walls, ceilings, and floors from paints and varnishes. Tests of the teachers in the school showed that they had slightly higher levels of only one PCB congener. In Germany, such contaminated schools are cleaned up either by painting over the PCB-containing paints or by removing the PCB-containing products from the building.[134]

PCBs and Wildlife

The Canadian study of Inuit people and the animals they use as traditional foods showed that PCBs multiply up the food chain. How do the PCBs affect the fish and animals? Just as people are exposed to more than one pollutant at a time, wildlife too is exposed to complex mixtures of pollutants. Residues of chlorinated pesticides, such as DDT, are also found at the top of the food chain, as are dioxins and furans. And methylmercury can also be found in many fish and in the wildlife that consume those fish.

Animal experiments in the 1970s showed that PCBs cause thinner eggshells and lower hatching rates in some kinds of birds. These effects even extend to the second generation after exposure is stopped. Lower fertility after exposure to PCBs has been found in dogs, mice, rats, rabbits, minks, and monkeys.[135]

In the 1970s in the Great Lakes, the frequency of thyroid goiters in Coho salmon was increasing as the levels of PCBs and other organochlorine chemicals increased.[136] Geese captured in a suburb of Milwaukee in 2001 had PCB concentrations above the consumption limit. They were feeding in a contaminated river.[137] The EPA says that pollution in the Fox River and Green Bay are implicated in declines in the populations of colonial birds, fur-bearing mammals, and waterfowl. Relatively high levels of deformities have been found in fish and fish-eating birds, and their reproduction is impaired.[138] Swedish researchers concluded that European otter populations have declined because of PCB exposure. Their inference is based in part on the finding that where PCB levels have declined, otter populations have recovered, but where PCB levels are still high, the otter populations remain low.[139] For killer whales, an endangered species, PCBs will pose potential health problems until 2060.[140]

PCB Pollution "Hot Spots" in the United States

PCBs are found in many Superfund hazardous waste sites in the United States. Where I live in Madison, Wisconsin, the sewage treatment lagoons are a PCB Superfund site. In the 1970s dikes broke sending sludge into a local river. There are many similar PCB hot spots around the country. In addition to discharge directly into waterways before they were regulated, PCBs were in industrial waste oil that was recycled by spreading it on dirt roads to control dust (see Chapter 6, which considers the effects that toxic waste dumps can have on communities).

Even though PCBs are no longer used in the United States, the European Union, and Japan, cleanup in these countries continues to be costly both to taxpayers and to corporations that are found to be responsible. As an example, GE reported that it spent approximately 200 million dollars on environmental remediation in 2006 and 2007, and expected to spend up to 300 million in 2008 and 2009.[141] That amount includes environmental cleanups of chemicals other than PCBs. But PCBs in the Hudson River are a major element of that expenditure (see the section on the Hudson River later in this chapter). For a huge corporation like GE, environmental cleanup costs are usually a very small percentage of their funds. In 2007 cleanup costs were only one-tenth of a percent of GE's total revenues, and only 0.9% of earnings. The former CEO of GE, Jack Welch, said that the controversy about cleaning up PCBs "is not about money. We'll pay whatever it takes to do the right thing."[142] Most corporate annual reports assure the stockholders that environmental costs (including litigation) will not materially affect the financial condition of the company. From that I conclude, with the former CEO of GE, that environmental cleanup should be about doing the right thing for future generations.

In this section I highlight just a few locations where PCB contamination has created local controversies, beginning with the Great Lakes. There are many other such situations in the United States and in most industrialized countries around the world.

PCB Pollution in the Great Lakes: One-Fifth of the World's Fresh Water   Lake Michigan is one of the five Great Lakes (Erie, Huron, Ontario, Michigan, and Superior), which form part of the border between the United States and Canada. The Great Lakes are uniquely valuable freshwater ecosystems that contain approximately one-fifth of our planet's freshwater. They also have the *world's largest* freshwater fishery.[143] Several Native American tribes also hold treaty rights to the fisheries of the Great Lakes. Almost 5 million people (about 8% of the population of the Great Lakes states) ate Great Lakes sport fish in 1993.[144] Approximately 25% of Canada's population and 10% of the U.S. population lives in the Great Lakes drainage area. Many large cities and industries are situated on the scenic shores of the Great Lakes in both countries. In Canada we find major cities like Toronto, Hamilton, and Windsor directly on the shores of Lakes Ontario and Erie. Buffalo, Rochester, Detroit, Toledo, Cleveland, Chicago, and Milwaukee all front the beautiful shores of the Great Lakes.

The Great Lakes have absorbed more than their share of pollution. In the past it was thought that the way to safely dispose of industrial wastes, including PCBs, was by diluting them. The Great Lakes allow dilution because they hold such a large volume of water. However, pollutants like PCBs (and other persistent organic pollutants, or POPs, as well as mercury) biomagnify as they move up the food chain. Biomagnification makes dilution ineffective. In addition, pollutants that enter the Great Lakes will be there for centuries before they are eventually flushed out to sea. And once the pollutants reach the Atlantic Ocean, biomagnification in sea life can occur.

The Great Lakes contain at least 362 kinds of chemical pollutants. The United States and Canada have jointly identified 11 chemicals as critical Great Lakes pollutants because they are known to be toxic to fish, wildlife, and people. PCBs are on that list.[145] Salmon[146] and trout in the Great Lakes contain relatively high amounts of PCBs because they are a relatively oily predator fish near the top of the food chain. Lake Ontario salmon and trout have almost 3 million times the concentration of PCBs as that found in the water.[147] This is true even though those measurements were taken more than 20 years since PCBs were last manufactured in the United States and most countries.

Yet Another Persistent Pollutant:
PBDEs

Researchers at the University of Wisconsin identified PBDEs (polybrominated diphenyl ethers) as a new persistent chemical pollutant in Lake Michigan salmon. PBDEs are flame retardants used in bedding and many household electronic products. The chemical structure of PBDEs is similar to that of PCBs. In Lake Michigan salmon, the higher the PCBs, the higher the PBDEs, and vice versa. The linear relationship between these two chemicals accounted for 82% of the differences in concentration of the chemicals among the fish. So if you know the PCB value, you can pretty much predict the PBDEs, and vice versa, at least for salmon sampled on the Wisconsin shore of Lake Michigan. PBDEs can also alter thyroid function in both animals and humans.[148] Because prenatal brain development is very closely tied to the levels of thyroid hormone, and because the concentration of PBDEs is so closely related to the concentration of PCBs in Lake Michigan salmon, these findings are worrisome.

Paper Manufacturing in the Fox River Valley and Green Bay, Wisconsin

The Fox River Valley in Wisconsin, home of the highest concentration of paper manufacturing plants in the world, drains into Green Bay, a part of Lake Michigan. The EPA listed this area as a Superfund site because of PCB and other organochlorine contamination. About a quarter of a million pounds of PCBs were discharged into the river during manufacturing and recycling carbonless copy paper with PCBs. The PCBs in sediments in the Fox River are gradually washed downstream into Green Bay and Lake Michigan.[149]

The Wisconsin fish consumption advisory for the year 2008 suggests eating no more than one meal a month of any species of game fish from the lower Fox River or Green Bay. Depending on the size and exactly where they are caught, some fish are designated as completely inedible, including walleye larger than 22 inches in the Fox River.[150] Walleye are an extremely popular fish among anglers in Wisconsin, both during summer and for winter ice fishing. And all anglers enjoy catching large fish. Mallard ducks and turtles from Green Bay have also been found to be high in PCBs. Other information on PCBs in the Fox River area is provided by the Appleton Public Library.[151] An environmental group has a Web site titled "FoxRiverWatch" at http://www.foxriverwatch.com/home.html.

Natural Gas Pipeline in Lobelville, Tennessee

PCB hot spots can occur in rural areas also. Lobelville is a relatively small community. It had 990 residents as of the 1990 census. A natural gas pumping station owned by Tenneco was built there in 1951. PCBs were used as lubricants in the pumps in the pipeline. PCBs sometimes escape from the pumps in leaks, in compressed air used to start the pumps, and in condensed water vapor. The EPA now has special rules for handling and recycling natural gas pipelines and their components because of the past use of PCBs. In 1992 Tenneco paid a $6.4 million fine for illegal use of PCBs after they were banned.[152]

The EPA declared the Lobelville pumping station and areas to which the PCBs had spread to be a Superfund pollution site. In Lobelville and Perry County, PCB-laden oils from the pipeline company were used on dirt roads to control dust from 1954 to 1992. In 1966 a hog farmer was reimbursed for the deaths of 18 hogs who were poisoned by PCBs. The PCBs that killed the hogs also leaked into a stream, Marr's Branch, near the pumping station.

In 1995, as part of litigation, the neurobehavioral and perceptual functioning of 98 adults who had lived adjacent to Marr's Branch for at least 5 years were tested (only four of the participants in the study were not involved in the litigation for damages). A comparison sample from another community not adjacent to the pipeline was also tested. The research staff who administered the tests were not aware of who was exposed and who was not (they were "blind" as to exposure).

Almost three-quarters of the Lobelville residents had visual field abnormalities, versus only 17% of the comparison group. These kinds of visual perception tests have not been done in other studies of PCB exposure. However, vision problems were noted in the rice oil poisonings. Of 28 tests (including both long-term and short-term memory, reasoning, sense of balance, and so forth), all but 6 showed significant differences between the groups. These results held after including age, gender, and educational background in the analyses. The researcher concluded that long-term exposure to PCBs, combined with the possibility of furans and dioxins, could account for the severity of effects seen in the Lobelville residents. Also, some residents were likely to have been exposed both prenatally and continuously as children. The investigator pointed out that there are 65 other pumping stations in other communities on Tenneco's pipeline. "The residents of these communities should have their neurobehavioral functions measured."[153] Of course this study could be criticized in many ways. I have included a brief description of it because there are not many studies that tested the neurobehavioral and perceptual abilities of PCB-exposed adults.

GE Capacitor Manufacturing Plant on the Hudson River, New York

GE used PCBs in its capacitor plant at Fort Edward on the Hudson River beginning in 1947. There are no records of the amount of PCB usage prior to

1966. In 1972 GE requested permission to discharge 30 pounds of chlorinated hydrocarbons per day from both Fort Edward and its Hudson Falls plant. By 1976 (when phaseout of PCBs nationwide was under way) the discharges were reduced to about a half pound per day. By 1977 only about four-tenths of an ounce, or 1 gram per day was discharged.[154] The EPA estimates that, over time, slightly over 1 million pounds of PCBs were discharged into the Hudson River.

There was a dam at Fort Edward until October 1973. The PCBs from the plant were discharged into the pool behind the dam. Removal of the dam allowed PCB-laden sediments to move downstream. A major flood in 1976 made matters even worse.[155] In 1983, the EPA declared the *entire* 200 miles of the Hudson River from Hudson Falls to New York City a Superfund site.[156]

The State of New York says that women of childbearing age and children less than age 15 years should not eat any fish caught between Hudson Falls and the Troy dam (Troy is near Albany).[157] The Hudson River used to have a profitable commercial fishing industry. Many kinds of wildlife also rely on Hudson River fish.

In 2006 the court approved an agreement between the EPA and GE for the Hudson River cleanup. GE fought for years against dredging the contaminants from the river. After years of delay and legal wrangling, in 2007 GE began construction of infrastructure for the cleanup. Dredging should finally begin in 2009.

CHAPTER 4

Neurotoxins Influence Neurodevelopment: Organophosphorus and Carbamate Pesticides

But the plague of locusts proved as certain as the seasons. All that grew above ground, with the exception of the wild grass, it would pounce upon and destroy; the grass it left untouched because it had grown here ere time was and *without the aid of man's hand....* A garment might be laid out on the ground to dry—a swarm would light on it, and in a moment only shreds would be left.—O. E. Rolvaag, *Giants in the Earth*

It is not my contention that chemical insecticides must never be used. I do contend that we have put poisonous and biologically potent chemicals indiscriminately into the hands of persons largely or wholly ignorant of their potentials for harm. We have subjected enormous numbers of people to contact with these poisons without their consent and without their knowledge....I contend, furthermore, that we have allowed these chemicals to be used with little or no advance investigation of their effect on the soil, water, wildlife, and man himself. Future generations are unlikely to condone our lack of prudent concern for the integrity of the natural world that supports all life. —Rachel Carson, *Silent Spring*

An ample, economically profitable, and safe food supply is the goal of pesticide use on food. Pesticides have been used in western European civilizations at least since the Roman empire. The Romans used sulfur on crops, and plant extracts were sometimes used as well. By the seventeenth century, tobacco was discovered to be an exceptional insecticide. Now it is known that nicotine is the active ingredient that makes tobacco an insect killer.[1]

In the United States there are more than 20,000 household pesticide products with more than 300 active ingredients.[2] Most Americans have pesticides of some sort in their homes. Look under your sink, in the bathroom cupboard, and in your basement, garage, or storage shed. Also look at your dog or cat. Flea and tick collars and flea soaps are insecticides. How many different kinds of pesticides do you have, how often do you use them, and how safe are they? What about pesticides in the foods you eat, water you drink, and air you breathe?

In this chapter I focus on two widely used categories of insecticides, organophosphorus (OP) and carbamates. These two categories of insecticides are similar to each other in their biological mechanisms. OPs and carbamates decay rapidly outdoors in the sun and do not bioaccumulate up the food chain. These features make them an improvement in some ways over organochlorine pesticides such as DDT. By focusing on these two categories, I do not intend to imply anything about their relative safety compared to pyrethrins, another major class of pesticides currently in use in the United States, or banned organochlorines.

We have already encountered one example of pesticide use gone awry. The Iraqi methylmercury poisoning described in Chapter 2 was caused by methylmercury applied as a fungicide on seed grain. It caused deaths and illness in adults who mistakenly ate the grain, and developmental delays in children who were prenatally exposed.

Pesticide misuses are still a problem. For instance, I play clarinet in the Madison Municipal Band. Before one of our summer concerts in a park near Madison, the park superintendent got over-zealous with the mosquito spray. A brass player ended up in the hospital with insecticide poisoning symptoms. No mosquito bites though! In Chapter 8 I explain how to protect against pesticide misuse. I also give some advice on how to reduce your overall pesticide exposure.

Research on how synthetic chemical pesticides affect children's neurobehavioral development is in its infancy. Here is an analogy to the history of lead. Until the mid-1940s, when Byers and Lord did their follow-up study of childhood lead poisoning victims, it was assumed that lead poisoning in children was only an acute illness. When the U.S. government provided money for lead exposure screening, researchers such as Needleman, Perino and Ernhart, and others began to find that IQ test scores were lowered by subtoxic lead exposure. The first studies were not long-term prospective studies. They simply measured lead and IQ test scores in children in a particular age group. After those initial studies showed that subpoisoning lead was tied to lower IQ and school performance, then long-term prospective studies were begun. Those studies tracked children's lead exposure and behavioral measures from birth all the way to adolescence.

Where are we now in the study of pesticides and children's behavioral development? Studies testing whether exposures early in life (prenatally or during infancy) relate to children's later behavioral functioning were just started in the late 1990s. Even though the potential for pesticides to harm children's psychological development is just now being explored, we do have some information on children's pesticide exposure. Exposure assessments by themselves are important for risk calculations that extrapolate from animal research. And there is a chain of evidence pointing to the possibility that OP and carbamate pesticides could harm child development.

Here's the chain of evidence casting suspicion on OP and carbamates. First, OP and carbamate pesticides alter the levels of certain brain chemicals that perform important signaling functions to direct the course of early brain development. Second, animal research shows that some OP or carbamate pesticides do alter aspects of brain development in a way that is similar to nicotine. Third, exposure to nicotine early in development (maternal smoking) is associated with altered development in children: low birth weight, a higher likelihood of learning disabilities, attention deficit disorder, lowered IQ scores, and behavior problems. In the United States the warnings about smoking during pregnancy have been on the cigarette packs for many years. Fourth, the evidence of a link between prenatal pesticides and adverse developmental outcomes is building. Fifth, exposure to OP and carbamate pesticides is quite widespread, not just in the United States but in most of the world. The metabolites of some types of these pesticides are found in detectable quantities in approximately 80% of Americans.

Before turning to the details, here is my tribute to Rachel Carson, the author of *Silent Spring*. She single-handedly changed the way we all look at pesticides.

Rachel Carson's *Silent Spring*

The *New York Times* named Rachel Carson one of the most influential people of her era. In 1964, her obituary appeared on the front page. An earlier editorial in that paper said, "If [Carson's book] helps arouse public concern to immunize Government agencies against the blandishments of the [pesticide] hucksters, and enforces adequate controls, the author will be as deserving of the Nobel Prize as was the inventor of DDT."[3]

When *Silent Spring* was published in 1962, it was an instant bestseller. Forty-five years later the book is still required reading for many college courses in environmental studies. *Silent Spring* described the disastrous effects of indiscriminate use of insecticides and herbicides on the world's ecosystems. Based on research, the book chronicled fish kills due to aerial spraying of forest areas, and massive deaths of robins, nuthatches, and chickadees on the campus of Michigan State University in the 1950s due to heavy DDT use to try to stop the insect that spreads Dutch elm disease. There were massive wildlife kills elsewhere in Michigan and Illinois when Aldrin (another organochlorine pesticide) was sprayed aerially to kill Japanese beetles.[4]

Carson was never opposed to limited and careful use of insecticides. Nevertheless, her book unleashed a storm of protest from representatives in industry and other skeptics. Robert White-Stevens, a spokesman for industry, was quoted in Carson's *New York Times* obituary as having said: "The major claims of Miss Rachel Carson's book, *Silent Spring*, are gross distortions of the actual facts, completely unsupported by scientific, experimental evidence, and general practical experience in the field."[5] But Carson had documented her book carefully. She was a professional biologist. The paperback edition of *Silent Spring* has 30 pages of references, mostly to scientific articles. Carson had earned her master's degree in biology at Johns Hopkins University in 1932, at a time when it was rare for women to pursue graduate degrees in science.

Stewart Udall, Secretary of the Interior in the Kennedy administration, wrote, " 'Silent Spring' was called a one-sided book. And so it was. She did not pause to state the case for the use of poisons on pests, for her antagonists were riding roughshod over the landscape. *They* had not bothered to state the case for nature. The engines of industry were in action; the benefits of pest control were known—and the case for caution needed dramatic statement if alternatives to misuse were to be pursued."[6]

The impact of *Silent Spring* was enormous. The book raised the consciousness of the American public (as well as readers in other parts of the world) about both the known and *unknown* toxic potentials of pesticides, to both people and wildlife. It also affected government. President Kennedy convened a committee to examine pesticide issues. In May 1963 the presidential advisory committee issued a report calling for more research on the potential health hazards of pesticide use. And the Committee urged *judicious* use rather than indiscriminate blanket spraying. Carson testified at Senate hearings on a bill to require labels on pesticides telling users how to minimize damage to wildlife. At that hearing, she advocated the creation of a commission in the executive branch of the federal government to coordinate environmental issues related to pesticides. It was not until 1970, during the Nixon administration, that the EPA was finally formed.[7]

Biological Processes Affected by OP and Carbamate Insecticides

The toxicity of OP compounds was discovered in 1932 by two German researchers at the University of Berlin, Lange and von Krueger. In 1936, another German, Gerhard Schrader, began searching for effective insecticides among organophosphorus compounds. Schrader is credited with inventing parathion, one of the most widely used OP insecticides. However, the German government interrupted Schrader's search for insecticides. He was ordered to develop chemical warfare agents. Schrader's lab synthesized three potent organophosphorus nerve gases, including sarin.[8] Sarin was used in a terrorist attack on a subway in Tokyo in 1995.[9]

Cholinesterase Inhibition

Organophosphorus and carbamate pesticides are poisonous because they inhibit an enzyme called acetylcholinesterase, or AChE. In order to understand how inhibiting AChE makes a substance poisonous, a few basic ideas about the nervous system are needed. Table 4.1 lists the key terms and abbreviations.

TABLE 4–1. Glossary of terms and abbreviations for chapter 4

ACh	Acetylcholine. A neurotransmitter, or chemical substance that is involved in communication between neurons.
Acetylcholine agonist	A substance that increases the effectiveness of the action of acetylcholine in the nervous system. Nicotine is an acetylcholine agonist.
AChE	Acetylcholinesterase. An enzyme that inactivates acetylcholine.
Apoptosis	Programmed normal death of neurons that occurs as part of the normal process of early development of the nervous system.
BChE	Butyrylcholinesterase. An enzyme found in blood plasma. Also called plasma cholinesterase or pseudocholinesterase.
Cholinergic	Sensitive to the neurotransmitter ACh.
Cholinesterase	A term that includes both AChE and BChE.
Cholinesterase inhibitor	Any substance that inhibits the action of cholinesterase. Organophosphorus and carbamate pesticides are cholinesterase inhibitors.
MUDDLES	Acronym for the symptoms of poisoning with cholinesterase inhibitors: Miosis, Urination, Diarrhea, Diaphoresis (profuse sweating), Lacrimation, Excitation of the central nervous system, and Salivation.
Myelin	The fatty sheath that surrounds nerve fibers which functions to increase the conduction rate of neurons.
NTE	Neurotoxic esterase. An enzyme that is not well understood, but that is believed to be involved in the neurotoxic effects of OP pesticides, hence its name.
OPIDN	Organophosphate-induced delayed neurotoxicity, a complication of exposure to OP chemicals that involves temporary or permanent paralysis. Also called OPIDP, organophosphate-induced delayed polyneuropathy.
OPIDP	Organophosphate-induced delayed polyneuropathy. Another term for OPIDN.
Synapse	The space between the membranes of two neurons that communicate with each other.

Neurons communicate with each other and with muscles by chemical signals called neurotransmitters. Acetylcholine (ACh) is one such neurotransmitter. When a neurotransmitter is released from one neuron, it can stimulate another neuron to transmit a nerve impulse (or "action potential"). Neurotransmitter chemicals are released into the microscopic space between neurons (the "synaptic cleft"). The chemical then affects receptors on the receiving neuron. The neurotransmitter ACh remains in the synaptic cleft until it is inactivated by AChE. (For other neurotransmitters, the excess is transported out of the synaptic cleft and back into the neurons.) If AChE is inhibited, as happens when a person, animal, or insect is exposed to an OP or carbamate pesticide, then the ACh builds up in the synaptic cleft. This excess neurotransmitter then results in an overstimulation of the neurons. If the overstimulation is enough, symptoms of poisoning will occur. If the overstimulation goes on long enough, then the cholinergic nerve cells quit working.

The symptoms of pesticide poisoning from AChE-inhibiting pesticides can be summarized with the acronym MUDDLES: miosis (pinpoint pupils), urination, diarrhea, diaphoresis (profuse sweating), lacrimation (tearing eyes), excitation of the central nervous system, and salivation.[10] If a severe overdose of an AChE inhibitor goes untreated, the person may stop breathing, and death will follow.

Organophosphate-Induced Delayed Neurotoxicity

Acute overdose of AChE inhibitors can be treated. The vast majority of victims seem to make a full recovery. But in a very small percentage of cases, about two weeks after exposure, a complication arises. The complication is temporary paralysis, called "organophosphate-induced delayed neurotoxicity" (OPIDN, or OPIDP, "organophosphate-induced delayed polyneuropathy"). Not all OP compounds cause OPIDN. It was first discovered for organophosphorus compounds that are *not* strong AChE inhibitors and are *not* insecticides. OPIDN usually begins with shooting pains and weakness in both legs about two weeks after exposure. The onset of symptoms is insidious and can be misdiagnosed as multiple sclerosis or encephalitis. After symptoms begin, they often continue to worsen for about 3 to 6 months, and then the paralysis stabilizes for a while. Six to 18 months after the onset of symptoms, improvement usually begins to occur. Recovery will continue for up to two years. But in severe cases, where the exposure has caused quadriplegia, complete recovery is unlikely.[11]

The most famous OP poisoning incident that caused OPIDN occurred in the United States in 1930 during Prohibition. A popular elixir was an extract of Jamaican ginger, called Jake, which was mixed with alcohol. So some people drank Jake for the alcohol, not for the ginger. Knowing this, the officials who enforced Prohibition raised the criterion for how much ginger extract had to be in a bottle of Jake. The new rule made it too costly to make a good profit. To cut costs, one bottler in New York City started adding the organophosphate tri-o-cresyl phosphate, or TOCP. TOCP is not an insecticide and does not cause cholinesterase inhibition. The result of using TOCP in Jake was more than 20,000 cases of partial paralysis known as "Jake leg." Other incidents of TOCP-induced paralysis have occurred in Morocco, where aircraft lubricants were mixed with cooking oil, and in South Africa, Fiji, Europe, and India.[12]

When organophosphorus insecticides were introduced, reports of paralysis following poisoning incidents began to appear. In 1953, an overdose of the insecticide mipafox caused paralysis.[13] Incidents of partial paralysis usually occurred after acute poisoning with full-blown symptoms of cholinesterase inhibition (MUDDLES). But there are cases in which OPIDN occurred without acute symptoms. Most victims of OPIDN recover at least partially.

Some Complexities of OPIDN   Researchers are not sure exactly what features of particular OP and carbamate compounds determine whether they will cause OPIDN. The safest pesticide would be one that would have low acute toxicity and also would not cause OPIDN or other long-term problems. Here are a few of the complex issues surrounding OPIDN.[14]

First, there are large species differences in susceptibility to OPIDN. This implies that it is not possible to test an OP or carbamate compound on just one species—such as mice or rats—and extrapolate to other species such as people, birds, or fish. Researchers are not certain why species differ in susceptibility to OPIDN. *Second*, there is also high variability among individual animals within a species in the occurrence of spinal cord damage as part of OPIDN. Some individuals may experience permanent effects, whereas others may have only temporary disability from OPIDN. *Third*, the age of an animal is also important. In some species young animals are *not* as likely to suffer OPIDN as adult animals. Studies on young chickens showed that a single oral dose of some OP compounds does not produce OPIDN, but repeated doses did. Adult chickens were susceptible to OPIDN from a single dose. Sensitivity to OPIDN increases with age in some strains of laboratory rats as well.[15] *Fourth*, the mode of exposure and whether exposure is repeated affect whether OPIDN occurs. In animal experiments, dermal exposure will sometimes produce OPIDN when oral exposure does not. Repeated exposure sometimes produces OPIDN when a single dose does not. Most substances are routinely tested for toxicity by feeding them to animals rather than testing them on skin. But people who apply pesticides, or work or live in areas that have been sprayed, are likely to absorb pesticides through their skin and lungs.[16]

The biological processes by which OP compounds cause OPIDN are not completely understood. It was originally thought that OPIDN occurred because the substance destroyed myelin. Myelin is a fatty sheath surrounding nerve fibers that increases their conduction rate. More recent work shows that the loss of myelin is a result of the deterioration of the nerve fiber, not vice versa. Prolonged inhibition of AChE was also a theory for OPIDN. But OP compounds that are only weak AChE inhibitors can still cause OPIDN, as with TOCP and Jake leg in 1930. Nevertheless, prevention of poisoning symptoms due to AChE inhibition is considered very important for preventing OPIDN, especially for insecticides.

Other processes may be involved in OPIDN, such as inhibition of another cholinesterase enzyme called butyrylcholinesterase (BChE) or of neurotoxic esterase (NTE). Some research has focused on changes in body substances called "protein kinases." If protein kinases are altered by an OP, they can in turn alter the concentration of calcium ions inside neurons, which ultimately can cause neurons to degenerate. Calcium is involved in the transmission of nerve impulses, so calcium balance is very important to the integrity of nerve cells.[17]

The occurrence of OPIDN shows that the processes by which AChE-inhibiting insecticides affect the nervous system can go beyond mere acute inhibition of AChE. For some OP substances that are not insecticides, AChE inhibition is not a prerequisite for the occurrence of OPIDN.

Cases of OPIDN in Children   In 1959 an Australian 18-month-old boy was afflicted with OPIDN. He was repeatedly exposed to malathion, a commonly used OP insecticide. Malathion had been sprayed in the family garden for about 6 weeks. The child was allowed to play in the garden. Three weeks after the spraying started, the boy became fussy, had a poor appetite and a bout of vomiting and diarrhea, and developed a fever. He had lost 7 pounds. These symptoms, though consistent with OP overdose, were not serious enough by themselves to prompt medical treatment. It just seemed like a gastrointestinal problem. But a few weeks later, when the little boy lost the ability to stand up, his parents took him to the hospital. Over the next 3 days in the hospital, the muscle weakness spread to his arms and trunk. After he had been treated in the hospital for 9 days, tests showed that his blood cholinesterase was normal. It took 4 weeks for him to recover the ability to sit up on his own.[18]

In 1993, a 3-year-old got OPIDN after playing with a bottle of Dursban (chlorpyrifos), an OP insecticide. When he was brought to the emergency room he was in a coma, with pinpoint pupils and twitching movements of the eyelids and limbs. He was treated with an antidote but had to be put on a ventilator. He recovered enough that the ventilator was removed on the third day, and his cholinesterase was in the normal range. But unexpectedly on the 11th day, his vocal chords became paralyzed, and he needed the ventilator again. By the 18th day of hospitalization, he was paralyzed totally. He finally went home after 52 days in the hospital and was able to walk again.[19]

Can Carbamate and OP Insecticides Have Long-Lasting Effects without Poisoning? Some scientists say "No." But there are case reports published in medical journals that dispute this conclusion. There are 8 cases of sensory (but not motor) neuropathy attributed to chlorpyrifos sprayed indoors. In 5 cases the patients had some minor symptoms consistent with OP poisoning: nausea, diarrhea, urinary urgency, dizziness, headaches, tearing eyes, and muscle cramps. No single patient had all of these symptoms, and none were serious enough for medical attention. About 2 to 4 weeks later, these patients lost feeling in their feet or hands (usually sensitivity to vibration was most affected). Five of the 8 people developed cognitive problems as well. Two of these were teenagers whose school performance declined. Psychological tests showed short-term memory problems. The doctors who reported these cases wrote, "We conclude that repeated environmental exposure to commercial chlorpyrifos (Dursban) can cause both reversible distal symmetric sensory neuropathy and mild cognitive dysfunction. This substance should be deployed with caution and its safety reassessed."[20,21]

The Roles of ACh, AChE, and AChE-Inhibitors in Brain Development

Anything that affects brain development has the potential to influence children's behavioral development. Because there is so little research on how exposure to pesticides affects children's development, we must look at the developmental processes that are influenced by OP and carbamate pesticides.

From the very earliest points in development neurotransmitters such as ACh serve as gene signalers. A chemical signal to the genes in a cell will change the ultimate development of that cell. Stimulation of a gene can result in cell proliferation, cell migration within the developing brain, cell death, or synapse formation. Which happens depends critically on the developmental timing of the gene stimulation. Even before the brain has become a distinguishable organ, neurotransmitters influence development of the neural tube (the embryonic structure from which the brain and spinal cord arise). Later neurotransmitters also influence brain development.[22] AChE and BChE work together to coordinate the growth of motor neurons in some species. One chemical is produced slightly before the other and precedes the other along a pathway to direct the nerve fiber to grow in that direction.[23] These developmental processes are exquisitely finely tuned. Anything that alters the concentrations of the chemical signalers has *potential* to alter the structure and later functioning of the nervous system.

But we know that ACh, AChE, and BChE are important signalers during neural development. And we know they are altered by exposure to insecticides. Could OP and carbamate insecticides alter brain development enough to show up in the behavior of children? Some animal research has shown that at least one OP insecticide does alter brain development in newborn rats, and evidence is starting to build about effects on children.

Nicotine and Chlorpyrifos Alter Early Brain Development and Later Behavior

I am going to make a case that OP and carbamate pesticides have potential to harm child development. Part of my case is based on research findings for nicotine. We have 3 decades of research showing negative effects of maternal smoking during pregnancy on children's development. And there is ample research showing negative effects of nicotine on animal development. The net effects of OP and carbamate pesticides are similar to nicotine. These chemicals all increase the excitation of cholinergic neurons. This does not imply that the differences among nicotine, OP, and carbamate pesticides in the nervous system are unimportant. However, the similarity in the effects of these substances on cholinergic neurons leads us to suspect similarities in how they affect brain development.

Maternal Smoking and Children's Development   Nicotine formerly was used as an insecticide, and there is a new class of insecticides in use called "neonicotinoids." Nicotine is a "cholinergic agonist." It increases the excitability of cholinergic neurons. Thus, its action in the brain and nervous system is similar to pesticides that inhibit AChE. Both result in more cholinergic activity. The evidence that maternal smoking during pregnancy harms children's behavioral development is quite strong and goes back at least to 1973.[24] Maternal smoking is associated with lower prenatal growth rates, lower birth weights, higher risk of fetal death and perinatal complications, a higher rate of sudden infant death syndrome, and lower growth rates during childhood. The behavioral effects include poor habituation to sounds during infancy[25]; lower scores on elementary reading and math[26]; lower scores on IQ tests, especially the verbal sections, from preschool to at least the end of grade school[27]; attention deficits and higher activity levels[28]; higher rates of aggressive behavior problems as adolescents; and a higher likelihood of violent criminal acts as adults.[29]

Figure 4.1 shows the results of part of a study by Naeye and Peters[30] done in 1984. They assessed pairs of siblings born to the same mothers. But these mothers smoked when pregnant with one child and did not smoke when pregnant with the other child. The siblings born when the mothers smoked had lower attention spans and higher activity levels than the siblings born when the mothers were not smoking.

Human epidemiological studies cannot draw cause-effect conclusions with certainty. Is it nicotine or other aspects of smoking that are harmful? Women who smoke during pregnancy differ on average from those who do not smoke,

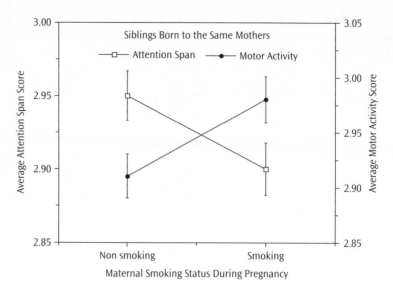

FIGURE 4–1. Attention span and motor activity of siblings born to smoking and nonsmoking mothers. *Source:* Data from Naeye & Peters, 1984.

including drinking and other drug use, socioeconomic status, marital status, their own mental health, self-esteem, dietary differences during pregnancy, and perhaps a genetic makeup that increases the chance of addiction to nicotine.[31] In addition, cigarette smoke contains many toxic substances other than nicotine. Smoking a cigarette also creates a short-term lack of oxygen and rise in blood pressure that can affect the fetus. Nicotine damps people's appetites. Pregnant women who smoke may not eat as well as those who don't smoke. Animal research can be used to find out if it is nicotine itself that is harmful to children whose mothers smoked during pregnancy, or something else. Another important issue is whether and how nicotine patches or gum should be used during pregnancy to help women avoid smoking. If nicotine itself is the problem, then patches and gum should be used during pregnancy only with great caution.[32]

Animal Research Shows that Nicotine and Chlorpyrifos Affect Brain Development At Duke University in North Carolina, Theodore Slotkin has a team studying how brain development is affected by the cholinergic family of chemicals ACh, AChE, BChE, and acetylcholine transferase. The group has been studying nicotine and the OP pesticide chlorpyrifos, sold under the trademarks Dursban and Lorsban.[33] Slotkin and his group found that nicotine in rats activates genes that cause cholinergic neural cells to die off during early brain development. Programmed cell death, called "apoptosis," is a normal part of brain development, but nicotine exacerbates it beyond the normal bounds. Fetal exposure to nicotine also affected how nerve connections formed and their sensitivity to neurotransmitters. Slotkin's procedures mimicked regular smoking. The nicotine was given to the animals at a steady state, so that sudden vasoconstriction and oxygen deficits were avoided. Slotkin concluded, "Our findings indicate conclusively that nicotine is a neuroteratogen, evoking cell damage and reducing cell numbers, impairing synaptic activity and behavioral performance, and eliciting these changes at doses commensurate with moderate smoking, below the level at which fetal growth is impaired. The underlying mechanisms are receptor mediated, accounting for selective effects on the brain at low-dose thresholds and for the involvement of brain regions and transmitter systems that have prominent cholinergic inputs."[34]

What about OP pesticides? Slotkin and his group studied chlorpyrifos because it is widely used on crops and seems not to cause OPIDN in animals at doses that are not lethal. Not long ago chlorpyrifos was also widely used as a home insecticide spray. Because OP pesticides affect cholinergic processes, the research team expected the effects of chlorpyrifos to bear a strong resemblance to nicotine. In addition to inhibiting AChE, chlorpyrifos also seems to be a direct ACh agonist like nicotine.[35]

First, Slotkin's team found that when chlorpyrifos was given to rats one day[36] after birth it reduced the number of brain cells even in regions of the brain that do *not* use ACh as a neurotransmitter. This nonspecific effect of chlorpyrifos was a surprise. These effects occurred without any signs of poisoning: AChE was reduced only 20%. Also, the effect of chlorpyrifos on rat brain development depended on developmental timing. Low doses of chlorpyrifos produced much larger effects early in development than larger doses did later on.[37] *Second*, when chlorpyrifos was given later in development (8 days after the rats' birth), the effects were specific to parts of the brain that rely on ACh as a neurotransmitter. This was what the researchers expected. *Third*, in rat embryo cultures in vitro, chlorpyrifos had an effect on cholinergic cell multiplication at very low doses. The other effects of chlorpyrifos on brain development required much higher exposure levels. All these effects occurred in the absence of any gross malformations in the brain. The effects of chlorpyrifos on brain development would not have been detected by a standard toxicology test in which the researchers look for frankly altered brain structure.

Slotkin concluded, "In the case of chlorpyrifos, our findings indicate that inhibition of cholinesterase, the standard biomarker for organophosphate toxicity, is inadequate to explain the effects of this compound on brain development. The uncovering of alternative mechanisms indicates the need for research on screening methods that emphasize unique attributes of developing systems such as DNA synthesis, cell acquisition, apoptosis [cell death], and cytoarchitectural modeling of specific brain regions."[38]

Another research group that reviewed the possible mechanisms of pesticides on early brain development agreed with Slotkin and drew this conclusion: "There is an ever increasing body of literature clearly suggesting that pesticides pose a particular threat to the fetal brain....The pesticides that function as cholinesterase inhibitors have been shown to affect [cell] proliferation, through mechanisms that are not well defined. Effects have been observed on markers of [cell] proliferation following exposure to organophosphorus compounds and, in particular, the insecticide chlorpyrifos."[39] This review was written by researchers working for the EPA.

Not Everyone Agrees   But another review came to a different conclusion.[40] Schardein and Scialli reviewed studies conducted by Dow Chemical as part of the EPA registration process for chlorpyrifos, as well as other studies carried out in university and government labs. They concluded that there was no effect on offspring's "regional brain DNA or protein levels in a study in which pregnant rat dams were given up to 5 mg/kg/day chlorpyrifos."[41] They also said that "chlorpyrifos did not adversely affect reproduction and was not developmentally neurotoxic or teratogenic, and no *selective* toxicity or sensitivity of the fetus or young animals was apparent in any guideline studies that were scientifically acceptable" (p. 12, emphasis added). Finally they argued that because the animal and human toxicology data on chlorpyrifos are quite complete, the additional uncertainty factor of 10 in the risk assessment to protect children should not be required. Schardein works for a private research lab, and Scialli is affiliated with Georgetown University.

But Schardein and Scialli used some rigid boundaries for their review. They very carefully qualified their conclusion as relying on "guideline studies that were scientifically acceptable" and said that no results showed "selective toxicity or sensitivity of the fetus." A "guideline study" is one that meets particular toxicity testing procedures that are mandated by federal agencies. Also, these reviewers excluded *nonselective* effects of chlorpyrifos on brain growth. Slotkin's team found nonselective effects when the animals were one day of age. Schardein and Scialli[42] did cite one study by Slotkin as evidence that newborn rats are more acutely sensitive to chlorpyrifos than adult rats. Slotkin[43] cited the same paper as evidence that acute doses of chlorpyrifos inhibit DNA synthesis with no regional selectivity. The inhibition of DNA was replicated with lower doses in two later studies not cited by Schardein and Scialli. One of those was published in 1997 and should have been available to Schardein and Scialli when they wrote their review.

Again, we see that the scientists are bringing different viewpoints to their interpretations of the same findings. Schardein and Scialli[44] set different boundaries on what studies they considered to be scientifically acceptable than did Slotkin.[45] Slotkin[46] concluded that the nonspecific effects of chlorpyrifos on brain development are an important mechanism of the chemical.

Schardein and Scialli ruled out nonspecific effects because they do not involve the major known mechanism of the chemical, cholinesterase inhibition.

Effects of Pesticide Combinations Are Complex   Pesticides don't hit us one at a time. When newborn rodents were exposed to DDT they showed less exploration, less habituation, and worse performance on learning tasks as adults if they were also exposed to either OP or pyrethroid pesticides as adults.[47] These researchers concluded:

> A short period of low-dose exposure to some pesticides during the neonatal period appears to be sufficient to increase the adult susceptibility to renewed, interventive exposure not showing any persistent effect in the neonatally untreated animals. The later adult exposure caused effects similar to those seen in neurodegenerative disorders associated with aging processes....Furthermore, these studies indicate that differences in adult susceptibility to environmental pollutants are not necessarily an inherited condition. Rather, they might well be acquired by low-dose exposures to environmental agents during perinatal life when the maturational processes of the developing brain and CNS are at a stage of critical vulnerability.[48]

What the Research Shows: Subacute Developmental Effects in Children

Yaqui Farming and Ranching Communities in Mexico

The homeland of the Yaqui of Sonora, Mexico, was the locale for one of the first studies to show a link between pesticides and child development. Sonora borders Arizona on the north and the Gulf of California on the west (see Figure 4.2). The research team from the University of Arizona and Instituto Tecnologico de Sonora chose children with different pesticide exposure histories. The 33 children with relatively high pesticide exposure came from a lowland intensive farming area near the mouth of the Yaqui River. The "green revolution" brought pesticide use into favor in the 1940s. Women's breast milk in this area was above the WHO and the U.S. FDA recommended limits for DDT metabolites and five other organochlorine pesticides. Farmers apply pesticides up to 45 times per crop, including aerial spraying. Household insect sprays were usually used daily.[49] The low-exposure comparison group of 17 children came from a foothill village of Yaqui where ranching without pesticide use is the major way of making a living. The people there garden with traditional methods of inter-cropping for pest control. Most people use fly swatters for household insects rather than getting out the spray can. Educational and health services were similar in both areas. None of the children in the study had started school yet. The lowland mothers reported more total problem pregnancies compared to the foothills mothers: premature birth, spontaneous abortion, birth defects, and death of an infant at birth or in the first two weeks of life. The children in the two groups did not differ in age or measures of physical growth. In both areas "the historical high degree of poverty has continued to exist."[50]

The high pesticide children scored more poorly than the low pesticide children on five out of the six research tests. Physical stamina, two tests of fine motor coordination, 30-minute memory task, and the Draw-a-Person test all differed between groups. The researchers concluded that "The role of pesticides on neuromuscular functioning and thought processes deserves such study."[51]

What could explain the results other than differences in pesticide exposure between the groups? The groups had similar cultural, language, and economic

FIGURE 4–2. Map showing Sonora, Mexico, and Yuma, Arizona.

backgrounds, and none of the children had started school yet. Both areas were sometimes sprayed with DDT to kill mosquitoes that carry malaria. The foothills children played more in groups compared to the valley children. Play is an important learning activity for children. It is also possible that some of the children had contracted malaria in the past. One type of malaria infects the brain and could therefore alter performance. The researchers pointed out that a limitation of their study is that they did not measure each child's pesticide exposure. Pesticide exposure presumably differs between groups because of location. But, pesticides are used in very complex combinations. Even if the researchers in the Yaqui study had tested for pesticide metabolites in urine samples, with only 54 children in the study it would be virtually impossible to sort out which particular substances were most strongly related to performance. I think of this study as analogous to studying children who live in housing with lots of lead-based paint versus those who live in lead-free housing. If the study showed a difference, it is certainly worth further exploration. Other researchers started looking at pesticides and child development starting in about the year 2000.

The Salinas "Chamacos"[52] Study

The Salinas valley of California is between Monterey and Santa Cruz, just a few miles from the Pacific Ocean. The novelist John Steinbeck grew up in Salinas, and he portrayed the area in his writings. The city calls itself the "salad bowl of the world" because agriculture brings in over $2 billion per year. The farm workers who help nurture your salad greens are exposed to pesticides on the job.

In 1998, Professor Brenda Eskenazi of the University of California at Berkeley assembled a research team to study the effects of pesticides on farm workers and their families. The project started with approximately 500 pregnant women who were receiving their prenatal care at the clinics that serve mostly farm worker families. Urine samples from the mothers and children were assayed for metabolites of OP pesticides. The women's blood was also tested for DDT, PCBs, and other OC pollutants. The research team was also working with the farm worker community on reducing pesticide exposure by measuring pathways of exposure.

The assessments of the infants have been very thorough. Newborns were tested with the standardized Brazelton Neonatal Behavioral Assessment Scale. Then the infants were tested with the Bayley Scales of Infant Development as 6-month-olds, and at 1 and 2 years of age. When they were 2-year-olds, mothers answered a standardized questionnaire about behavior problems, the Child Behavior Check List. These are all excellent tests that are the gold standard for measuring child functioning.

The Salinas study is starting to show some links between prenatal pesticide exposure and worse infant outcomes. *First*, higher prenatal OP pesticide exposure increased the likelihood of abnormal reflexes in the newborn.[53] This result held after including covariates that were even weakly ($p<.15$) associated with the reflex scores. The chance of an infant showing 3 or more abnormal reflexes increased almost 5 times for each 10-time increase in prenatal pesticides. *Second*, for the 2-year-olds, higher prenatal pesticides predicted lower mental development scores on the Bayley test. Prenatal pesticides were not significantly related to motor development at any age, and were not related to mental development before 2 years of age. *Third*, prenatal pesticides also predicted worse scores on the Pervasive Developmental Disorder (PDD) scale of the Child Behavior Checklist. This scale measures behaviors that might be symptoms of autism such as body rocking, failing to make eye contact, and not being responsive to affection. Child OP exposure measured at the time of tests also predicted higher PDD scores. Finally, a puzzling finding for the 2-year-olds was that higher child pesticide exposure at the time of tests predicted better mental development scores. The research team speculated that perhaps 2-year-olds with higher mental function explore more and come into contact with more pesticides. Another possibility is that a recent exposure to OP insecticides has a positive short-term effect on cognition. Cholinesterase inhibiting drugs are used to alleviate some cognitive symptoms of Alzheimer's disease. Remember that all the other findings of the Salinas study associate worse scores with higher pesticide exposure.

New York City Prenatal Pesticide Study

Are the findings of the Salinas study supported by other research? A study in New York City began by measuring prenatal exposure for 200 infants and has followed children up to age 3.[54] The study measured prenatal exposure to just one pesticide, chlorpyrifos. The study also used the Bayley tests and the Child Behavior Checklist. At 3 years of age higher prenatal chlorpyrifos predicted a higher chance of scoring in a developmentally delayed category on the mental and motor scales of the Bayley test. The 3-year-olds exposed to the highest prenatal chlorpyrifos were also significantly more likely to have attention problems and actual ADHD symptoms. At 1 and 2 years of age there had been no significant effects of prenatal chlorpyrifos on the children's development.

The New York study findings are not identical to the Salinas study, but they are similar. These are the first two prospective studies of pesticides and child development. At this point we don't know exactly what accounts for the differences. Both studies show some effect of prenatal pesticides on outcomes in toddlers. The specific outcomes differ. But so do the cultural backgrounds of the samples. The Salinas participants are from Chicano farm worker families. The New York participants were either African Americans or immigrants from the Dominican Republic. The New York study measured metabolites of only one pesticide. In Salinas the researchers measured metabolites of 80% of the OPs used in the nearby fields. The New York study selected women who weren't smokers, while the Salinas study included smoking as a covariate. Both studies incorporated a good set of family background variables in their analyses.

The most disturbing findings are the increases in symptoms of pervasive developmental disorder in the Salinas study and of ADHD symptoms in the New York study. These kinds of symptoms interfere with a child's overall school and social functioning. Follow-up of these children is very important. The Salinas team is testing their sample for a genetic trait that determines the body's ability to detoxify OP pesticides. This particular genetic trait was associated with autism in another study.[55]

Exposure to Pesticides: How Much and How?

It was 30 years after Rachel Carson's book when the United States got serious about pesticide residues in food. The 1993 National Academy of Sciences (NAS) committee on pesticides and children recommended that "estimates of expected total exposure to pesticide residues should reflect the unique characteristics of the diets of infants and children and should also account for all nondietary intake of pesticides."[56] That committee also recommended better monitoring of pesticide residues on foods and the effects of food processing on pesticide residues. Pesticide residues are usually measured one at a time. But pesticides come to us in unpredictable combinations through residues in our homes, schools and workplaces, and through food, drinking water, and air. Scientists measure exposure externally or internally. External exposure estimates measure the amount of the substance outside the person that has potential to enter the person. Internal exposure is a measure of how much of the substance actually entered the person or has been metabolized and excreted by the person. For lead, mercury, and PCBs, the best studies used internal exposure estimates—blood lead, tooth lead, lead in the umbilical cord, hair mercury or mercury in cord blood, and PCBs in blood or milk samples. The Salinas and New York studies used internal estimates—pesticides in blood or urine. Pesticides in air, dust in the home, dust on children's hands, residues on farm workers' clothes, food samples, and drinking water give us external exposure estimates. Just because a pesticide is in a child's immediate environment or food does not necessarily imply that it is absorbed by the child, although it might be. In order to figure out how pesticides get inside people, we need both external and internal measures.

External Exposure Estimates

Outdoor Air Exposure   Remember the brass player in the band who got sick from the mosquito spray in the park? It happened outdoors. For those who live in proximity to agricultural areas, pesticide exposure can be an every-day event. One of my students grew up in the middle of an apple orchard in Wisconsin. Her parents asked the owner of the orchard to phone them to let them know when he was going to spray. But the phone call rarely came. Instead they would hear the air-blast sprayer coming down the rows of trees. They'd run indoors and shut all the windows, sometimes too late. But what pesticide measurements do the scientists find outdoors?

In the summer in Jacksonville, Florida, 95% of outdoor air had detectable levels of the OP insecticide chlorpyrifos. In Springfield, Massachusetts, 52% of air samples had chlorpyrifos.[57] In the vegetable-growing Central Valley of California, air pollution enforcement officers found that the air in a school had a soil insecticide (called Telone) more than 320 times the level of EPA's "safe" cutoff.[58] The insecticide had never been used inside the school. It came from nearby farms. Partly as a result of this incident, California officials canceled the use of Telone.[59] OP pesticides may drift from the Central Valley to the Sierra Nevada mountains.[60]

But risk assessments for pesticides in air usually conclude that exposure from outdoor air is safe. Those risk assessments examine exposure for one chemical at a time, and extrapolate the risks from animal research. The wide presence of pesticides in outdoor air also indicates the potential for wildlife exposures that extend to nontarget animals in the fields during application.

Indoor Air Exposure: The Florida and Massachusetts Air Exposure Study Indoor concentrations of pesticides are sometimes much higher than the outdoor concentrations. Diazinon (an OP pesticide that was an ingredient in home ant poisons, now phased out by the manufacturer from home use) was more than 30 times higher in indoor air compared to outdoor air in the summer in Jacksonville.[61] In the Springfield area in winter, diazinon was about 3 times higher outdoors than inside, but in the spring diazinon was about 6 times more concentrated indoors than out. Geography and climate influences how the pesticides are being used and, therefore, the exposures of people. The Jacksonville area has a subtropical climate where pests such as roaches, ants, and fleas can be a year-round problem. The Springfield area has a distinct winter season with below freezing temperatures. Many household pests disappear in the winter.

The Florida/Massachusetts study compared estimates of the average air exposure to pesticides to estimates of the exposure to the same pesticides from food. The research team based this on a survey of what the residents said they ate, combined with FDA estimates of pesticides in food. Exposure to pesticides in food is normally greater than air pesticides for the average adult. The exceptions are termiticides and home pesticide products. There are a lot of assumptions behind these calculations, and the authors noted that "because these estimates of exposures via the food route are more indirect than the estimates of the exposures via the air route, they are subject to considerably greater uncertainty."[62]

The researchers also tested carpet dust in a few homes only in Jacksonville. More pesticides were found in carpet dust than had been found in the air of the same homes. One carpet that was 18 years old had very high levels of a number of pesticides. The researchers estimated how much pesticide a child could ingest from carpet dust versus air: "In the worst case scenario involving the 18-yr-old carpet, potential dust ingestion would outweigh air route exposure by a factor of nine for chlordane and four for chlorpyrifos."[63]

These investigators also crunched the numbers for a formal risk assessment. They assumed a 154-pound person (an average man). They used EPA data on how dose relates to cancer risk ("slope factors"). For some of the pesticides there is no estimate of cancer risk, but there is an estimated reference dose, the dose that should not be exceeded in one day. They summarized their risk assessment this way: "For pesticides other than cyclodiene termiticides . . . as compared to thresholds commonly used by the EPA (10^{-6} for cancer risk and one (1) for noncancer hazard), the risks for nonoccupational airborne exposures all appear to be negligible. Of course, the risk estimates are based *exclusively* on estimated *average* concentrations via the air route of exposure, assuming exposure at this level throughout a 70-year lifetime. Moreover, the reference dose and slope factors are subject to considerable uncertainty."[64]

Notice three features of this risk assessment. *First*, it assumed an average adult man, not a child. Children's respiration is higher per body weight than adults. Higher respiration yields higher exposure from air for children. *Second*, the risk analysis used the *average* levels of air exposure to pesticides, not an estimate of the highest 10%, 5%, or 1% of the exposure values. A risk assessment based on the average exposure will usually understate the risk for half of the population. *Third*, the risk assessment was calculated for each of the pesticides separately. These decisions have a dramatic impact on the results of the risk assessment.

What if we just assume that the pesticide risks are additive? No one knows whether cancer risks are additive, and very little is known about the acute or chronic effects of pesticide combinations. For the Florida/Massachusetts study, I summed the cancer risks and the hazard index values. For Jacksonville, the excess cancer risk due to the average daily air pesticide concentration is 4 in 10,000. Most of the risk is due to the banned (but environmentally persistent) organochlorine termiticides. For Springfield, the cancer risk is 1.1 in 10,000, mostly due to the termiticides. The EPA says a cancer risk that exceeds 1 in 10,000 is hazardous.[65] Both locations exceed this value, and that is for *average* exposure in air.

The hazard index includes risk other than cancer, in principle. If a hazard index is less than 1.0, the EPA considers it "negligible." The sums of the hazard indices were 1.53 and 1.03 for Jacksonville and Springfield, both above 1.0.

In Jacksonville, the hazard index without termiticides is 0.11, and in Springfield it is 0.4, both well below 1.0. But I have already pointed out that these risk assessments are based on the average daily concentrations, the adult respiration rate, and no consideration of other sources of pesticides. The authors of the study concluded that exposure from air was lower than exposure from food. If we added exposure to pesticides in food, the picture would probably change pretty dramatically. And these risk estimates for pesticides were calculated prior to any studies of long-term risks to children's behavioral development. As you will see below, other researchers compared the exposures for the top 5% of the population to the EPA reference doses, rather than using the average exposure. These different decision standards embedded in the research make a difference in the conclusions.

Internal Exposure Estimates

There are two ways to measure internal exposure to OP and carbamate pesticides: cholinesterase inhibition and pesticide metabolites in urine. Normal cholinesterase activity, measured from blood samples, varies widely among individuals. Unless someone has been acutely poisoned, cholinesterase inhibition isn't very useful. Nevertheless, cholinesterase inhibition is the main measure of poisoning in farm workers. One issue with measuring metabolites in urine is that several different pesticides produce the same metabolite. Another issue is that urine varies considerably in concentration. The concentration of pesticide metabolites found in urine must be standardized in some way, and different ways of standardizing may yield slightly different results. Plus, one substance in urine used to standardize (creatinine) seems to vary both seasonally and among ethnic groups, perhaps due to diet or liquid intake.[66] Another problem is that laboratories' tests for the metabolites of pesticides in urine differ in their sensitivity, especially at the lower limit, or "limit of detection." We saw this same issue in the PCB chapter: different labs, as well as the evolution of the technology over time, make the PCB levels difficult to compare across studies. Some measures of pesticide exposure, especially the percentage of people in whom the pesticide is detected, will vary between studies just because the sensitivity of laboratory tests differ.

Arizona Children's Exposure to Pesticides Survey   A team of researchers from the University of Arizona and from the EPA measured the urinary metabolites of pesticides in children 2 to 5 years old in Yuma County, Arizona.[67] Yuma is an agricultural area that borders California and the northwest corner of the state of Sonora, Mexico (see Figure 4.2). Crops are grown all year, and pesticide drift from the fields is a concern. To be eligible to participate, the children had to be toilet trained. The participation rate was high: 140 of 150 families contacted agreed to participate. All participating families were Hispanic, and all but a few preferred to answer the questionnaires in Spanish. The urine was tested for 6 metabolites of OP pesticides with the limit of detection at 25 ppb (or 25 μg/liter).

One-third of the children had detectable levels of urinary metabolites of at least one OP pesticide. The child in each age bracket (3-, 4-, and 5-year-olds) in their study with the highest levels of OP pesticide metabolites had exposure between 25 and 589 *times* the EPA's daily reference dose for methyl parathion (an insecticide commonly used on many kinds of crops). The calculation assumes that all of one metabolite came from methyl parathion. Because different OP pesticides can create the same metabolite, this might be an overestimate. Nevertheless, it is shockingly high. For malathion, only the 5-year-old with the highest metabolites exceeded the reference dose (by about 6 times). For diazinon and chlorpyrifos, one 5-year-old in the study exceeded the reference dose by 4 times (chlorpyrifos) or 126 times (diazinon).

The team of scientists also estimated the exposure for the upper 5% of children living in this area. This gives a more conservative estimate of overexposure than just looking at the single most highly exposed child. Estimating the upper 5% exposure implies we want to protect at least 95% of the children. Looking at only the average exposure implies we want to protect the average child. The upper 5% estimates still showed daily dosages of two pesticides that exceeded the EPA's reference dose: diazinon (10 to 24 times) and methyl parathion (11 to 49 times the reference dose). Where is this high exposure coming from? Parents who work in the fields can track home pesticides on their work clothes and possessions. Another possible source is pesticide drift from the fields.[68]

What are the assumptions behind these results? First, the research team assumed that *all* the metabolites of a pesticide would come out in the child's urine. If not all of the metabolites are excreted in urine, then the calculations would underestimate exposure. Second, it was assumed that a given metabolite was produced by only one particular pesticide. Most of the urinary metabolites tested were not specific to single OP pesticides but common to several. The researchers describe this kind of urinary testing as "screening" and acknowledge its limits for inferring the dose of any particular pesticides. But, on the other hand, the EPA has been criticized for setting reference doses for individual OP pesticides, rather than for a class of pesticides that have similar effects on our bodies.[69] Two OP pesticides that create the same urinary metabolite are likely to be rather similar in their biological effects.

Minnesota Children's Pesticide Exposure Study Children from 3 to 13 years of age were sampled in Minneapolis, St. Paul, and in two counties to the south of the Twin Cities. The data were collected during the summer. To measure pesticides in food, the families submitted a plate of food that was a duplicate of that eaten by the child. Metabolites of four pesticides were also measured in urine. One was a herbicide (metabolite of atrazine), one was the metabolite of a major ingredient of mothballs, and the other two were metabolites of the OP pesticides malathion and chlorpyrifos. Both of these insecticides are used on food crops and were used in household pest products. (Chlorpyrifos is now restricted from indoor home use except by licensed pesticide applicators.) Ninety-three percent of the sample had detectable metabolites of chlorpyrifos, and 37% had detectable malathion. Chlorpyrifos was higher in urban children than in nonurban children. This is because chlorpyrifos was a widely used indoor roach spray. Malathion exposure also showed a slight trend toward higher levels in the urban children. Nonwhite children had higher levels of malathion metabolites than white children. Compared to studies that have measured urinary metabolites in adults, the children were more likely to have higher levels of both malathion and chlorpyrifos.[70]

Children of Orchard Workers in Washington   Apples are the targets of a variety of insect pests. Orchards are often sprayed more than 15 times per year with between 9 and 16 different pesticides, mostly insecticides.[71] A University of Washington research team is measuring pesticide metabolites in the urine of children in agricultural and nonagricultural households near orchards. The agricultural children had approximately 5 times the OP pesticide metabolites of children from nonagricultural families. The homes of children in agricultural families also had higher levels of OP pesticides in house dust. The closer the family lived to a treated orchard, the higher the exposure to OP pesticides. Approximately 17% of the agricultural children had detectable pesticide residues on their hands. None of the nonagricultural children had pesticide-laden hands.[72]

The University of Washington team also estimated total doses of OP pesticides. Fifty-six percent of the agricultural children's total OP dose would exceed the chronic reference dose for azinphos methyl and 9% would exceed the chronic reference dose for the pesticide phosmet. Compared to the acute reference doses, 35% and 7% of the agricultural children are overexposed. The research team concluded: "The data presented here demonstrate that OP pesticide exposures among children in agricultural communities fall into a range of regulatory concern and require further investigation."[73]

The Washington researchers also pointed out that the pesticide exposures that agricultural children receive from being close to the orchards are over and above their pesticide exposures from food, water, and normal home pesticide use. OP pesticides are normally considered one chemical at a time, but the doses of all OP pesticides are cumulated by the human body because they have the same mechanism of action: cholinesterase inhibition. Figure 4.3 shows a diagram similar to one created by that team of scientists to capture these ideas.

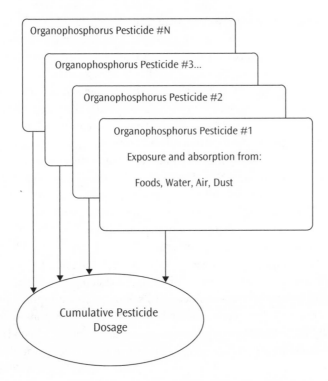

Figure 4–3. Cumulative exposure from different organophosphorus pesticides. *Source:* Adapted from Fenske et al., 2000, *Environmental Health Perspectives*, 108, 515–520. Reproduced by permission.

Pesticides in Food: Is Certified Organic Better?

In 2000 the U.S. Department of Agriculture's organic standards rule became effective. The rules for certified organic crops prohibit the use of the OP and carbamate pesticides, along with other chemical insecticides and herbicides. Organically grown food should be lower in pesticide residues. Is it? The University of Washington team recruited families who said they fed their 2- to 5-year-old children organically grown food. They compared them with families recruited from the same grocery store who said they fed their children conventionally grown food. The organically fed children had only one-ninth of the OP pesticide metabolites of the conventionally fed children. This was true for the metabolites of the OP pesticides that are most commonly used on food.[74] Even though the sample was only 43 children, a difference of the size found would occur by chance only 2 in 10,000 times.

But the families that fed their children organic versus conventional food did report some differences in using pesticides in their homes. To rule that out, the Washington team did an experiment with families eating conventional food. The children in these families gave urine samples for 15 straight days. During a 5-day streak, the research team provided organically grown foods from a shopping list chosen by the family. The metabolites of chlorpyrifos and malathion both fell to undetectable levels when the children were eating organic foods. Metabolites of other OP pesticides also fell.[75] Because this study compared the same children when they were eating different diets, the results must be due to the dramatically different pesticide residues in organically versus conventionally grown food. The research team concluded that OP pesticide exposure of suburban children is primarily through food. "We conclude that organic diets provide a protective mechanism against OP pesticide exposure in young children...Such protection is dramatic and immediate."[76]

Other scholars have questioned whether organic foods are safer. All plants have some natural toxins in them. Some of these toxins are produced when the plant is damaged by an insect. Pesticides are intended to reduce insect damage. Therefore, pesticides might make plants healthier.[77] The Washington research team replied that organically raised plants have higher concentrations of some positive chemicals such as antioxidants.[78] Tomatoes grown organically have higher concentrations of antioxidant flavonoids than conventional tomatoes.[79] The Washington researchers also pointed out that organic food production has other benefits. Land is used in a way that retains and enhances fertility. Farm worker pesticide poisonings are avoided.

The Problem of Acute Poisoning by OP and Carbamate Pesticides

DDT had the advantage of low *acute* toxicity to mammals. OP and carbamate insecticides have much higher acute toxicity. For this reason, their wide use has been accompanied by human poisoning incidents, especially as organochlorines were phased out. Users were not always aware of the higher acute toxicity of OP and carbamates.

Poisoning Symptoms in Children May Differ from Adults

Remember that the symptoms of OP or carbamate poisoning are summarized by the acronym MUDDLES: Miosis (pinpoint pupils), urination, diarrhea, diaphoresis (profuse sweating), lacrimation (tearing eyes) excitation, and salivation. In younger children the symptoms may differ. Unless the source of the child's exposure is known and told to medical personnel, life-threatening delays in treatment can occur. An Israeli study of insecticide-poisoned children found that the major symptoms are stupor or coma and muscle weakness. But usually the children didn't show muscle twitching (fasciculations) that is seen in older poisoning victims. Gastrointestinal symptoms occurred in about 50% of children, salivation and tearing in about 60%. One of the classic signs of poisoning with an AChE inhibitor is sweaty, cold skin. Only about 50% of the children in the Israeli study showed that symptom. About 75% showed pinpoint pupils, but a few of the children had dilated, nonreactive pupils.[80]

The problem of diagnosing pesticide poisoning in children can be exacerbated by cultural differences between the medical staff and the families of the victims. The doctors in Israel said that parents who brought the poisoned children to the hospital told about the poisoning in only three cases. Some parents strongly denied that the child could have been poisoned, apparently because the particular pesticides were illegal in Israel. Of the 25 cases that were studied, only 1 child was Jewish. The others were of Bedouin or other Arab cultures. This exacerbated communication problems with the doctors.[81]

Pesticide Poisoning: A Global Problem

Pesticide poisonings of people, whether accidental, suicidal, or homicidal, are a problem in most of the world. Pesticides are not very well regulated in many industrially developing countries, and this factor in combination with low literacy rates can be disastrous.

WHO (the World Health Organization of the United Nations) estimated approximately 1 million accidental pesticide poisonings and 2 million suicidal pesticide poisonings in the world.[82] That means approximately 1 out of every 6,000 people are poisoned each year. The poisoning rate and causes of poisoning vary across different countries. The 3 million global poisonings are estimated to cause 220,000 deaths, and almost 99% of the deaths are thought to occur in industrially developing nations.[83]

In the United Kingdom between 1982 and 1991 researchers estimated that between 1,500 and 1,850 children were sent to emergency rooms because of suspected pesticide poisoning. Between 350 and 450 were admitted to the hospital each year, but there were no deaths. More than 95% were children under 6 years of age. Rodent poisons were the most frequent causes (41%); insecticides were responsible for about 35% of cases; herbicides for 13%, and other pesticides (mothballs, creosote) for the remainder.[84] In the United Kingdom studies, pesticide poisonings accounted for about 6% of all children's emergency room visits due to incidents that occurred at home. In the United States in 1993 approximately 70,000 pesticide "exposures" of children less than 6 years old were reported to poison control centers.[85]

What kinds of pesticides are involved in poisoning incidents depends on the part of the world. Organophosphate pesticides accounted for almost 80% of poisonings in China, 70% in Sri Lanka, and 50% in Malaysia.[86] In Israel, about 7% of all children's poisonings are due to OP and carbamate pesticides.[87]

Case Reports of Pesticide Poisonings

Here are some examples of medical publications of pesticide poisoning case reports of children, as well as adults. These incidents help us understand the variety of situations in which poisonings can occur.

Children and Pesticide Containers in North Carolina   A 1975 paper reported eight child pesticide poisonings in North Carolina.[88] In the year the paper was published, 70% of North Carolina farmers disposed of empty pesticide containers by throwing them in the woods or leaving them in the fields. A seemingly empty 5-gallon container can easily contain about half an ounce (about 14 grams) of parathion residue.[89] The authors of the North Carolina study argued for better packaging and labeling of pesticides to prevent accidental contact, especially by children. Here are some cases from that paper.

A 7-year-old girl needed hospitalization after playing with "empty" pesticide containers in the yard. The containers had collected rainwater. The girl made "mud pies" with some of the water the day she became ill. A few days earlier, the family dog died suddenly, probably from drinking out of a discarded container.

Two boys, aged 1 to 2 years, became gravely ill after playing in a 55-gallon drum that was empty. It was empty except for pesticide residue.

An 8- and 11-year-old were poisoned by cleaning out the trunk of a car in which a glass container of the OP pesticide methyl parathion had broken. The 8-year-old died.

A 4-year-old boy and his two cousins were poisoned by playing in powdered insecticide. The 4-year-old put the insecticide powder on his face and hands. He was in a coma by the next morning, suffered respiratory arrest in the hospital, and had to be put on a ventilator. He recovered after treatment. No follow-up on his behavioral function was reported.

Aerosol roach spray became a toy to two toddlers. They took turns spraying it in each others' mouths. When taken to the hospital they were comatose but recovered with treatment. No follow-up on their behavioral development was reported.

Improper Indoor Insecticides Poisoned Newborn   In 1980 in San Francisco, an 11-day-old newborn boy was admitted to the hospital with pinpoint pupils, limpness (no muscle tone), heavy salivation, and no response to pain. He was put on a ventilator. After the doctors discovered that his home had been professionally sprayed for roaches and termites, they treated the baby for pesticide poisoning. After eight days in the hospital the infant recovered. Inspection of the home by the California Department of Food and Agriculture found the pesticide chlorpyrifos (Dursban) on dish towels, the infant's clothing, and on food preparation surfaces. The U.S. Centers for Disease Control (CDC) commented that chlorpyrifos has a long half-life indoors (greater than 30 days) and that it is illegal to use it on food preparation surfaces. Further, the CDC said that the case punctuates both the higher susceptibility of infants and young children to pesticides compared to adults and the importance of using pesticides carefully in the home.[90]

Methyl Parathion Indoors   Methyl parathion is an OP pesticide registered for outdoor use on agricultural crops. Because it is cheap and effective, it attracts illegal indoor use.

In 1984 Mississippi investigators found that a home had been sprayed with parathion. Two weeks later seven siblings were taken to the local doctor with diarrhea and abdominal pain. It just seemed to be a gastrointestinal illness. But within 2 days two of the children stopped breathing, and both died after being taken to the hospital. The adults living in the home did not experience symptoms. An investigation showed that an "empty" container of methyl parathion had been obtained from a farm by one of the adults. The adult mixed the residue with water and sprayed to control spiders. The mix was about 3 times stronger than would normally be used outdoors on crops. Air in the home had 100 times the concentration of parathion that would be found in the air in a field during spraying. The home did not have running water. The family obtained its drinking water from a neighbor's well and kept it in an open ice chest on the back stoop. The ice chest was the source of water to mix the pesticide, too. The water in the ice chest was found to contain parathion residue. The public health officials concluded that the contaminated drinking water was the source of poisoning. The adults in the home had been drinking mainly bottled soft drinks, while the children had been drinking water. My own conclusion is that poverty was a contributor. Had there been running water in the home, the poisoning would have been much less likely despite the illegal pesticide use.[91]

The EPA has found other incidents of illegal indoor use of methyl parathion: 232 homes in Ohio in 1994, four homes and a homeless shelter in Detroit in 1995, more than 1,100 homes in Jackson County, Mississippi, and approximately 1,000 homes in Chicago. Symptoms of low-level chronic exposure to methyl parathion or other OP pesticides are flulike, including weakness, loss of appetite, nausea, malaise, and headache. Sometimes the EPA ordered emergency relocation of the residents. Decontamination and renovation of the illegally sprayed homes can cost $45,000 to $85,000 per home. In the Jackson County case, the two individuals who did the spraying were charged with 30 federal counts of mishandling pesticides. The EPA said, "The full scale of the problem of indoor methyl parathion use is not known."[92]

Pesticide-Caused Food Poisonings

Pesticides can get into foods from use of the wrong pesticide on a crop, use of pesticides too close to harvest time, contamination during shipping, and contamination during food storage or preparation. Cases occur all over the world.[93]

In 1985 Aldicarb caused illnesses through its use on watermelons. Aldicarb, which is not licensed for use on watermelons, was found at the level of 2.7 ppm in the watermelon that caused illness in one family. After three days of investigations, the California Department of Health Services destroyed all the watermelons in the California supply chain. Laboratory tests showed that 4% of the melons were contaminated. six hundred ninety-two cases of illness, including 17 hospitalizations, 6 deaths, and 2 stillbirths occurred in people who ate the affected watermelons.[94] Many of the people who got sick ate only a single slice of watermelon. Similar incidents occurred with aldicarb in cucumbers in 1978 in Nebraska, and in 1987 and 1988 in California. The California researchers concluded that "it appears possible that illness may be caused by produce with contamination below the detection level of screening assays that are used by regulatory agencies."[95]

In Singapore, 105 people were seen in hospital emergency rooms in a sudden outbreak of gastroenteritis. The poisoning was traced to OP and carbamate residues on *gai-lan* vegetables from Malaysia. The remainder of the shipment of vegetables was seized and destroyed.[96]

In 1998, 60 men in India all became ill after they ate at a community lunch. Four had to be hospitalized, and 1 died 9 days later. The kitchen had been sprayed with malathion, an OP insecticide, the morning of the lunch. The flour and other foods were in open jute bags in the kitchen during the spraying. Tests showed malathion in some of the food, and no botulism. The authors commented, "In developing countries, the widespread use of organophosphates has been accompanied by an appreciable increase in the incidence of poisoning with these agents.... The increasing and *indiscriminate use* of organophosphates as agricultural and household insecticides without any accompanying public education about their storage and safe use increases the potential for more outbreaks of food poisoning."[97]

Incidents of Agricultural Worker Pesticide Poisoning

People who work in agriculture are sometimes accidentally exposed to high levels of pesticides. Even though pesticide illness of agricultural workers must be reported by law, the frequency of reporting such incidents is low.[98] Reports of poisonings help estimate how much of a public health problem such incidents are. Reports help protect workers and their family members. Here are a few examples of agricultural pesticide illnesses.

In 1981, 44 workers harvesting iceberg lettuce started having symptoms of AChE inhibition after about 2 hours of work. The field had been aerially sprayed with mevinphos (trade name Phosdrin) that morning even though the spray order had been canceled. Thirty-one workers and 3 produce inspectors went to the hospital. They were told to undress and were "hosed down." Two were hospitalized. The rest were told to wash their clothes after going home. They were sent home wearing their contaminated clothes. Several were unable to go to work the next day.[99]

In 1993, the Washington Department of Health (WDH) received 26 separate reports of people who became ill from applying mevinphos to kill aphids in apple orchards. Most of them were using an airblast system in which a tractor pulls a sprayer through the orchard. All required medical treatment, and four required intensive care. All of the workers were men, and approximately 70% were Hispanic. In 78% of the cases, the full protective equipment required by the label was not used: gloves, goggles, or respirator were removed at some point during handing, or leather boots were worn rather than rubber ones. But in the other 22% of the cases, the required protective procedures were carried out.

In 1998 California workers were sent into a cotton field 2 hours after it had been sprayed aerially with a mix of pesticides including carbofuran (a carbamate pesticide that is an AChE inhibitor). The reentry time for carbofuran is 48 hours after application. After about 4 hours of work, the 34 workers were moved to another field. The other field had been sprayed 2 days earlier with a mix of 3 pesticides (a pyrethroid insecticide, an organochlorine, and a growth regulator). The reentry time for the second field was well within compliance (the regulation reentry time was 12 hours, and more than 48 hours had passed). After about a half hour the workers began to feel ill with nausea, vomiting, muscle weakness, tearing, and headache. Most of the workers were sent to hospitals. All but four were released that day. Initial blood samples failed to show cholinesterase levels below normal. The lab had not put the specimens on ice for transportation to the analysis location. Other blood samples drawn by a different clinic showed below-normal cholinesterase. The CDC commented that "several of these pesticides have been associated with adverse effects in animals, but reliable data for humans are lacking. The toxicity related to combined exposures to pesticides remains unresolved and requires further research."[100]

In 2006, twenty-seven workers in a California vineyard became ill when cyfluthrin (a pyrethroid) was sprayed in the citrus orchard nearby. Workers were decontaminated by the HAZMAT team. Symptoms were headache, nausea, eye irritation, muscle weakness, anxiety and shortness of breath.[101]

Greenhouse workers are also at risk. In 1995 a Texas greenhouse worker became ill after using an OP fumigant, Sulfotepp. The worker was wearing all recommended protective equipment, including the right type of respirator. The Texas Department of Health sent an observer to the greenhouse. The next time Sulfotepp was used, two workers became ill even though they too used the correct protective equipment. Approximately 80% of greenhouses in Texas use fumigants, and 7% of the companies using fumigants reported that a worker had been ill during the process. The Texas Health Department recommended to the EPA that the label on Sulfotepp be modified to indicate better respirator protection.[102] Pesticide labels are legal documents. Violating the label is against federal law.

Non-agricultural Worker Pesticide Poisonings

In 2007 California's Department of Health Services received 17 reports of insecticide illness in flight attendants in one year. Pyrethroid insecticides were "fogged" in the cabin of the airplanes in Sydney, Australia, prior to departure for Los Angeles. In five cases the flight attendants had heart palpitations, nausea and itching. Almost all had headaches and all sought medical care. The report concluded that applying pesticides in an airline cabin poses a human health hazard.[103]

Over 300 retail workers were poisoned in a 5-year period in 6 states. There was one death. Insecticides were responsible for 44%.[104] How did these accidents happen? Sometimes the clerks or customers dropped a bottle of pesticide, or containers broke during shipping and receiving. Bakery and deli workers are at risk from spills and splashes of disinfectants such as ammonia and chlorine. Disinfectants are classed as pesticides.

Are There Long-Term Neurobehavioral Effects on Workers?

Whether chronic low-level pesticide exposures have negative effects on adults has been controversial for several reasons. First, the research has yielded inconsistent results. Second, a sound study of pesticide exposure is difficult to accomplish. One difficulty is that pesticides usually contain solvents or other carrier substances. These carriers can also have effects on people's mental functioning. It is also difficult to track people's exposure to pesticides because they are used in complex combinations. Also, for some highly exposed populations, such as field workers, it is difficult to find a good comparison group. A good comparison group would be equivalent in age, education, income, and cultural background. Here I will describe just one study of pesticide applicators as an example.

Professional pesticide applicators in North Carolina who used the OP pesticide chlorpyrifos to treat termites participated in a battery of tests. Eight of the 191 applicators said they had experienced poisoning symptoms while using chlorpyrifos, and one of them had sought medical treatment in the past. The pesticide applicators and nonexposed[105] comparison people took tests of visual acuity, color vision, sense of smell, sensitivity to vibrations, postural sway (ability to stand without swaying), fine motor dexterity, eye-hand coordination, nerve conduction velocity, self-reported moods and symptoms (such as memory trouble, irritability, anger, depression, lack of coordination, and so on), vocabulary test, short-term memory, attention, and neurological abnormalities. The researchers also collected a urine sample to test for metabolites of chlorpyrifos.[106]

The termiticide applicators reported more fatigue and tension, more dizziness, confusion, memory difficulties, irritability, uncoordination, weakness in the limbs, and headaches compared to the nonexposed group. The applicators performed slightly worse on the vocabulary test, even after education was included in the analyses. The two groups also differed in their fine-motor skills, with the applicators performing more poorly. Postural sway was significantly related to the concentration of pesticide metabolites in urine. The groups did not differ significantly on the other measures. The researchers concluded, "We found few exposure-related effects for most tests....However, the exposed subjects did not perform as well as the nonexposed subjects on pegboard turning tests and some postural sway tests. Furthermore, exposed subjects reported more symptoms than nonexposed subjects; this is a cause for concern because previous studies lend some support to this finding."[107]

Self-report symptom questionnaires can be a very good window on people's behavioral well-being and quality of life. Your quality of life is affected when you *feel* ill, and only you know how you really feel. Symptom questionnaires are sensitive to psychological stress and other events that affect our lives (see Chapter 6 on pollution disasters). Furthermore, much of medical diagnosis is based on self-report. For example, "Doc, I have a splitting headache." There is as yet no objective headache monitor. Mental illnesses are also diagnosed primarily by self-report using interviews and questionnaires. Symptoms of depression such as poor sleep, irritability, loss of appetite, loss of interest in sex, feeling sad, feeling like life is pointless, and so on, are determined by self-report. The drawback is that self-report measures can be faked if the person decides he or she wants to appear to be ill.

Home Safety: Are Labeling
Requirements Effective?

The label of a pesticide container is federal law. Do people read and follow the safety warnings on containers? Household pesticide containers are now labeled, "Do not re-use empty container. Wrap container and put in trash collection. Do not contaminate water by disposal of waste." Other cautions are printed on the label. But on an old container of flea powder I found at home, the price tag was placed right on top of the cautions! I had not peeled it off to read the warnings.

Pesticide manufacturers argue that pesticides are safe when they are used in accordance with the label. Many of the actions taken by the EPA involve changing the labeling of a pesticide. For example, Dow Chemical, the manufacturer of chlorpyrifos (often sold under the name Dursban), agreed voluntarily in 1997 to relabel the product so that it would not be used as a broadcast treatment in homes.[108] But as the indoor use of parathion shows, pesticides are sometimes misused.

How do labels affect pesticide users? What level of visual acuity and reading ability is necessary to read the label? Do people read the labels on pesticide containers? Do people say they follow the instructions on the labels?[109] To answer the first question, the researchers examined the labels of a variety of pesticide products. The visual acuity needed to read the labels was greater for general-use pesticides than for restricted-use pesticides, and insecticide labels required better vision than herbicides or fungicides. Approximately 25% of Americans would not have sharp enough eyesight to read the labels on 15% of the containers. On average, reading ability at the 11th grade level is needed. About 40% of Americans would have difficulty understanding the label. The research team also surveyed a sample of adults, approximately half of them registered pesticide applicators, about whether they read and understood pesticide labels. Nearly one-half of the sample said they always read all the label on a pesticide, and 57% said they always read at least part of the label. More than 10% said they do not read the entire label because they do not understand it. Forty percent said that the information on the label was "irrelevant." Thirty percent said they already knew how to use the product. About two-thirds of the sample said they always followed the instructions on the label, and 29% said they "usually" followed the instructions. Among those licensed to use restricted pesticides (mostly farmers and ranchers), 72% said that they "always" followed the label directions. The rest said they "usually" followed label directions. In the general population, a slightly lower percentage, 60%, said they always followed directions.[110]

Social Justice Issues

Involuntary exposure to pesticides, exposure of agricultural workers, exposure of residents of substandard housing in the inner city, and pesticide use in industrially developing countries with high illiteracy rates and poor worker training are social justice issues.

Inner-city residents and families who work in agriculture are most likely to have the highest pesticide exposures. Children living in agricultural areas can be routinely exposed to pesticides at levels that exceed current government standards. Just as poverty in America is associated with childhood lead exposure, children in low-income families in large cities are also likely to be exposed to higher levels of pesticides. In the state of New York, the heaviest use of pesticides occurs in Manhattan and Brooklyn. The New York City Housing Authority routinely sprayed all its apartments with insecticides monthly.[111] Pesticides that are sprayed in crevices to poison cockroaches also accumulate on absorbent surfaces in the rest of the home, such as carpets, textiles on furniture, and toys with soft fabric. Children are likely to receive higher exposure than adults from such pesticide "fallout." Children crawl on the floor, put their hands and other items in their mouths, and hug their toys.[112] Recall that the Minnesota pesticide exposure study showed that children in the Twin Cities had higher pesticide levels than nonurban children. Inner-city children also have a high incidence of asthma, and asthma is sometimes associated with allergy to cockroaches.[113] There are also reports in the medical literature of an association between asthma and exposure to organophosphorus and carbamate insecticides.[114]

Pesticides and Wildlife

OP and carbamate pesticides are nonspecific "biocides" with the potential to kill any animal, not just insects. In *Silent Spring* Rachel Carson documented the wildlife effects of indiscriminate use of pesticides in the United States in the 1950s to the 1960s. That was when DDT was widely used. What are the current problems with pesticide uses? There is no doubt that many nontarget animals are affected negatively by pesticide use on farms, forests, parks, and golf courses.[115] Pesticides can affect nontarget animals directly, by making them ill, killing them, or inducing OPIDN, or indirectly, by affecting their reproduction, the availability of food, or other aspects of habitat and behavior. One pesticide, Diazinon, was withdrawn from broadcast use on residential lawns because it caused bird die-offs.[116]

Some wildlife researchers think the negative effects of pesticides on birds are dramatically underestimated.[117] *First*, humans must be present in a sprayed area to see birds killed. But it is illegal to enter sprayed fields immediately after a pesticide application. *Second*, dead birds are scavenged very quickly by predators. Within 24 hours predators remove at least half of bird carcasses in a field. *Third*, when more birds are killed, more predators are attracted and more carcasses removed. Therefore, the numbers involved in dead birds will be underestimated. *Fourth*, when birds are sickened by pesticides, they seek a hiding place. If they die in their hiding places, they will not be counted in the toll of a pesticide kill. *Fifth*, birds that are not killed but become ill may be more easily captured by predators and not counted as pesticide victims. *Sixth*, birds that do not live in the area that was sprayed may also be killed by visiting the field to feed there. If they leave the area to die, they will not be counted. *Seventh*, people searching for dead birds after a pesticide kill find only 30%–80% of those that are present. Researchers discovered this by placing bird dummies in a field, and having searchers look for them. *Eighth*, search intensity affects carcass recovery. But when wildlife authorities are notified of a pesticide killing, they rarely have more than a couple of staff members to help with the investigation. Many of these principles apply to estimates of the effects of pesticides on nontarget insects and animals other than birds as well. The upshot is that the effects of pesticides on animals are very difficult to estimate.

Does the fact that unknown numbers of animals are killed by pesticides imply that the use of pesticides should be decreased? The answer will depend on your own ethical standards and your own decision standards for what constitutes *unreasonable* harm to the environment.

One argument is that it is unethical to harm *individual* sentient animals. By that criterion, being vegetarian is not enough. Eating food grown through conventional agriculture that uses pesticides, or purchasing a Christmas tree grown conventionally,[118] indirectly harms sentient animals. Pesticides also affect behavioral functioning and "quality of life" in birds. Abandoning territory, and a decrease in the amount of time spent singing, gathering food, and feeding nestlings are behavioral changes that can occur due to pesticide.[119]

Another point of view argues for protecting natural resources, including animal populations, rather than individual animals. Are *populations* of animals and biodiversity being harmed by pesticide uses in agriculture, forestry, homes, and recreation? This is a much more difficult question to answer. There is great uncertainty in the estimates of deaths due to pesticides, as I described above. If scientists cannot accurately estimate the effects of specific pesticide applications on localized populations of birds and animals, how can they estimate the effects on larger populations or biodiversity? Some argue that the use of pesticides in conventional agriculture has actually helped stop global species loss. If the chemicals allow higher crop yields per acre, this would reduce the loss of other habitats, such as forests, prairies, and wetlands. The other side of this argument is that conventional chemically intensive agriculture damages the environment and people in ways that are not included in its "costs." For example, insect resistance to chemicals leads to more concentrated or more frequent applications of chemicals.[120] Certified organic agricultural methods can produce yields as large as chemically intensive methods.[121]

At least one study has shown greater biodiversity of bird species in organically managed as opposed to conventionally managed orchards.[122] Ornithologists have concluded that some species of birds are declining globally or in some areas.[123] The contributions of pesticides and habitat loss as causes of the decrease of bird populations are unknown. Effects of pesticides on beneficial insects and small mammals are also largely unknown.

Conclusion

The research on the effects of OP and carbamate pesticides on children's psychological development is just beginning. The findings so far show some negative effects on infants and toddlers. Because OP and carbamate pesticides alter the levels of ACh and cholinesterases, there is reason to believe that these categories of pesticides do harm early brain development. Animal research supports this. Estimates of exposure to OP and carbamate pesticides show that most children are more highly exposed than adults. Families near agricultural areas and in urban settings are more highly exposed than are others. Children's exposure to some OP pesticides sometimes exceeds the EPA's reference dose. The reference dose itself is based on animal studies. But the reference dose does not include studies of animal brain changes or the recent behavioral studies in children.

The good news is that two OP pesticides widely used in residences, diazinon and chlorpyrifos, have been phased out for some uses. But what will replace these substances: other less thoroughly tested substances, or least-toxic alternatives? The bad news is that no studies have been started on how children's behavioral development might be influenced by herbicides, neonicotinoids, or pyrethroid insecticides.

CHAPTER 5
Noise: A Barrier to Children's Learning

Everyone knows that noise that is loud enough can damage a person's hearing. This includes iPods and rock concerts. But those of you who live near loud factories, airports, busy highways, or train tracks know that noise can affect your whole life. A factory about a mile from my house has a ventilation system that makes it too loud for people who live close to sleep during the night. One neighbor wrote this noise diary in the on-line chat: "Tuesday: 1 AM thru 9:45 AM, LOUD, lowered volume at 8 AM.... Friday: Yippee...quiet all day and all night!"

What are the effects of living or going to school in high noise? In this chapter I concentrate mainly on the "nonauditory" effects of noise, or how noise affects well-being aside from its effects on our hearing organs. The nonauditory effects of noise in children fall in two main categories: reading and other aspects of cognitive performance, and stress-related responses such as annoyance, blood pressure, secretion of stress-related hormones, and mental health. These two categories are also linked to the auditory effects of noise. A very important auditory effect of noise is interference with the intelligibility of speech. Hearing other people talk is critical to children's early language development. But trying to hear someone talk in a noisy environment can also impair memory and other aspects of cognition. Cognitive performance can be impaired because of the extra effort required to decipher speech.[1] Mood can also become more negative, and the person feels tired or stressed.[2] Keep in mind that the auditory and nonauditory effects of noise are often interlaced.

Living or going to school in high noise can affect children's reading. Poor reading can cascade into poor overall academic performance and lack of motivation in school. One important principle of developmental psychology is that negative impacts early in life, if combined with other negative influences, can affect the direction of the rest of the child's development.[3] Poor reading is a negative factor that can lead to other academic and social problems. So anything that affects children's early reading needs careful consideration. As you will see, the research regarding noise and reading is not unequivocal. Researchers do not agree about why noise can impede the process of learning to read. Some think noise affects children's ability to discern important speech sounds, an auditory effect of noise. Others think noise affects children's ability to concentrate or persist at difficult tasks, a nonauditory effect. Both of these processes could be involved. But regardless of the specific cause, poor reading can be the first in a chain of events that ultimately affect a child's later quality of life—educational attainment, employment and job performance, enjoyment of literature, exposure to a wide variety of ideas through books, and so on.

Another type of cascading effect on later development can begin with exposure to excessive noise prenatally or during early infancy. We have seen that prenatal exposures to PCBs (Chapter 3), mercury (Chapter 2), and pesticides and nicotine (Chapter 4) can alter a child's later functioning. Well-designed long-term studies of prenatal exposure to noise and later human development have not really been done yet. But, evidence from both animal and human research suggests that prenatal noise may have negative auditory and nonauditory effects on development.

After reviewing the research on how noise affects children's behavioral health and well-being, I examine U.S. noise policy and how it fails to consider the effects of noise on children's development. Before addressing the effects of noise on development, first I cover a little bit of technical information about sound and how it is measured.

How Sound Is Measured

Noise is the term for sounds that we do not want to hear. Sounds are alternating compressions and expansions (waves) transmitted through the air, water, or solids. Two characteristics of sound waves are important for understanding this chapter: frequency and amplitude.[4]

The *amplitude* or intensity of a sound is measured in *decibels*. This influences how loud a sound seems when we hear it. There are different decibel scales. The dBA scale weights the different frequencies of the sound in a way that takes the average sensitivity of human hearing into account. The distinction between amplitude and loudness is important. Amplitude in decibels is the scientist's measure of the physical intensity of the sound. Loudness is our perception of it. Loudness does not increase linearly with sound amplitude.

Decibel scales are log (base 10) scales of sound amplitude. In a log scale, each increase of 1 unit represents a multiplication of the previous unit by 10. In the dB scale an increase of 3 units reflects a doubling of the physical amplitude of the sound, or the amount of energy in the sound. But doubling the amplitude of a sound in dB does not double the perceived loudness. It takes about a 9 to 10 dBA increase in order for a sound to be perceived as twice as loud.[5,6] To give you a feel for sounds measured in dBA, Table 5.1 shows some common sounds and their approximate dBA values.

The *frequency* of a sound is measured in Hz (Hertz), the number of oscillations in the sound wave per second. Frequency determines the pitch that we perceive. Pitch is analogous to the notes on a piano. Higher notes on a piano are higher-frequency sounds. Middle C is 256 Hz. Human speech sounds are in the range from approximately 100 Hz to 6,500 Hz but mostly below 1,000 Hz. Some speech sounds, such as *th*, *f*, and *s*, are in the upper range of frequencies. Sounds that have frequencies that overlap with the sound frequencies of speech can mask speech, making it difficult to hear what someone is saying.

Measures of Noise Exposure

Ambient noise is regulated in the United States by calculating the average noise level over different time periods. The FAA (Federal Aviation Administration) uses a noise measure called DNL (for "day and night sound level"). DNL expresses the decibel level of the noise over an average 24-hour

TABLE 5–1. Approximate dBA values of some everyday situations

Situation	Approximate dBA
Setting off firecrackers	125–160
Operating a chainsaw	103–115
Jet flyover at 1,000 feet	100
Diesel locomotive at 50 feet	87–103
Apartment next to freeway in Los Angeles (in 1970s)	87 (DNL)
Operating a power lawn mower	85–94
Waterfall in a small canyon at 25 feet	85
Operating a vacuum cleaner	77–86
Large office (10 or more people)	67
Automobile at 50 feet	60–90
Older residential area in Los Angeles (in 1970s)	60 (DNL)
Small office (1 or 2 people)	58
Normal speech at 10 feet	55
Wooded residential area in San Diego (in 1970s)	51 (DNL)
Tomato field in rural California (in 1970s)	44 (DNL)
Quiet wilderness area	20–30
Threshold of hearing	0

Sources: Compiled from Chen & Charuk, 2001; EPA, 1974; Department of Consumer and Employment Protection, Government of Western Australia, 2002
Note: DNL means "day-night average level," with a 10 dB penalty for sound between 10 p.m. and 7 a.m.

TABLE 5–2. Key terms and abbreviations in Chapter 5

dBA	The A-weighted decibel scale, a log scale measure of sound amplitude that weights sounds of different frequencies according to the sensitivity of human hearing.
decibel	A measure of sound amplitude. Decibel scales are log scales so that an increase of approximately 3dB is a doubling of the sound amplitude.
DNL	A measure of daily noise exposure in dBA that expresses the average amplitude of noise occurring in a 24-hour period, but that treats noise between 10 p.m. and 7 a.m. as if it were louder than it actually is. The FAA calculates DNL from annual average noise exposure.
FAA	United States Federal Aviation Administration
FHWA	United States Federal Highway Administration
Frequency	The number of compressions and expansions in a sound wave that occur in a specified period of time. Normally measured in Hz (Hertz). A sound with a frequency of 256 Hz would have the same perceived pitch as middle C on a piano.
GAO	United States General Accounting Office. An independent agency of the U.S. government that evaluates government operations, at the request of members of Congress or the president of the United States.
Leq24	A measure of noise exposure for a 24-hour period that expresses the average intensity of sound in a 24-hour period.

period, and then averaged over a year. DNL is calculated so that noise during sleeping hours (10 p.m. to 7 a.m.) is treated as if it were louder than it actually is. When noise is loud enough to disturb sleep, people find it much more annoying than daytime noise (see Table 5.2 for a summary of terms).

Other noise measures are also used, both in the United States and in Europe. Most are averages of various sorts. "24 Leq" is the decibel level averaged over a 24-hour period. This is sometimes used in Europe. It differs from DNL in that it does not apply a penalty to night noise. In the United States, highway noise is measured as Leq for 1 hour, or as the "L10," the level of noise that is exceeded 10% of the time, on average.[7]

After reviewing some of the research on how noise affects children, I also examine the decision criteria that agencies in the United States are using to regulate noise.

How Does Noise Influence Children Learning to Read?

Learning to read is exceedingly important. Children who are behind in reading often perform poorly in other school subjects, on other cognitive tasks, and fall further behind their classmates over time.[8] There are many possible reasons for this. Here is how Stanovich[9] explained that early poor reading could affect the rest of a child's academic development. First, children who have more difficulty learning to read will read fewer words each day in school. Second, they also read less on their own. Third, less reading experience results in poor skill at decoding words, and also lower reading comprehension. Poor readers have to spend their mental effort figuring out how to translate the letters into words. They have less mental energy left for understanding the meaning of a story. Also, reading helps children build their vocabularies. Those who read less learn fewer vocabulary words. Slower vocabulary development then impedes reading comprehension. As children advance through school, higher reading comprehension is required to deal with more complex materials. Such effortful reading becomes a handicap in other academic subjects and can also damage children's overall academic motivation.

Because poor reading is linked to poor overall academic performance, any condition that hampers children's early reading should be taken very seriously. Exposure to high noise levels is one such condition.

The Manhattan Bridge Apartment Study

The Bridge Apartments are a set of four high-rise apartment buildings that span Interstate 95 in Manhattan, New York City. In 1973, an excellent study of the effects of noise on children's reading was carried out with the children in the Bridge Apartments.[10] Because the apartments span the interstate highway, people living there were exposed to high traffic noise. The traffic noise was highest on the lower floors of the building and lowest on the upper floors, where distance from the highway was greater.

Just how noisy were the Bridge Apartments? When overlooking I-95 beside the buildings, the researchers found 84 dBA. That's the same as the sound of a city bus at close range. In the hallways (windows closed) the noise levels diminished toward the higher floors of the buildings: 55 dBA on the 32nd (top) floor, 58 dBA on the 26th floor, 60 dBA on the 20th floor, 63 on the 14th floor, and 66 on the 8th floor. Because dBA is a log scale in which an increase of 3 is a doubling of the physical intensity of noise, the lower floors had approximately 6 to 12 times the noise intensity of the upper floors. Also, the dBA levels in the hallways of the lower floors of the Bridge Apartments are louder than normal conversation. It would be difficult to hold a conversation without raising one's voice. Very few of the apartments had air conditioning at the time of the study. Air conditioning creates its own noise but would have allowed the residents to keep their windows shut in the summer.

If noise in the home interferes with how children learn to read, then we would expect the floor of residence to be related to the children's reading scores. The children in the building all attended the same public school and they were of approximately the same socioeconomic status. Because the Bridge Apartments were partly funded by New York State, family income had to fall in a certain range to be eligible for residence. These two factors are important because schools do vary in how effective they are in teaching children to read. And, in the United States reading scores are related to both socioeconomic status and ethnic background. The researchers mailed Bridge Apartments parents a questionnaire that requested information about parental education, length of residence in the apartments, and number of children in the family.

The research team recruited children in the 2nd to 5th grades who lived in the Bridge Apartments. Fifty-four children completed all the testing for the study, did not have a significant hearing loss, and did not have any problems with spoken English. The researchers tested the children's auditory word discrimination, and the school provided reading scores from the New York City Metropolitan Achievement Test (with parental permission). The reading test was given by the school in the children's regular classrooms within a few weeks of the rest of the testing. In the auditory word discrimination test, each child listened to 40 pairs of words on a tape recorder and said whether the pair of words were the same or different. Ten pairs of words were identical, but 30 pairs were similar except for one sound feature (which speech scientists call a "phoneme")—for example, "nap/map," or "thick/sick." An audiometrist also tested the children's hearing as part of the study.

The researchers thought that noise might interfere with word discrimination. Children living in high noise might "tune out" the details of the soundscape around them. In fact, there are large differences among children in their ability to attend to speech when there are distracting noises.[11] If the children can not attend to speech carefully enough, then the children would have trouble discriminating between words that sound similar. Failure to discriminate among similar-sounding words would then interfere with learning to read. If you cannot hear the difference between "peer" and "fear," how can you learn to distinguish them in reading? Other research has shown that the ability to discriminate among speech sounds is a key factor in initial reading.[12] Some schools use tests of phonological awareness to classify children as at risk for reading problems.[13]

Major Findings The results depended on how long the children had lived in the Bridge Apartments. For those children who lived in the Bridge Apartments for 4 years or longer (34 children), lower floor of residence in the Bridge Apartments was related to both worse auditory discrimination and worse reading test scores. Approximately 20% of the differences among children in their reading scores could be predicted from floor of residence in the Bridge Apartments. This was also the case for auditory word discrimination. But floor of residence was also related to mother's education and father's education. When parental education was included, the relationship between auditory word discrimination and floor of residence remained significant. Floor accounted for 19% of the differences in auditory discrimination among the children. The relationship between reading test scores and floor just missed the standard cutoff of 5 in 100 chances of being due to chance alone. But floor still accounted for about 10% of the differences in reading scores among the children.

I have outlined the framework of this study in Figure 5.1. Here are the steps of logic. First, long-term residence in a noisy environment impairs children's ability to discriminate among similar-sounding words. Second, auditory word discrimination is important in learning to read. Therefore, residence in a noisy environment impairs learning to read. Even though it is a correlational study with a small sample, this study is particularly powerful. The study controlled three important factors that can influence children's reading scores. Parental education levels were similar, income was within a relatively narrow range, and the children all attended the same school. The study measured a key variable that is known to affect reading—auditory word perception.

Alternative Interpretations and Further Questions   The main criticism of this study is that the children took the reading test in a noisy environment, the regular school classroom. Perhaps children who live in noisy homes are

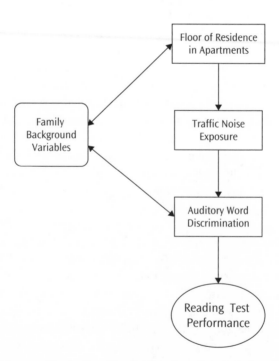

FIGURE 5–1. Diagram illustrating the Bridge Apartment study of Cohen et al., 1973.

unable to concentrate as well in a noisy environment. This could be handled by giving reading tests individually in a quiet environment. But children are normally given standardized tests in the classroom. The current trend in education is for more and more standardized testing. Also, if living in a noisy environment damages children's ability to concentrate in normal school testing circumstances, then we would still conclude that high noise is bad for children's school performance.

Could differences in family income account for the results? The rents did vary with floor, but by less than 20%: $185 to $219 for a two-bedroom, and $235 to $250 for a three-bedroom. Residence in the building was restricted to families with income in a certain range. The researchers included parental education in their data analyses, and education is related to income. Also, if income differences accounted for the results, we would not expect the relationship between floor of residence and reading to be strongest for the children who had lived in the building for 4 years or longer. If income were the primary influence on the relationship between floor and reading, then we would expect the correlation to be the same regardless of the length of residence.

Could it be that children living lower in the Bridge Apartments were exposed to higher levels of carbon monoxide (CO) than those living on the higher floors? CO is a poisonous gas produced by cars, trucks, and buses. CO is often high near busy urban streets. If the level of CO is high enough, it impairs oxygen absorption, cognitive functioning, and can cause brain damage in adults as well as children. Senator Robert F. Kennedy had asked New York air pollution officials to measure CO inside the Bridge Apartment buildings in 1967. The data for the reading study were collected in 1972. In 1967 CO in the apartments was 13–15 ppm, well above the current U.S. federal pollution standards of 9 ppm for 8 hours of exposure. But carbon monoxide was *not* related to floor of the building. The whole building was polluted at approximately the same level.

Bottom Line   The 1973 Bridge Apartment study by Sheldon Cohen and his colleagues is an excellent naturalistic study of how long-term noise affects children's reading. As you will see, the results are stronger than those of some other studies. I think one reason this study's results are robust is that it avoided many of the confounding variables that are troublesome in the other studies. The children all lived in the same apartment complex. Their family incomes fell in the same range. They attended the same school and took their achievement tests in the same school. The conclusion I draw from this study is that long-term (4 years or more) exposure to road traffic noise at home does interfere with the development of auditory word discrimination. Auditory word discrimination is related to how quickly children learn to read and to their proficiency on standardized tests of reading.

LAX Aircraft Noise in the Late 1970s

Sheldon Cohen assembled a team of collaborators for a two-year study of children going to school near Los Angeles International Airport (LAX). The flight corridor over the noisy schools had approximately 300 flights per day. That's an airplane about every 2½ minutes during school hours. During the first year of the study the flight pattern classrooms had 79 dBA noise, while the comparison schools had an average of 56 dBA. The maximum noise levels were 95 dBA and 68 dBA, respectively. The L.A. school district had sued the airport because of noise. Part of the settlement provided some money for sound insulation of the noisy schools. Some of the classrooms in the noisy schools were sound insulated, but others were not.[14] The sound insulation reduced classroom noise to 63 dBA, below the FAA's target value of 65 dBA. But the noise levels are still high enough to interfere with understanding speech.

In the first year of the study, 142 3rd and 4th grade children in four schools that were in the flight pattern of LAX were compared to 120 children in three schools not in the flight pattern.[15] The second year, the research team retested as many of the children as possible: 83 students from the noisy schools and 80 from the quiet schools.[16] As in the Bridge Apartment study, the researchers gathered reading scores from the school records and gave the children the auditory word discrimination test. In addition, the researchers also measured blood pressure, ability to attend to a task in the presence of a distracting noise, persistence in solving puzzles (a measure of motivation), and annoyance with noise from airplanes. As before, they excluded the data of children who showed hearing losses of greater than 25 dB on an audiometric test. Aircraft noise on the playgrounds of the schools very close to LAX exceeded the guidelines for adult exposure to noise at work. It was loud enough on the playgrounds to damage children's hearing.[17,18]

In the year between testings some children moved away. If the children who moved away were the ones who were most sensitive to noise, that would obviously bias the results. The children who moved away from the noisy schools did have higher average blood pressure than those who were retested in the follow-up. Because of this selective attrition, the results from the first year of the study are the easiest to interpret, and that is mainly what I summarize here. But notice that if the families with the most noise sensitive children moved away, it would make it more difficult to find a negative effect of noise.

Over the 2 years of the study, the children in the noisy and quiet schools differed on almost all the measures *except* reading and auditory word discrimination. The reading and math scores were from standardized tests the children took when they were 2nd and 3rd graders, a year before the rest of the tests were given. The results did show that auditory word discrimination was related to reading scores, regardless of whether the children were in the noisy or quiet schools. Children who had attended the noisy schools for 3½ years or more took longer to solve a puzzle than the children in the quiet schools. They were also more likely to fail the puzzle because they gave up.[19] And, of course, children in the noisy schools reported that classroom noise bothered them more.

Children in the four noisy schools also had higher average systolic and diastolic blood pressure than the children in the quiet schools. The difference was largest for those children attending their schools for 2 years or less. I have graphed these results in Figure 5.2.

The differences in blood pressure suggest higher chronic stress for the children in the noisy schools.[20] But the lack of an effect of noise on the children's reading scores is puzzling. The Bridge Apartment and LAX studies involved the same team of investigators and some of the same tests. Let's look at the differences between the studies in more detail.

Differences between LAX and Bridge Apartment Studies There are at least four key differences between these two research projects that could account for the different findings with respect to whether noise affects children's reading: the ethnic mix of students, the ages of the children when they took their reading tests, the sources of noise, and intensity of noise in the home.

First, in the LAX study the noisy and quiet schools were matched for the percentage of families receiving Aid to Families with Dependent Children (AFDC, informally called "welfare"). But the quiet schools had more children

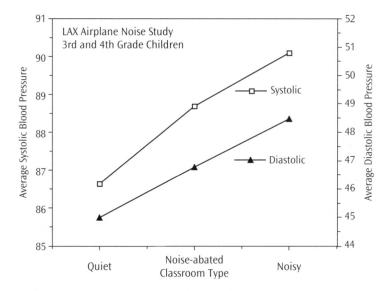

FIGURE 5–2. Blood pressure in LAX noise study. *Source:* Data from Cohen et al., 1981 (note different scale for diastolic and systolic blood pressure).

who were likely to be bilingual Spanish-English speakers. There were equal percentages of white children (approximately 30%). The noisy schools had 32% African American and 33% Chicano (Mexican American) children. In the quiet schools 50% of the children were from Mexican American families, and 18% from African American families. In the LAX study the research team did not mention whether they excluded children who did not speak English fluently. In the Bridge Apartment study they did exclude those with poor English. The effect of the ethnic differences in the LAX research could erase any negative effect of noise on children's reading. The quiet school scores might have been lowered by a larger percentage of bilingual children. In the United States, both Hispanic and African American students lag behind white students on average in reading scores.[21] This is unfortunately true at present and was also true at the time the LAX data were collected. Being bilingual can be an advantage in a multicultural society such as Los Angeles, but it presents separate issues for learning to read. Spanish-speaking children generally lag behind English-speaking children in learning to read *English*, the language in which standardized tests are given in the United States.[22]

A second difference is that the reading tests in the LAX study were given when the children were 2nd and 3rd graders, about a year before the research tests were done. In the Bridge Apartment study the children's reading tests were given in school within a few weeks of the other tests, and the children were 2nd to 5th graders. Because children who are poor readers often fall further behind with age, the older children in the Bridge Apartment study may have been more likely to show the damaging effects of noise exposure.

The third difference was that the LAX study dealt with airport noise, while the Bridge Apartment study dealt with highway traffic noise at home. But airport noise was also significantly related to reading scores in the New York City area,[23] in Munich, Germany, and possibly London (see below). I do not know of any studies that have directly compared children who are exposed to road traffic noise with those exposed to airport noise.

Fourth, the *home* noise of the children in the LAX study is not known. The children in the quiet schools could have been living near expressways or other sources of high noise. If so, this would have made it more difficult for the study to find effects of the airport noise in the schools.

Which of these differences between the studies is enough to make a difference in the results? We can't be certain. Here are other studies of how noise affects children.

Other Studies of Noise, Children's Reading, and Stress Reactions

New York City Aircraft Noise in the 1970s   Schools differ in their effectiveness at teaching reading. Most of the studies of aircraft noise have tested children at only a few schools. This study is a welcome exception. A team of researchers from New York University tallied the number of students in each school who scored 1 year below grade level in reading or 2 or more years below grade level. They did this for *all* 362 elementary schools in Brooklyn and Queens.[24] The investigators also gathered other information about school characteristics: ethnic makeup of the student body, absentee rates, turnover of students during the school year, teacher experience and advanced education, percentage of students eligible for free school lunch, and the student-to-teacher ratio. For each school, aircraft noise from Kennedy and LaGuardia airports was estimated using the official FAA noise contour maps. Noise contour maps are the FAA's official method for estimating aircraft noise in residential areas. Because most children attend elementary school in their neighborhoods, the noise estimate is for airport noise both at home and at school.

The results showed a dose-response relationship between airport noise and reading. I have graphed it in Figure 5.3. The relationship between school noise and reading is especially noticeable for 4th graders reading 2 years or more below grade level. The other grades also show a general trend toward worse performance in higher noise schools. The research group concluded that "an additional 3.6% of students in the noisiest schools read at least 1 yr. below grade level.... The dose response relationship suggests that the percent reading below grade level increases with increasing noise level."[25] The upper and lower limits of their estimate were that between 1.5% and 5.8% more children read below grade level due to airport noise.

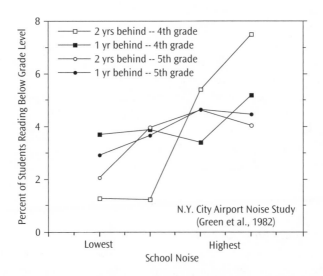

FIGURE 5–3. Percentage of students reading below grade level in New York City aircraft noise study. *Source:* Data from Green et al., 1982.

One way to interpret such statistics is to take the "magic bullet" view. If someone had a magic bullet to raise the reading performance of 3% of the worst readers, it would probably be worth giving it a try, even though 3% of the worst readers is not very many children. That magic bullet might just be lower noise exposure.[26]

New York City Elevated Trains and Noise Abatement in the 1970s   At Public School 98 in New York City, approximately 80 elevated trains passed the school each day during school hours. About 10% of instructional time was lost to train noise for those classrooms that faced the tracks. The train noise was measured in the classroom at 89 dBA. That's enough to make most of us cover our ears. Children whose classrooms were near the tracks scored lower by 3½ months on reading tests compared to children in classrooms on the other side of the same school.[27]

Since 1972 parents at PS 98 had been asking the transit authority to quiet the noisy trains. Finally in 1978, rubber pads were installed on the tracks, and noise-absorbing material was added to the ceilings of some classrooms. Train noise was reduced from approximately 89 dBA to 81 to 86 dBA (measured with the windows open). The children's reading scores on both the noisy and quiet sides of the school were measured in 1978, before noise abatement, and in 1980–1981, after noise abatement.[28] Classes were chosen that were expected by the assistant principal to be approximately equivalent in intelligence, teaching method, and overall achievement. Bilingual classes were excluded. Even this minimal reduction in train noise significantly reduced the differences in the children's reading scores on the two sides of the school, compared to the previous testing.

Why would noise abatement have virtually an immediate effect on children's reading? One possibility is that the noise abatement made more instructional time available. The teachers were at least able to shout over the train noise, rather than having to stop entirely. Perhaps the children were able to concentrate a bit better during the standardized tests after noise abatement. Adults have more difficulty concentrating in high noise.[29] It is also possible that after noise became a hot issue, the school administrators were more careful to avoid bias in assigning teachers and students to the noisy and quiet sides of the building. The researcher also mentioned the possibility that noise abatement boosted morale in the school among children, parents, teachers, and administrators. Finally, perhaps the noise abatement reduced everyone's stress.

Airport Noise in Munich, Germany, in the 1990s   The Munich airport has been the subject of a long-term study of the effects of aircraft noise on children's functioning. A new airport opened and the old airport closed in 1992. Before the old airport closed, researchers tested 135 3rd and 4th grade children living near the old airport. Children living in a similar urban neighborhood not affected by airport noise but matched in socioeconomic status served in the comparison group.[30] The noisy neighborhood averaged 68 dBA over 24 hours, with 80 dBA peak noise. In the quiet neighborhood the noise average was 59 dBA, with a peak of 69. Children with a hearing loss were excluded from the data analyses.

The children took a set of research tests in a soundproofed trailer at their schools. The tests were reading, long-term memory (remembering a story 24 hours later), short-term memory (remembering a string of consonants in order), three attention tasks, motivation (solving a puzzle after having failed a different one), noise annoyance, a questionnaire measuring four aspects of quality of life (physical, psychological, social, and functional), and blood pressure. In addition, the children's parents collected a first morning urine sample from the children. The urine was tested for three different hormones related to stress: cortisol, epinephrine, and norepinephrine. Children from the noisy and quiet areas differed significantly on most of the tests: reading scores, long-term memory, systolic blood pressure, epinephrine, norepinephrine, motivation, psychological quality of life, and noise annoyance. The authors concluded: "Our results reflect a general pattern of adverse psychological stress reactions associated with chronic exposure to noise among elementary-school-aged children.... These data are sobering when one considers that more than 10 million American schoolchildren are exposed to comparable noise levels and that worldwide population exposure to noise is escalating exponentially with accompanying industrial development."[31]

The same research team also tested 4th grade children near the new airport before and after it opened, and near the old airport before and after it closed.[32] Closing an airport is the ultimate noise abatement. The 24-hour noise decreased from 68 to 49 dBA, and the peak noise level (exceeded 1% of the time) decreased from 80 to 59 dBA. After the new airport opened, the 24-hour average noise rose from 53 to 62 dBA, and the peak noise from 63 to 73 dBA. The noise levels near the new airport are about 6 times more intense than prior to the construction of the airport. For the quiet area, the noise did not change when the new airport opened (53 and 64 dBA average and peak). The researchers tested the children 6 months before the opening of the airport, and at 6 and 18 months after it opened.

For the children living near the new airport, stress-related hormones increased (epinephrine and norepinephrine) after the new airport opened. There was also no evidence for habituation to the noise. The stress-related hormones stayed high in the children living in the noisy areas when they were tested 18 months after the new airport opened.[33] The children in the noisy area near the new airport scored slightly worse than the quiet-area children on a reading test.[34] The reading test results just missed the $p = .05$ cutoff. Eighteen months after the airport closed, the children living in formerly noisy neighborhoods performed as well as the quiet neighborhood children on reading tests. These results suggest that noise abatement could have positive effects on children's school performance.

Three Country Road & Airport Noise Study   The gateway airports to Europe, Schipol, Heathrow, and Barajas, are among the busiest airports in the world. What about the children who live and go to school in the flight paths? An international research team recruited 89 schools that were high or low in aircraft noise and either high or low in road traffic noise.[35] The high- and low-noise schools were matched for socioeconomic status in each country. The 9–10-year-olds in each school took a reading comprehension test, three memory tests, and a test of sustained attention. The children took the tests in school in groups. Parents also answered questions about their children's behavior problems and family background such as maternal education, ethnicity, language spoken at home, employment, length of time in the same home, and home crowding.

Higher aircraft noise was associated with both lower reading comprehension and lower recognition memory scores. This result held when the confounding variables were included. A puzzling finding was that school road traffic noise was associated with higher recall memory scores, and was not linked to reading scores. Behavior problems reported by the parents were not related to either kind of noise.

The study had over 2800 children in three different countries. The researchers estimated a dose-response curve and concluded that 5dB increase in aircraft noise predicts a 1- to 2-month delay in reading.

Developmental Processes behind the Effects of Noise on Reading

Children's Speech Perception   One way in which noise affects children's reading is by disrupting speech perception. This is shown by the association between poor auditory word perception and noise exposure. How much noise interferes with the intelligibility of speech depends on a variety of aspects of the situation: the skill of the speaker, the skill of the listener, and the familiarity of the words being spoken, as well as the exact characteristics of the noise itself and whether the noise is variable or constant.[36] Standardized tests of speech and hearing show that children's speech perception is more easily disrupted by a masking noise than the speech perception of adults is. The ability to selectively attend to a chosen sound or speech while ignoring other sounds is still developing even up to age 16.[37] This selective attention phenomenon is called "informational masking." For both children and adults there are large individual differences in informational masking by an unexpected sound.[38] The long development period for these listening skills, combined with large individual differences, has important implications for education. As early as

1975 one reviewer wrote, "Data...suggest that levels of noise which interfere minimally with the performance of adults on tests of speech intelligibility (speech perception) can interfere substantially with the performance of children."[39]

Children with Minor Hearing Problems   The conclusions above are for children and adults with normal hearing. Children with mild hearing losses have much more trouble understanding speech against a noisy background. Noise poses a serious learning barrier to such children.

How common are minor hearing losses in children? An NHANES (National Health and Nutrition Examination Survey) survey of more than 6,000 children between 6 and 19 years of age showed that at any given time approximately 15% have at least a 16 dB hearing loss in one or both ears.[40] This implies that approximately 1 in every 7 children in any classroom is not hearing fully. Those children will be more severely affected by noise. The study did not distinguish between temporary or permanent hearing loss. Children may have poor hearing because of colds or middle ear infections when they were tested. Over half of middle ear infections last a month or more.[41] The authors of the NHANES study concluded: "Unilateral hearing loss in children impacts speech perception, learning, self-image, and social skills. Slight hearing loss affects children in classrooms and other reverberant listening environments in which a child with transient auditory dysfunction can have difficulty perceiving and understanding speech sounds....Because the decibel scale is exponential, even a slight decibel change in a child's hearing threshold at any frequency can significantly affect the child's ability to hear."[42] Children with a slight hearing loss have a higher chance of being held back a grade than their peers with normal hearing.[43] Noise is a barrier to learning for such children.

What Teachers Say about Aircraft Noise   In a 1978 survey of schools near San Diego International Airport, 50% of teachers reported that aircraft noise interfered with speech four to five times per hour. Sixty-five percent said that noise interfered with their own silent reading, writing, and mental work. Seventy-two percent said building vibrations due to the noise were annoying. Ninety-one percent said aircraft noise interfered with playground and sports activities. One-hundred percent said the noise interfered with use of audiovisual teaching aids.[44] The noise in these schools averaged 70 DNL, with a peak of about 87 dBA. In Hong Kong, even in schools with noise of approximately 55 DNL, 20% of the teachers said that aircraft noise interfered with classroom speech very much. Over 40% of teachers said they had to pause very often. In schools with a very high noise level such as 80 DNL, almost 75% said that they had to pause very often while teaching.[45] Near London's Heathrow Airport, disruption of speech in the classroom by aircraft noise was the number one problem named by the teachers. More than 70% of the teachers in the noisiest schools (DNL approximately 80) said that aircraft flyovers caused them to abandon a lesson.[46]

What about Air Conditioning?   Where aircraft noise is damaging the school learning environment, does it help to install air conditioning in the classrooms? The answer is a bit complex. In Florida the primary background noise in classrooms was from air conditioning and heating systems.[47] A study of schools in Singapore found that the noise from air conditioners was higher than the noise coming from outside. This was true even with a brand-new central air system. The Singapore researchers concluded: "If not carefully implemented, air-conditioning a school may result in higher noise levels in classes than would occur in its absence....In the case of aircraft noise, where the sources are intermittent and can be very high in magnitude, the closed building envelope may reduce the aircraft noise by 15 dBA or more."[48] They also commented on the potentially negative effects of spread of diseases through the air in air-conditioned schools, as opposed to those with natural ventilation.

Overall Conclusions

Noise is a barrier to children's learning. The weight of the evidence is that children's reading performance is negatively affected by chronic noise. We have studies covering three decades. Higher noise yields a greater likelihood of lower reading scores. Children's speech perception is more easily disrupted by noise than is adults'. Children with hearing losses are likely to be the most severely affected by noise. Teachers say that noise from outside the classroom interferes with instruction. The Health Council of the Netherlands (1999) and the United Kingdom Institute for Environment and Health both concluded that noise damages children's performance in school.[49] Other studies have also found that children are negatively affected by exposure to noise, as measured by stress responses[50] or academic performance.[51] Also, all studies that have examined stress-related hormones or blood pressure have found that noise-exposed children show higher levels of physiological stress than those exposed to lower noise.

Other Effects of Noise

Annoyance, blood pressure, and stress-related hormones, such as cortisol, epinephrine, and norepinephrine are all affected by noise.[52] Whether the stress children experience will affect their functioning later in adulthood is not known. Animal studies of prenatal exposure to noise stress show that it can have lasting negative effects on offspring.[53] Also, there is some evidence, both in humans and animal research, that prenatal noise exposure can harm hearing.

In 1997 the American Academy of Pediatrics issued a statement on the hazards of noise to the early development of infants and newborns.[54] First, prenatal noise and noise in the hospital intensive care unit can cause hearing loss. Hearing loss, especially if undiagnosed, is important for enjoying the soundscape in which we live. But it also is a risk factor for poor performance in school. Children with even a slight hearing loss (>10 dB) have a higher likelihood of performing poorly in school or having peer problems than children with normal hearing.[55] Second, high prenatal noise exposure is correlated with low birth weight and premature delivery. Third, for preemies in the hospital, noise in the infant care unit can disrupt normal growth and development. These last two findings may be due to the fact that high noise is a stressor, and they are corroborated by animal research, as I describe below.

Prenatal and Early Postnatal
Noise-Induced Hearing Loss

It has been known since 1927 that the human fetus responds to loud noise during the third trimester.[56] Fetal hearing is relatively well developed by about the 24th week of pregnancy.[57] Noise can only affect the hearing of the fetus directly if it penetrates the womb. Knowing how much sound is reduced as it is transmitted to the fetus is important for protecting the fetus from damage. Researchers summarize the transmission of sound to the fetus by saying that hearing music in the womb would be like listening with the bass turned up and the treble turned down.[58] Below approximately 250 Hz (about middle C on a piano), sounds are damped very little (about 10 dB).[59] This implies that low-frequency noise is likely to be more of a hazard to the hearing of the fetus than high-frequency noise. In addition, vibrations that are below the human hearing threshold (e.g., 10 Hz) have potential to create harmonic vibrations within the human hearing range.[60] Also, the organs of the inner ear in young animals are more susceptible to noise damage than those in adult animals. For this reason, high-noise exposure of infants and children should be avoided.

Canadian Study of Work Noise during Pregnancy A group of Canadian researchers tested the hearing of children 4 to 10 years old whose mothers worked during pregnancy in high-noise jobs.[61] The researchers contacted manufacturing plants that employed large numbers of women. Based on detailed interviews and assessments of the manufacturing plants, the researchers categorized the female employees into three pregnancy noise exposure groups: 65–75 dB, 75–85 dB, and 85–95 dB. Their children were given audiology tests to determine hearing thresholds. When a person's hearing has been damaged by noise, the hearing loss shows up best at 4,000 Hz. Audiologists call this the "4,000 Hz notch" in hearing threshold. The Canadian investigators found a dose-response relationship in the percentage of children with a 10 dB or greater hearing loss at 4,000 Hz. The higher the mother's noise exposure during pregnancy, the more likely the child was to have a hearing loss. I have graphed these results in Figure 5.4. The research team also found that the hearing threshold of the child was most likely to be affected when the pregnant mother worked in low-frequency noise. Other risk factors also increased the likelihood of a hearing loss, including neonatal jaundice and staying in an incubator after birth. The researchers recommended that the Canadian government adopt temporary noise exposure standards of 85 dBA for each 8-hour period, until further research could be done, instead of the 90-dBA-per-8-hours standard.

Noise in Incubators and Hearing Loss   The American Academy of Pediatrics committee stated that "numerous studies have documented the continuous noise exposure of infants, without intervening periods of quiet, associated with neonatal intensive care units (NICUs).... Many studies have documented hearing loss in children cared for in the NICU."[62] One important way of avoiding damage to hearing from noise exposure is to get out of the noise and give your ears a rest. But a preemie in an incubator cannot do that.

How high is the noise in most NICUs? The motors of most NICUs create 60 dBA of noise, bubbling in the ventilator tubing can transmit 70 dBA to the infant, tapping the top of an incubator is about 80 dBA, and closing the plastic porthole can cause 100 dBA of noise to the newborn. If one of the medical staff strikes the side of the incubator to startle an infant who momentarily stops breathing (and some say this is done with some regularity), it can create a jolting 130 to 140 dB noise.[63] In the 1980s, one research team found

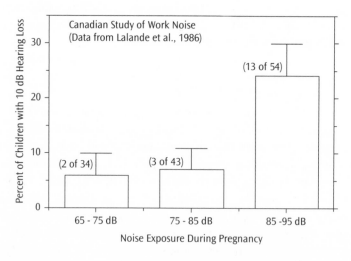

FIGURE 5–4. Effects of prenatal noise on hearing loss in Canadian study. *Source:* Data from Lalande et al., 1986. Bars are +/– 1 estimated standard error.

that the average noise in the NICU ranged from 66 to 109 dB.[64] Another issue is that the effects that the medications and other health issues of such fragile newborns might have a synergistic effect on hearing loss. The research is incomplete on this point, but some antibiotics are known to be toxic to the inner ear. Such fragile newborns should not be subjected to the added hazard of high noise.

The Academy of Pediatrics committee recommended simple approaches to reducing newborn noise exposure. Do not tap on incubator hoods, do not use the incubator hood as a writing desk, close incubator lids and portholes very gently, wear soft-soled shoes in the nursery, talk in soft voices, and incorporate sound control as a criterion for new equipment purchases and remodeling.

Animal Research on Prenatally Induced Hearing Loss In animals ranging from sheep to hamsters, the amount of damage that noise causes to the cochlea depends on the particular time during pregnancy of the noise exposure, as well as the frequency and intensity. (The cochlea is the part of the inner ear that converts sound to nerve impulses.) A recent review of animal research concluded: "Numerous studies have shown that exposure of young mammals to noises at levels that would not produce damage in adult animals can cause severe high-frequency hearing loss and histologic damage of the cochlea. The period of increased susceptibility corresponds to the final stages of morphologic and functional development of the cochlea. It has not been determined if and when this period of susceptibility occurs in human fetuses and neonates.[65] Most of the animal studies have used noise of 110 dB or higher. There are large differences among studies in the length of exposure. Some use 8 hours per day, others used short bursts. Some studies measured damage to the hearing of the offspring by postmortem exam of the cochlea. Others have used brain responses to sound called "auditory evoked potentials."

Vibroacoustic Testing of Human Infants Late during pregnancy, obstetricians sometimes use a vibroacoustic device (an "electronic artificial larynx") on the abdomen of a pregnant woman to create a noise that will startle the fetus.[66] The American Academy of Pediatrics committee on noise as a hazard wrote that "1 to 4 seconds of 100 to 130 dB of 1220–15,000-Hz sound is used as a stimulus to document the well-being of the fetus."[67] Questions about the safety of stimulating the fetus in this way have been raised. It can cause tachycardia (racing heart rate), unusual heart rate patterns, increases in baseline heart rate, and disorganization of fetal sleep and activity cycles.[68] Studies of fetal sheep found that sound pressures between 129 and 135 dB inside the womb were created by the electronic artificial larynx.[69] No one has done a follow-up study of children after vibroacoustic fetal testing.

Noise Exposure as a Prenatal Stressor

Prenatal stress is associated with a variety of negative outcomes in humans.[70] In animals, loud noise as a prenatal stressor has been studied in species ranging from mice to monkeys. Pregnant rhesus monkey females were exposed to just three short blasts of an air horn once a day (1300 Hz, 115 dB when measured 1 meter from the monkey), 5 days per week. This treatment increased cortisol, a stress-related hormone, in the pregnant females.[71] The offspring of the stressed monkeys scored lower than control monkeys on infancy tests of motor activity, orientation to stimuli, and motor maturity. The prenatally stressed animals were slightly smaller at birth. They showed more abnormal behaviors, were slower to learn a task to earn food treats, and showed higher levels of cortisol and ACTH when separated from their mothers for weaning.[72,73] In rodents, noise stress during pregnancy yields fewer litters, litters with a smaller number of animals, and litters in which the animals weigh less, as well as offspring with altered immune function.[74]

Based on the animal research, one might expect noise during pregnancy to have similar effects on people. Some studies have found that women who lived in high airport noise had a higher risk of either shortened pregnancy or decreased birth weight. But not all research finds this.[75] One study in Denmark found lower birth weight when aircraft noise was between DNL 60 and 65 dBA.[76]

Is vibroacoustic stimulation of the fetus by obstetricians a prenatal stressor similar to the noise stress used in the monkey studies? The doctor's purpose in using the vibroacoustic device is to make sure that the fetus is healthy. There are many differences between the noise stress in the monkey studies and the vibroacoustic devices used by some obstetricians—the frequencies of the sounds, and how often they occur (once or twice in the doctor's office versus every day for the monkeys). Whether one or two such events are harmful to development, either because of fetal stress or damage to hearing, is not known. But the topic has not been carefully researched. One study measured adrenaline and noradrenaline in fetuses exposed to a vibroacoustic stimulus. They found no change in these hormones, but they tested only 13 pregnant mothers.[77] This topic merits research.

Noise Sensitivity and Other Effects of Noise on Adults

Do children differ in their psychological sensitivity to noise? There is not much evidence on this question. In one laboratory study, children adjusted the intensity of white noise (a hissing sound) played to them through headphones. Children with more extroverted personalities adjusted the noise to a higher level than those who were more introverted. Boys made the noise louder than girls did. The children were 10–11 and 5–6 years old.[78]

Adults who say they are more sensitive do react more to noise, both physiologically and when answering questionnaires.[79] Noise-sensitive college freshmen were more disturbed by dormitory noise than those who were less sensitive. As the school year went on, the noise-sensitive students became more disturbed by noise in their dorms, whereas annoyance stabilized in those who were less sensitive.[80] Another project on traffic noise by the same author also cast doubt on the idea that people habituate to irritating noise.[81] The issue of habituation to transportation noise is a controversial area with conflicting findings.[82]

Noise-sensitive people are more likely to have their sleep disturbed by traffic noise, take longer to fall asleep, and rate their sleep quality as worse than people who are less noise-sensitive. Noise-sensitive people also showed more body movements, and their heart rates sped up more in response to noise even while they slept. But performance on a reaction time task and self-ratings of mood deteriorated equally in noise-sensitive and nonsensitive people after sleeping with the simulated traffic.[83]

Noise sensitivity is related to some aspects of personality and well-being. German researchers found that noise sensitivity in university students correlates slightly with depression, stress, the trait of anxiety, and the trait of anger.[84] The correlations accounted for a maximum of 8% of the differences among people.[85] Other studies in Europe have also found that both noise sensitivity and noise annoyance are related to personality and mental health variables such as neuroticism[86,87] and depression.[88] General somatic health complaints such as colds, flu, aches and pains, nervous symptoms, heart symptoms, headaches, and fatigue are also related to noise annoyance and noise sensitivity.[89] There is a controversy about whether living in noisy neighborhoods increases the likelihood of psychiatric problems.[90] One team of researchers in Britain concluded that "hypersensitivity to noise should be considered among the risk factors for psychiatric illness."[91] This conclusion is based on the finding that people who are noise-sensitive tend to have other characteristics that may make them more susceptible to stressors such as noise.

Both the 1999 Health Council of the Netherlands and the 1997 United Kingdom noise reports listed sleep disturbances, mood after noise-disturbed sleep, annoyance, and ischemic heart disease as documented health effects of noise. These international panels of scientists judged that the evidence was "limited" or "inconclusive" for the link between noise and low birth weight, psychiatric disorders, and poor performance after noise-disturbed sleep. The Dutch report concluded that hypertension (high blood pressure) is linked to noise exposure, but the British report judged the evidence inconclusive. A meta-analysis of 43 studies of the effects of noise on cardiovascular functioning concluded that both occupational and aircraft noise exposure are associated with hypertension.[92]

Decision Criteria in Noise Policy

The evidence that noise affects children's speech perception, reading, and stress responses can inform policy on noise exposure. In this section I describe acoustical standards for schools and classrooms.[93] The standards are based solidly on the research evidence that children need to hear well in school to have equal access to education. Second, I describe how America's current regulations and voluntary guidelines for highways and airports fail to provide sufficient noise protection. Third, I summarize research on noise annoyance, the controversies surrounding the annoyance research, and why I think the annoyance research is both over-emphasized and misused in setting noise policy.

Acoustical Standards for Schools

In 2002 the United States joined France, Germany, Italy, Portugal, and Sweden by adopting noise standards for schools.[94] Standards adopted by the American National Standards Institute (ANSI) are normally incorporated in local building codes. But a prerequisite for a quiet classroom is a quiet location in the community: "Site selection is a vital concern, and high environmental noise levels must be avoided."[95]

The standards are: (1) noise in unoccupied regular classrooms should not exceed 35 dB, even when the ventilation system is on, (2) the signal-to-noise ratio of instruction should be at least 15 dB at the student's ears, and (3) sound reverberations (long time-lag echoes) in classrooms should not exceed 0.7 seconds (echoes can degrade the intelligibility of speech). When external sound exceeds 65 dB, it is recommended that special building techniques be used to reduce the amount of noise that enters the school building.[96] Standards specify that if the 1-hour exterior sound exceeds 75 dB, the location should not be used for a school.

Current Noise Policy in the United States

How do regulations for highway and traffic noise compare with the proposed classroom standards? Ambient noise is regulated in the United States (as well as most other countries) by calculating the average noise level over different time periods. The FAA's current cutoff for noise abatement is 65 DNL or greater. In special circumstances the FAA will consider abatement when noise is above 60 DNL.[97] The FHWA (Federal Highway Administration) criteria for noise abatement are that highway noise must have a 1-hour average *above* 52 dBA or must be above 55 dBA for 10% of the time when measured *inside* a residence or school. The FHWA will also consider noise abatement if a new highway increases existing noise levels "substantially."[98] According to FAA land use recommendations, it is acceptable to locate schools wherever the DNL from an airport is below 65. And if the community decides that a school is needed where DNL is between 65 and 70, the FAA recommends that the school construction reduce external noise by at least 25 dB (regular construction decreases interior noise by about 20 dB).[99] The ANSI standards would require at least a 30dB decrease in external noise.

The FAA's cutoff of DNL 65 (outside noise) is incompatible with the classroom acoustical standards, and so is the FHWA's standard of 52 dBA for interior traffic noise. Interior highway traffic noise of 52 dBA is 17 dB above the acoustical criterion of 35 dBA for unoccupied classrooms! The FAA's land use guidelines, which influence local zoning rules, also conflict with the recommended standards for school noise. According to FAA land use guidelines, it is acceptable for noise from outside to be 40–45 DNL inside schools. This is 5–10 dB higher than the ANSI standard. But also remember that because DNL is a 24-hour average, a DNL of 65 can be created in many ways: 500 daily flights at 87 dB, 100 flights at 94 dB, or 50 flights in a day at 97 dB.[100] When DNL is 65, each and every aircraft event (takeoff or landing) will interrupt speech inside a school. Even with the quietest airplanes, a building that reduces external noise by 25 dB will only dampen the aircraft sound to about 62 dB.

The FAA reaffirmed DNL 65 or greater as the criterion of significant noise impact and says that "DNL is the only metric backed with a substantial body of scientific survey data on the reactions of people to noise."[101] The FHWA's latest documents also maintain the criteria of a highway noise hourly average of either 52 dBA indoors, or 55 dBA for 10% of the time, as significant noise impacts for schools and homes.[102]

The noise criteria of the FAA and FHWA are based almost exclusively on noise annoyance, not the other negative effects of noise discussed in this chapter. To predict noise annoyance, the FAA uses the "Schultz curve." The Schultz curve goes back to 1978. It was based on a synthesis or meta-analysis of surveys of noise annoyance. It is an amalgam of studies of aircraft (both commercial and military), highway, and train noise annoyance.[103] Over the years, the FAA has reaffirmed the use of the Schultz curve "to determine community noise impacts."[104] The FAA says that according to the Schultz curve, approximately 12% of people will be highly annoyed when aircraft noise is DNL 65 dB.[105] In its latest policy documents, the FAA[106] uses the terms "annoyance" and "noise impact" virtually interchangeably.

Because of the prominent role of the Schultz curve for noise annoyance in U.S. noise policy, and because the curve totally neglects children's learning and stress reactions, in the next section I expose some of the scientific controversies that have surrounded it.

Decision Criteria Embedded in the Schultz
Curve for Noise Annoyance

Schultz[107] combined 11 different noise annoyance surveys. In noise annoyance surveys, people are asked to rate their degree of annoyance using a numerical category scale. For example, numbers might be labeled with terms such as "not at all annoyed," "a little annoyed," or "moderately" or "very" or "extremely annoyed." The 11 different studies were conducted in different years and different countries and different languages. The annoyance scales used by researchers varied considerably. For each survey Schultz chose a cutoff for what to count as "highly annoyed." He then graphed the percentage of highly annoyed people for each noise level in each survey. Then he fit a curve to the graph. This was the origin of the Schultz curve that the FAA still uses for estimating community noise impact.

What decisions did Schultz have to make? First, he decided to use only the category "highly annoyed." Second, he had to decide how to define "highly annoyed." These decisions provided a good starting place. But later they were treated as if they were the only way to go. Schultz was originally cautious about his assumptions: "given the survey data as published, the largest uncertainties in the results of this study are associated with the judgment as to who is counted as 'highly annoyed.' "[108] Schultz argued that when people are highly annoyed, the relationship between intensity of noise and people's reaction will be the strongest. Schultz cited studies by the EPA that established the precedent of examining only the "highly annoyed" category. Finally, he argued that "'percent highly annoyed' carries a commonsense import that is clear, even when it is not precisely defined."[109]

Schultz also looked at surveys that asked people if they were disturbed by noise during conversations, while listening to radio or television, and sleeping. He concluded: "These data do not cluster so closely as the curves concerning annoyance, for reasons having to do with who was counted as seriously disturbed; but nevertheless it is possible to draw meaningful averages from the data."[110]

Schultz's graph is reproduced here as Figure 5.5. The results showed that at DNL 65, approximately 28% say noise disturbs both conversation and TV/radio. In order to protect children's speech perception and language learning at home, it seems more logical to set noise policy based on the curves for conversation disturbance than on the annoyance curves. But U.S. government agencies chose to use Schultz's annoyance curve, *not* the speech interruption curves.

In spite of the caveats Schultz included in his original paper, the Schultz annoyance curve quickly became the center of U.S. noise policy. And partly because of that, a controversy erupted over it a few years later.

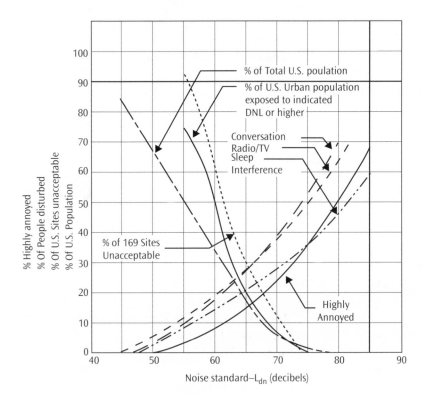

FIGURE 5–5. The original graph from Schultz (1978) showing the "Schultz Curve" for noise annoyance. *Source:* Reproduced with permission from Schultz, 1979, *Journal of the Acoustical Society of America*, 64, 371–405.

The Controversy over the Schultz Curve for Annoyance Two questions fueled the controversy over the Schultz curve. First, should aircraft noise be considered separately from road and rail traffic noise? Second, what level of annoyance should be used in public policy: high annoyance alone, or other levels of annoyance, such as moderate or slight annoyance?

Karl Kryter is a scientist from California who was serving as a consultant in several lawsuits against noisy airports. He looked at the same noise surveys that Schultz had used, plus some others. But Kryter graphed the results separately for aircraft noise, road traffic, and trains. Kryter concluded that aircraft noise is much more annoying than road and rail traffic.[111] Instead of approximately 15% of people showing high annoyance at DNL 65, Kryter said it was 28%. Kryter also noted that aircraft noise complaints begin at DNL 55. Community groups organize and start legal actions at about DNL 65. The GAO[112] also found that at least half of airport noise complaints come from people exposed to *less* than DNL 65. Also, because the percentages of people saying they are moderately and slightly annoyed are as clearly related to noise as are the data for highly annoyed, Kryter concluded that there is nothing magical about using only the highly annoyed.

Schultz[113] penned an argumentative reply to Kryter's[114] 1982 paper. Schultz attacked not only Kryter's published paper but also a previous draft of it. Schultz charged that many of the details of Kryter's analysis were wrong. Schultz objected to the cutoffs Kryter used for "highly annoyed." Schultz claimed that Kryter included only surveys that would support the conclusion that airport noise is more annoying than road and rail noise. Schultz defended his own use of "highly annoyed" by claiming that it provides information about *relative* noise impact." Schultz ended his paper with a sharp accusation of scientific bias: "A comparison of the first and second versions of his [Kryter's] paper suggests that he is determined to reach a preselected goal by ad hoc means."[115] Schultz mentioned twice that Kryter was serving as an expert witness against airports in several different lawsuits and added, "It will be shameful if his testimony prevails."[116]

Kryter published not one but two rejoinders to Schultz's commentary.[117] It is exceedingly rare for the editor of a scientific journal to allow two rejoinders. That alone tells us the arguments were heated. Kryter[118] claimed that his opponent modified his criticisms after seeing Kryter's rebuttal of them, something that would be "dirty pool."

In his rebuttals, Kryter[119] stuck to his guns. He advocated criteria other than the "highly annoyed" category: "For example, revealing only that 5% (according to Schultz) of all people exposed are 'highly annoyed' by an aircraft noise environment at an Ldn [or DNL] of 55, and not, at the same time, indicating that 50% of all these people are 'moderately' to 'extremely annoyed,' could be misleading to some decision makers or persons less than expert in attitude survey techniques and findings."[120] Kryter[121] also pointed out that only a single survey made a direct comparison of road traffic and aircraft noise annoyance. That survey also concluded that aircraft noise is more annoying than road traffic noise. Kryter's[122] parting shot was this: "It is truly unfortunate that [Schultz's] 'synthesis' curve has found its way into various textbooks and guidelines for noise control."[123]

Why did the decision criteria of these scientists differ so dramatically? First, why did Schultz object to separating aircraft and other sources of noise? Second, why use only the "highly annoyed" ratings? Third, why did the controversy center on *annoyance*, not on interruption of conversation and sleep?

Status Quo Science in Noise Annoyance   The predominant theory at that time was that different noise exposures of the same DNL should yield equal reactions from people. Some acoustical scientists summarize this by saying, "Noise is noise."[124] Kryter[125] questioned this theory by concluding that aircraft noise is more annoying than highway or railroad traffic noise at the same

DNL. Schultz and his collaborators held to the holy grail that equal DNL values should result in equivalent reactions. In 1991, Schultz was a coauthor on another review of noise and annoyance.[126] That paper completely neglected interruption of conversation. The 1991 paper also combined aircraft noise with road and rail traffic noise, and did not mention the question of whether it is sensible to combine them. Finally, the 1991 paper did not reference any of Kryter's publications, not even the arguments with Schultz.

The graphs in the 1991 paper[127] show that the majority of the aircraft noise annoyance data fall above the Schultz curve (80 data points above and 34 below the curve). The road and rail traffic data fit the Schultz curve much better (29 above and 44 below).[128] The conclusion that air traffic is more annoying than road traffic was also supported by another research team.[129] More recently, two Dutch researchers calculated curves for annoyance and concluded that "if DNL is used as a descriptor of noise exposure, different curves have to be used for different modes of transportation."[130] The aircraft noise annoyance curve derived by the Dutch team agreed very closely with Kryter's.[131] Approximately 28% of people are highly annoyed at DNL 65. Of course, the conclusion that aircraft noise is more annoying than road traffic does not imply that road traffic noise should not be taken seriously. The Bridge Apartment study shows how negative the effects of road traffic noise can be. Further, as outlined below, I think policy should move beyond its obsession with the category of "highly annoyed."

The "Highly Annoyed" Fiction, Misuses of the Schultz Curve, and Neglect of Speech Interference   You may also be surprised to learn that only a few noise surveys actually used the phrases "high annoyance" or "highly annoyed." Of the surveys in the 1991 review,[132] only one used the term "highly annoyed" as a response category. In the 11 surveys reviewed by Schultz[133] only a Swedish survey used a phrase that translates as "highly annoyed." In light of this, and Schultz's[134] caveats about what to count as "highly annoyed," it is extremely disturbing to find that the concept of high annoyance is the central tenet of noise policy in the United States.[135] This stands in stark contrast to European countries, where a wide range of health impacts are considered.[136]

Even if Schultz's criterion of high annoyance is accepted, the Schultz curve underestimates the proportion of people who are highly annoyed by aircraft noise. The proportion is much higher, and the FAA's continuing problems with community objections to aircraft noise stand as evidence of that fact.[137] The situation is compounded by the fact that the Schultz curve is commonly misinterpreted. The GAO[138] summarized the Schultz curve as showing that "when…sound levels exceed 65 dB, individuals report a noticeable increase in annoyance."[139] The Schultz curve in Figure 5.5 shows that there is no discernible threshold at DNL 65 dB, and there is no particular upturn (or point of inflection) in the slope of annoyance at DNL 65 dB.

The Federal Interagency Committee on Noise (FICAN) "decided not to recommend evaluation of aviation noise impact below DNL 60 dB because public health and welfare effects below that level have not been established."[140] This is astonishing to me. First, a glance at Schultz's original graph (reproduced as Figure 5.5) shows that 18% report conversations are interrupted at 60 DNL. Second, the Hong Kong school study found that 40% of teachers had to pause "very often" at DNL 55! Third, at DNL 55, more than half of the people are *at least* moderately annoyed. The category "moderately annoyed" could just as easily qualify as a noise impact on public welfare, depending on one's particular ethical standards for involuntary exposure. Fourth, the FAA has not examined the impacts of the number of peak noise events or the amount of time that noise exceeds different levels.[141] Schultz himself[142] strongly advocated measuring "the occurrences (levels and numbers) of maximum...noise levels outdoors. These might be associated with identifiable events, such as a fire truck siren, an aircraft flyover, or a train or heavy truck passage. These noise events are the *only* candidates likely to intrude indoors with sufficient intensity to attract the subject's attention and thus generate annoyance."[143] There is also very little research examining how short-term high-intensity noises alter stress hormones, blood pressure, mood, sleep, conversation, school performance, and other indicators of well-being.

The original goal of combining annoyance surveys was to provide an index of "what constitutes a 'suitable living environment'."[144] The question of what constitutes a suitable living environment with respect to noise can be informed by the results of scientific surveys. But the question is an ethical one that can not be decided by science.

Some Good News   FICAN held a symposium on how noise affects children's learning in February 2000, a few months before the FAA issued its draft noise abatement policy.[145] Unfortunately, the draft noise abatement policy continued to neglect the effects of noise on children's speech perception and reading. The symposium included presentations by researchers from the Munich and Three Country research teams. In 2007 FICAN also studied 35 schools in Illinois and Texas where aircraft noise decreased. Twenty schools were near an airport that completely closed, and 15 schools were sound insulated. Noise reduction was associated with a significant improvement in test scores. For high schools, noise reduction decreased the percentage of students failing.[146] Perhaps these findings will convince the FAA that children's school performance is of national importance.

Another bit of good news is that aircraft engines are not as noisy as they used to be. Airlines in the United States were required to have quieter engines on 75% of their large jets ("Stage 3" aircraft) as of the start of the year 2000. According to the FAA, the number of people who are negatively affected by aircraft noise dropped from 2.7 million in 1990 to only 440,000 people in

2000.[147] It would be more honest for the FAA to say that now there are 2.2 million fewer people being subjected to DNL of 65 or greater than there were in 1990. Most of those 2.2 million still are subjected to a DNL of at least 60, a level that creates moderate annoyance in more than 50% of people and interferes with conversation for about 20% of adults. Stage 4 engines are required to be approximately 10 dB quieter, but those criteria will take a long time to be phased in.

Social Justice Issues in Noise

Noise affects quality of life. More than 30% of people spontaneously mentioned noise or quiet in an interview about neighborhood dislikes and likes.[148] No one has claimed that high ambient noise is good for children or adults. If it were, wealthy families would be scrambling to relocate to high-noise areas. Instead, research has found that home prices are lower where noise is higher. In southern Ontario, Canada, house sale prices dipped by $254 for each decibel increase in noise from an arterial highway, and by $312 per decibel from expressway noise.[149] This implies that a new highway that increases noise 15 to 20 dB[150] decreased a homeowner's property value by $4,700 to $6,200 in 1980 dollars. Kryter[151] concluded that real estate prices decrease between 0.5% and 1.0% *per decibel*, depending on the type of housing.

Areas with very high noise are often pockets of poverty. The Heathrow researchers in England commented that "areas of high environmental noise exposure are often also areas with high levels of social deprivation."[152] They said it was difficult to find low-noise schools that matched the high-noise schools on socioeconomic and ethnic variables. The New York University research team that studied airport noise in the mid-1970s noted that, "Transfers [between schools] of minorities for purposes of racial balance have also generally been from noise-exposed to nonexposed areas."[153] A 1998 study of New York City aircraft noise randomly sampled residents of two communities that were in a high- or low-noise impact area. The two communities were selected to be comparable in socioeconomic status. It turned out that 23% of the high-noise survey respondents identified themselves as black or Hispanic, whereas only 3% of those in the low-noise area said they were of those racial or ethnic groups.[154] In the LAX study, the four noisy schools were in a "poverty area"[155] and had a majority of "minority" children—Hispanics and African Americans.[156] A study of all kindergarten to grade 12 schools in California found that the heavier the road traffic near the school, the higher the proportion of African American and Hispanic students. Heavy road traffic schools also had a higher proportion of low-income students.[157] Heavy road traffic brings higher air pollution, an asthma trigger. Airports also are high air-pollution areas.[158]

Hearing Loss, Income, Race/Ethnicity, and
Lead Exposure

In the NHANES survey children from low-income families had more than twice the risk of a high-frequency hearing loss as children from high-income families, and 1.5 times the risk of a low-frequency hearing loss.[159] Mexican American children have a higher chance of hearing loss than either non-Hispanic whites or African Americans.[160] NHANES also found a correlation between hearing thresholds and blood lead in children aged 4 to 19 years.[161]

The Whole Picture

First, most of the studies of airport noise around schools ended up with samples that were predominantly low-income, with a high proportion of minority group members. Second, the U.S. government's airport and highway noise criteria are not adequate to protect children's speech perception in school, even for those children who hear normally. Third, lower-income children have a higher chance of slight hearing loss than high-income children. Fourth, low-income children have a higher chance of having high lead exposure than high-income children. Lead exposure affects hearing, intelligence test scores, and behavior. Add these four findings together and it paints a rather ugly portrait of educational opportunity in America.

Noise and Wildlife

I will mention five categories of research here: noise and marine mammals; highway noise and birds; effects of ecotourism on monkeys in the Amazon; effects of aircraft noise on ungulates and waterfowl; and the need for natural soundscapes in natural areas. Most of the research that I have read on these topics seems to take the point of view that "the jury is still out" with respect to whether human-created noises actually harm animals. Some researchers assume that the mere presence of animals in noisy areas indicates that they are not being harmed. What conclusions one draws from the research will depend on what decision criteria one applies for determining whether the animals are being harmed.

Ecotourism in the Amazon

In Ecuador, South America, Stella de la Torre and Charles Snowdon of the University of Wisconsin, along with Monserrat Bejarano, of an Ecuadoran university, studied the effects of tourists on a small species of monkey, the pygmy marmoset. These animals use intricate trills to communicate with one another when they are foraging. During heavier tourism there were fewer trills by the animals, more use of the upper branches in the trees, and use of a larger home range area. The research team recommended that "tourist groups should always be led by trained guides working in coordination with the Reserve authorities to control and stop the capture of wild animals, to reduce the use of motor boats, which significantly increased the levels of ambient noise, and to create a sense of awareness in the tourists about the potential negative impacts they may be causing on the ecosystem."[162]

Underwater Noise and Marine Mammals

Many species of whales and dolphins have "exceptional hearing and sound production capabilities."[163] Marine mammals vocalize to communicate among themselves, but some species also vocalize in order to search for prey, and may even use very loud vocalizations to disable prey.[164] There are many studies of the effects of underwater noise on marine mammals such as whales, dolphins, seals, and sea lions.[165]

One study tested the hearing thresholds of dolphins and white whales by training these clever animals to whistle when they heard a tone.[166] Once the animals were well trained, the research team could test the effects of loud noises on their hearing. The tests involved loud sounds that were similar to sonar "pings" in frequency, amplitude (dB), and their impulsive characteristics. After hearing a loud ping, the researchers then presented a series of softer tones to see if the animal could hear them. This allowed the researchers to graph the hearing thresholds of the animals. After hearing the loud pings (180 dB) the animals had a temporary hearing loss between 6 and 17 dB. The louder the pings were, the more likely the animal was to show irritation. One dolphin attacked the station where the loud noise originated. Others slapped their tails and popped their jaws. And the animals increased the time they took to swim from the location of the loud ping noise to another location they had been trained to go to listen for the hearing test tones. These behaviors show that the sonar-type pings were unpleasant and stressful to the animals.

Icebreaker ships make noise greater than 190 dB. This much noise would be audible to beluga whales from at least 35 km (21 miles) away and would probably cause the whales to avoid the area.[167] A temporary hearing loss could occur when the whales are within about 1 to 2 miles (about 1 to 4 km) of the ship for 20 minutes or longer. But icebreakers are not the only noise in the arctic. There can be seismic exploration, oil drilling, helicopters, sonic booms, and factory fishing vessels. The whale researchers said, "Summed noise levels could be very high and ongoing for long durations and cover large areas such that animals might either be permanently scared away from critical habitat or be adversely affected because they have nowhere to flee to . . . projects focusing on critical locations and addressing a large variety of man-made noise are needed."[168]

Aircraft Disturbance of Wildlife

Low-altitude military practice flights are often over wildlife refuges and other areas where only a few people live. In the United States, the military must file the appropriate paperwork with the EPA whenever it proposes to change its low-altitude operations (either an environmental assessment or a full environmental impact report). Studies of how low-altitude military flights affect animals are funded by the military because the data are needed for environmental impact reports.

One research team summarized the effects of aircraft on ungulates by saying, "Resource management agencies have been addressing wildlife and habitat disturbance for many years and have identified a variety of problems regarding aircraft disturbance of ungulates, including (1) movement from the area of disturbance and changes in habitat use by mountain sheep...and mountain goat..., (2) decreased foraging efficiency of desert bighorn sheep..., (3) increased heart rate in mountain sheep...and desert mule deer..., (4) panic running..., (5) decreased frequency of nursing..., (6) overt behavioral responses by barren-ground caribou...and woodland caribou..., and (7) decreased calf survival of woodland caribou."[169]

Another group of scientists summarized the research literature as showing that geese are particularly sensitive to aircraft disturbance, including small planes and helicopters.[170] There is controversy in the research about whether animals habituate to aircraft flyovers and noise, and not all studies find negative effects of aircraft on wildlife.[171] The criteria for what constitutes harm varies greatly across studies.

Highway Noise and Birds

There are thousands of scientific studies of the effects of roads on wildlife.[172] Highway traffic noise has the potential to mask the calls of birds, making communication more difficult. A German scientist found evidence supporting the masking theory. Forest bird species whose song pitch does not overlap with the traffic noise were less impacted by a road. Species whose song pitch was masked by the traffic noise were less abundant closer to the road.[173]

Noise in National Parks and Wilderness Areas

The National Park Service has been criticized for promoting high-noise visitor activities such as snowmobiling and "flightseeing." How much noise annoyance should be tolerated in National Parks such as Grand Canyon, Denali, and Yellowstone? Among acoustical scientists there is a debate over this topic that echoes some of the issues regarding noise annoyance. FICAN is now measuring noise in national parks. Another issue is how noise affects wildlife in those areas. As noted above, research is quite incomplete on the latter point.[174]

CHAPTER 6

It Isn't Fair: Environmental Pollution Disasters and Community Relocations

We all contribute to pollution. Pollution would be equitably distributed if, whenever you purchased an item or turned on the lights, you also received a packet of wastes that you had to keep, the by-products of whatever you consumed. But it is not that way. Instead, the waste products pile up in certain places closer to some people than to others. At the Chernobyl nuclear plant disaster, the people who lived nearby suffered varying degrees of exposure to radioactivity so that a large region of the former Soviet Union could be supplied with electricity. People close to the Nevada Test Site received more than their share of fallout so that the United States could develop the most destructive nuclear arsenal on the planet. People at Love Canal in New York were subjected to toxic chemicals in their basements and schoolyard so that others could use herbicides, insecticides, and marvels of the modern chemical industry.

In this chapter I describe the psychological aftermaths of going through a pollution disaster. In each of the pollution crises, people were uncertain about how much pollution exposure there was. But they ended up having to evacuate their homes temporarily or permanently, voluntarily or by force.

Policy decisions in pollution crises are always controversial. There is a pollutant present, but its exact nature, the extent of human exposure, and the severity of the hazard it poses are not well understood in the heat of the crisis. In the disasters I describe, the extent of the threat was initially uncertain. It became evident only over time. Because of the many uncertainties, government agencies and communities can easily become entangled in a web of conflict and mistrust. Government officials face a dilemma. A pollution crisis seems to pit the inequity of risk to one group against the rest of the society. Should the government spend tax money or require a corporation to pay to relocate a polluted neighborhood? Should a community have the right to refuse to accept a hazard such as a chemical disposal facility, a nuclear power plant, or a bomb test site? How much compensation do people deserve if they are irreparably harmed by such a facility?

Communities in the midst of a pollution crisis are often fractured along lines corresponding to perceptions of the risk. People within the community have different opinions about the severity of the situation. Opinions differ about what ought to be done. People living through a pollution crisis face double jeopardy. The pollution itself can affect their psychological functioning, physical health, or both. And stress comes from the knowledge that the pollution exists, the uncertainty about exposure and the effects of the pollution, when or whether the government will act to help, and the evacuation or lack of an evacuation.

Government agencies often cannot answer questions about the severity of the hazard. Exposures usually have to be reconstructed retrospectively. These estimated dosages can be helpful, but they are not direct measures of internal exposure. The exposure estimates themselves sometimes lead to disputes. This happened at Love Canal and also at the nuclear accident and bomb test sites.

I examine two facets of pollution crises and disasters. The pollutants themselves can directly affect the health and psychological well-being of children and their families. But the indirect effects of the psychological stress are equally important. Keep in mind that the environmental crises in this chapter concern unjust suffering. The people at Love Canal, Chernobyl, and Three Mile Island were not at fault for their pollution crises. It could just as easily happen to you or me. The unfairness of pollution crises can add to the psychological stress.

Some pollution disasters have led to greater knowledge and a better world for the rest of us. The mercury poisonings in Minamata and Iraq and the PCB poisonings in Japan and Taiwan are examples that led to knowledge that helped the rest of us. The benefit of the knowledge from these disasters does not heal the individuals who suffered. I have written most of this book with a focus on hard-nosed empirical science. However, science cannot give us a heartfelt appreciation of the agonies, anger, and moral outrage experienced by those who are victims of pollution crises.

Part I: Radioactivity: Three Mile Island, Chernobyl, and the Legacy of Nuclear Bomb Testing

I don't think they knew what they were doing.—a U.S. veteran describing his experience in military maneuvers during a nuclear bomb test (quoted in Garcia, 1994, p. 654)

Do we know what we are doing now? Radioactivity is one of the most dreaded types of pollution,[1] and also one of the most controversial. Scientists have argued about how hazardous radiation is from the time it was discovered. We start our atomic journey quietly in Pennsylvania, proceed to the former Soviet Union, and end with a thermonuclear blast in the tropical paradise of the Republic of the Marshall Islands.

The hazards of chronic radiation were not understood very well initially. I work in a building named for Wilfred Brogden. Prior to becoming a professor at the University of Wisconsin, Brogden studied the effect of x-rays on hearing thresholds in animals for his Ph.D. at the University of Illinois. He severely damaged his right foot with radiation while doing that research in the 1930s. What is presumed safe at one point in history is found to be harmful later. Brogden personally suffered the consequences of the presumed safety of ionizing radiation.

The Nuclear Reactor Accident at Three Mile Island

What Happened

The Three Mile Island (TMI) nuclear power plant is located near Middletown, Pennsylvania. It is on the Susquehanna River, just about a dozen miles south of the state capital, Harrisburg. Metropolitan Edison's two reactors began producing electricity in 1974 and 1978. The second reactor had been running for only 3 months when, at about 4 in the morning on Wednesday, March 28, 1979, an accident began. The result was a release of radioactivity and a partial meltdown of the fuel core. By 7:30 a.m. the TMI plant declared a general emergency. The radiation monitor in the stack vent set off the alarm. Radiation exceeded the maximum that the stack monitor could record during the early part of the accident. The Pennsylvania Bureau of Radiation Protection was not notified that radiation was detected off the plant grounds until 10 a.m.. Metropolitan Edison's field crews had found excess radiation on the west shore of the Susquehanna River at about 8:30 a.m. Kunkel School, approximately 6 miles to the west-northwest of the TMI reactors, had the highest radiation reading (13 mR per hour) at 11:30 a.m. on the first day of the accident.[2]

At the same time that staff started monitoring radiation readings around the local area, Metropolitan Edison officials held a press conference. They assured the public that there was no danger.[3] From the citizen's perspective the events of that Wednesday did not seem alarming. But conflicting reports began to emerge. At 4 p.m. the mayor of Middletown was finally told that a slight radiation release had occurred. The mayor appeared in public with a Geiger counter. The mayor of Goldsboro, a town just 1½ miles to the west of the plant, went door to door talking with people about the possibility of an evacuation.[4]

Meanwhile, "the radiation releases from the plant continued."[5] The U.S. Department of Energy sent a helicopter to sample the air in the vicinity of the plant. By Thursday morning, March 29, the TMI accident had become a major media event. Governor Thornburgh's press conference that day emphasized that there was no danger to the public.[6] But at 8 a.m. on Friday the radiation in the stack vent rose unexpectedly. This led some Nuclear Regulatory Commission members in Washington to recommend an evacuation.[7] Hydrogen had built up inside the plant, and there was a chance that the hydrogen would explode again. In the worst case scenario if the hydrogen exploded it could break the containment building and allow larger amounts of radiation to spread. It was not until shortly after noon on Friday, March 30, that Governor Thornburgh recommended a partial evacuation. He urged pregnant women and those with preschool children within 5 miles of the nuclear plant to evacuate the area. The partial evacuation was not called until more than 48 hours after the accident began.

After the advice to evacuate many communities sounded their emergency sirens.[8] The mayor of Middletown had issued police a "shoot to kill" order in the event of looting of evacuated homes, adding the feeling of a state of siege.[9]

More than 60% of people within 5 miles of the plant evacuated, and more than three-quarters of those with preschool children left.[10] Some families were divided in their opinions about evacuating, creating added stress.[11] By Saturday, March 31, the hydrogen inside the plant had dissipated. Everyone breathed a sigh of relief. On Sunday, April 1, President Jimmy Carter[12] and his wife Rosalynn toured the damaged reactor. But most schools within 5 miles of the plant remained closed for the following week. The advisory to evacuate was lifted on April 9.[13]

How much radiation was released? After the TMI accident it was estimated that 13 to 17 curies of one substance, radioactive iodine, were released.[14] (Table 6.1 gives a list of some radioactive isotopes and their half-lives, and Table 6.2 provides a brief glossary of terms related to radiation.) For comparison, the Chernobyl accident released approximately 7 *million* curies of radioactive iodine.[15] Health officials try to keep a close eye on radioactive iodine, because it enters the food chain and can be passed into cows' milk and consumed by children (as well as adults), where it can damage their thyroids. Radioactive iodine was found to be higher in the thyroids of wild meadow voles trapped about a mile from the plant compared to those captured about 8 miles away in early April.[16] This finding met with controversy.[17]

But the TMI accident also released a plume of the radioactive gases xenon (half-life of about 5 days) and krypton (half-life of more than 10 years).[18] The gas plume contained 2.4 to 13 million curies of radiation.[19] The initial radioactive plume was detected in the air 225 miles away in Albany, New York, on Thursday and Friday of the accident. The weather was "rather stagnant" the first day these radioactive gases were released, so they would have remained

TABLE 6–1. Half-lives of some radioisotopes

ELEMENT	HALF-LIFE
*Ba-140	13 days
C-14	5,720 years
*Cs-137	30.1 years
Ce-144	32 days
Co-60	5.2 years
*I-131	8 days
*Kr-85	10.8 years
*Pu-238	86.3 years
*Pu-239	24,360 years
*Pu-240	6,575 years
*Pu-241	13.1 years
Ra-226	1,622 years
Rn-222	3.8 days
*Sr-89	53 days
*Sr-90	27.7 years
Th-232	14 billion years
U-235	713 million years
U-238	4.5 billion years
*Xe-133	5.3 days

Sources: Adapted from Ginzburg & Reis, 1991, Table 1; Smith, 1969, Table 23.1
*Released in Chernobyl reactor accident

in the TMI area before being blown away as they decayed.[20] On Saturday, March 31, the highest radiation reading of 38 mR per hour was recorded just to the northeast of the plant. At about the same time, the EPA began installing additional radiation dosage monitors (called TLDs, or thermoluminescence dosimeters) around the area, as did the Department of Energy, the Nuclear Regulatory Commission, and the U.S. Department of Health, Education, and Welfare. All these agencies deployed a total of 333 additional TLDs. But only 20 radiation detectors were present when the accident began.[21] And only five TLDs were in inhabited areas within 5 miles of the plant.[22] Ten dosimeters were on the island occupied by the plant or on other nearby islands in the Susquehanna River. This means that the radiation dosages in approximately a 19-square-mile area occupied by about 35,000 people had to be estimated from 5 dosimeters, and the plant radiation detectors, along with the wind and weather conditions during the accident. Remember the stack vent went off-scale during part of the accident. Radiation from the accident continued to be released unpredictably until April 4, as well as during the later cleanup operations. The 333 extra dosimeters were not in place in time to be used in most dosage estimates for the accident itself. Nevertheless, government reports issued soon after the accident gave estimated exposures of between 20 and 70 millirem for people on the east bank of the river, and less than 20 millirem

TABLE 6–2. A glossary of radioactivity terms

alpha radiation	Nuclei of helium that have a positive charge. This type of radiation is emitted by radium, plutonium, thorium, and uranium.
background radiation exposure	Internal and external exposure to radiation that comes from cosmic rays and natural terrestrial sources such as potassium-40, uranium, radon, and their decay products in soils and rocks. For current information on estimated background radiation see UNSCEAR (2000).
beta radiation	Decay of a neutron into an electron and a proton. The electron is emitted from the atom, and a gamma ray may also be present.
Curie	A measure of the emission of radioactive particles in disintegrations per second. 1 Ci = 3.7 × 10E10, or 3,700,000,000 disintegrations per second.
gamma radiation	Electromagnetic radiation emitted when an emitted particle does not contain all the energy available from nuclear decay.
Gray	A measure of radiation energy transfer. 1 Gy = 1 joule/Kg; 1 joule = 4.1868 calories. The Gray replaced the rad as a measure of absorbed radiation dose.
half-life	The time that it takes a substance to release one-half of its radioactivity. If an isotope has a half-life of 30 years, then after 30 years it will have 1/2 of its original radioactivity, after 60 years it will have 1/4, and after 90 years it will be 1/8 as radioactive as it was originally.
LD50	The amount of exposure to any toxic substance that is lethal to 50% of the species exposed at that level. For humans, the LD50 for radiation is approximately 4 Gy whole body radiation.
nuclear fission	Splitting of the nucleus of an atom. Fission reactors involve bombarding U-235 or plutonium with slow neutrons. Fission is a self-sustaining reaction that breaks the original atoms into smaller atoms accompanied by the release of large amounts of energy and radioactive isotopes.
nuclear fusion	A reaction of two small nuclei that produces a larger nucleus and releases a large amount of energy. Hydrogen bombs (thermonuclear) involve a fission and fusion reaction. The earth's sun produces heat by fusion reactions.
rad	A measure of absorbed radiation dose. 100 rad is approximately equal to 1 Gy.
rem	A measure of absorbed radiation dose. *See* Sievert.
Sievert	A measure of absorbed radiation dose that is weighted by a factor expressing its linear energy transfer. Alpha radiation is normally weighted by a factor of 20. 1 millisv is approximately equivalent to 100 millirem. The Sievert replaced the rem as a measure of dose equivalent.
x-radiation	Electromagnetic radiation that is emitted when electrons are removed and other electrons are rearranged in the atom.

Sources: Compiled from Amdur et al., 1991; Oakley, 1972; Warner & Kirchmann, 2000

for others living within 2 miles of the plant.[23] There were many problems in estimating dosages. The dosimeters did not record beta radioactivity, the wind and weather records did not allow the scientists to predict where the plume of radioactivity came near the earth, the exhaust stack release rate was never directly measured, and radioactive gases such as xenon and krypton were not directly measured.[24,25]

Controversy Continues over TMI's Health Effects

Whether the radiation from the accident was enough to cause health problems itself remains controversial. This book focuses on the *psychological* effects of pollution. Psychological effects can result from exposures that are directly toxic to the biological substrates of behavioral functioning or from indirect effects. However, before turning to the psychological effects, I give a brief synopsis of the controversy over cancer in the TMI area.

Official pronouncements soon after the TMI accident said that the likelihood that radiation had caused immediate health effects was virtually nil. Also for the long term they said, "its potential carcinogenic, mutagenic, and teratogenic effects combined add up to only about a one-in-a-million risk of death."[26] Jacob Fabrikant,[27] the director of the Public Health Safety Task Force of the President's Commission on the Accident at Three Mile Island, wrote: "We can conclude, therefore, that since the total amount of radioactivity released during the nuclear reactor accident at Three Mile Island was so small, and the total population exposed so limited, that there may be no additional detectable cancers resulting from the radiation."[28] For reproductive effects Fabrikant[29] said: "We can conclude, therefore, that no case of developmental abnormality can be expected to occur in a newborn child as a result of radiation exposure of a pregnant woman from the accident at Three Mile Island."[30]

After the accident some residents testified in sworn statements that they had experienced symptoms consistent with radiation poisoning, such as red skin, hair loss, and vomiting. Others testified that pets had died.[31] There were also reports that cows nearby died unexpectedly[32] and that some farm animals had had miscarriages. Farm animal deaths were not investigated as systematically as would be desirable.[33] Some residents attributed the sudden deaths of trees to radiation. In court, plaintiffs brought in a former Soviet scientist who testified that killed trees appeared very similar to radiation-killed trees in areas of the former Soviet Union where radiation releases had occurred.[34] The court ruled that plaintiffs had not provided sufficient evidence that the TMI radiation releases were causally related to their illnesses.[35]

Researchers have found an increase in all cancers, and leukemia, in the area. The scientific controversy is over whether the amount of radiation exposure could be responsible for the increased cancer, or whether the results are attributable to factors such as stress or other confounding variables.[36] One group of researchers interpreted the data as showing that "overall, the pattern of results does not provide convincing evidence that radiation releases from the Three Mile Island nuclear facility influenced cancer risk during the limited period of follow-up."[37] Using the same data, another set of researchers concluded that "cancer incidence...increased more following the TMI accident in areas estimated to have been in the pathway of the radioactive plumes than in other areas.... Causal interpretation is further strengthened by the observation that...higher and lower dose study tracts are all within 10 miles of the source and differ in exposure only as a function of weather conditions at the time of the accident."[38]

Biases in the interpretation of results have been implied in the commentaries.[39] The reanalysis of the data by Wing and colleagues[40] was funded by a grant to the University of North Carolina from attorneys for approximately 2,000 TMI-area residents suing for damages.[41] The original data collection was funded by money from the utility company administered by the court. During the litigation, the court set restrictions on the radiation dose estimates such that the estimated health effects could not exceed 1%.[42] The court also required that any dose estimates be agreed to by the nuclear industry representatives.[43] Wing writes, "The studies themselves were funded by the nuclear industry and conducted under court-ordered constraints, and a priori assumptions precluded interpretation of observations as support for the hypothesis [of radiation-induced cancer]" (Wing, 2003:1816). The controversy is likely to continue: "Despite a century of research since Roentgen's discovery of X-rays, fundamental disagreements exist over biophysical mechanisms, dose-response assumptions, analytical strategies, interspecies extrapolations, and the representativeness of studies of select human populations."[44]

In the research on disasters involving radioactivity, the retrospective dosage reconstructions are sometimes interpreted as if they were direct measures of individual exposure. The absolute estimated exposure values are often used to decide whether or not differences in disease or psychological functioning should be attributed to the effects of radiation.[45] Let me make an analogy to lead exposure. Suppose that a researcher studied children in an area in which leaded gasoline is used. Using data on air pollutant dispersion from major highways, the researcher could construct a model of how much lead exposure was received by children at different distances from the highways. These estimates could then be used *as if* they were measures of lead exposure. Suppose that the researcher also found a significant relationship between estimated lead exposure and IQ scores (including appropriate confounding variables). But now imagine that the researcher concluded that the lower IQ scores *cannot* be attributed to lead because the estimated lead exposures were too low to affect IQ. This is analogous to some of the TMI research. Estimated radiation exposures from a model of air flow are said to be too low to be responsible for illnesses. The assumption about how much exposure is needed to yield a particular effect is being given primacy over an association between *relative* exposure and outcomes.

The Psychological Impacts of the TMI Accident

In contrast to the official estimates that there would likely be no detectable increase in cancer or birth defects among people living close to TMI, the effects of stress on psychological well-being have been readily acknowledged. A point I have made repeatedly in this book is that the psychological effects of pollution are real effects that have enormous impacts on our daily lives. This was certainly true of the TMI accident.

Technological failures can pose unique psychological problems to the victims.[46] Technological failures involve a loss of societal control and a loss of trust in authority and experts. People who lived close to TMI went through a disturbing crisis in which accurate information was unavailable.[47] There was the risk of exposure to radioactivity in uncertain quantities. There were also the trauma and stress of temporary evacuation. There was the uncertainty of the crisis's outcome, and the uncertainty about how much exposure to radiation from the accident had already occurred.

The consensus among social scientists is that this very stressful event had relatively long-lasting consequences for psychological well-being.[48] The TMI accident was well researched by social scientists. People from the TMI area fared worse than comparison groups on measures of stress and emotional functioning almost 5 years after the accident.[49] It is important to keep in mind that the possibilities of large-scale evacuation and exposure to uncertain quantities of radiation are inherent in nuclear power. Wherever there is a nuclear power plant or nuclear weapons facility, there must be an evacuation plan and a radiation monitoring program. This is even more important now in the age of terrorist threats. The potential for evacuation and the *psychological* impacts of an accident should be included as a social impact in weighing the benefits of nuclear power. As a result of the TMI incident, social scientists know more about those impacts than they did before the accident.

Mothers with Preschool Children Mothers with preschool children experienced the most stress of any demographic group. One researcher compared TMI mothers of preschool children to others near TMI and to people living near a coal-fired electricity plant or near a different nuclear plant (the Shippingport plant near Pittsburgh).[50] The study assessed mental health, beliefs about whether the TMI plant was dangerous, and distance of residence from the plant. The assessments were done 5 times after the accident (9 months after; at 1 year, 2 years, and 3 years after; and approximately 6 years later, in 1985, when the plant was restarted).

As time passed, differences between mothers who did and did not believe that the plant was dangerous became greater. Three and a half years after the accident, mothers who thought TMI was a hazard had 3 times the risk of depression or anxiety during the previous year, compared to TMI mothers who did not believe the plant was dangerous. In contrast, plant workers had fewer long-term psychological adjustment problems than the mothers, and the workers' problems disappeared after the first year.[51] The authors concluded, "The TMI accident has had a long-term adverse effect on the mental health of the mothers of young children, particularly those living within 5 miles of the plant when the accident occurred and those continuing to perceive TMI as dangerous years later."[52]

A 10-year follow-up of TMI mothers found that approximately one-third of them had recurrent psychological distress. At the time of the accident the women with recurrent distress believed more strongly that the power plant was dangerous, and they were more likely to have evacuated. Their symptoms were highest at key points later such as the restart of the undamaged TMI reactor and the 10-year anniversary of the accident. This fact led the authors to conclude that, "... at least some portion of their distress was attributable to this chronic stress situation."[53]

Research on fathers of preschool children at TMI was apparently not carried out. Studies of other disasters (floods and chemical pollution in Missouri) suggest that men and women often differ in their reactions. One study found that in both genders there was an increase in depression and somatic symptoms 11 months after a disaster. But for males there was also an increase in alcohol abuse and these effects varied depending on social support from others. The researchers concluded: "Men are more adversely affected by personal exposure to disaster than are women....Only when exposure [to disasters] is accompanied by heavy demands for nurturance—an obligation traditionally associated with the female role—does it have a negative impact on women's mental health."[54] Given these findings, it is unfortunate that fathers at TMI were not studied as thoroughly as the mothers.

Children and Youth at TMI   A study of teenagers (7th, 9th, and 11th graders) approximately 2 months after the TMI accident found that girls remembered being more distressed than boys. Youth whose families evacuated remembered more psychological distress than those who did not evacuate. Two months after the accident, teenagers with a preschool sibling reported more psychological distress than others. In the 7th and 9th graders, there were more somatic symptoms (headaches, stomachaches, and so on) than in the older students.[55]

Three and a half years after the accident, researchers interviewed children 8 to 16 years old and their mothers. There were four groups: those who lived near the TMI plant, those with a parent employed at the TMI plant, those who lived near another nuclear plant, and those with a parent employed at the other plant.[56] Mothers completed a questionnaire about the children's social competence, behavior problems, and the mother-child relationship. The children answered a fear survey, a self-esteem questionnaire, and were interviewed about the TMI accident and knowledge of nuclear power. The results showed no significant differences among the four groups of children in how upset they were, although the averages were in the direction of the TMI children having slightly worse psychological adjustment. More fine-grained analyses showed that how upset the TMI children were, or their mothers said the children were *during* the accident, was related to the children's fearfulness 3½ years later. These correlations accounted for a maximum of 10% of the differences among children in their fearfulness. Better mother-child support was related to better child self-esteem, fewer behavior problems, and better social competence. The research team concluded that children adjust well "over time when faced with the stresses caused by...man-made events" but that "children who initially were upset by the accident may continue to be more vigilant and unable to deny the situation's severity."[57]

This study was based on retrospective interviews with the children and their mothers. But I included it because there is so little data on how children react to pollution disasters. As the researchers noted, the results of any retrospective study can be partly due to reporting bias. Those children and mothers who are not doing as well could reconstruct the past to be consistent with their current functioning. Also, the study compared the TMI children with children who lived near the Shippingport reactor, the first commercial reactor built in the United States. This comparison may underestimate the impacts of the accident on the children. The Shippingport plant had also been the subject of controversy, and hearings on a plan to build another reactor there had been held.[58] Almost all nuclear power plants were controversial in some way during this time. The public was becoming more skeptical of nuclear power safety even before the accident.[59]

Risk Perception, Stress, and Coping at TMI

The research on the TMI accident found that mothers of preschool children living within 5 miles had a higher likelihood of long-lasting negative effects on their psychological well-being than other adults. This problem was exacerbated among those who believed the plant was dangerous. How children adjust to most stresses is linked to how the rest of the family reacts.[60] For this reason, it is important to look at the adult coping strategies at TMI. Also, the same general principles of stress and coping seem to apply to children under stress, depending on how old they are and how much they can understand about a situation.

What people think and do about threatening situations can affect the amount of stress those situations create.[61] There are three major variables:

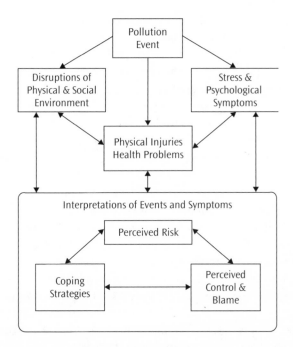

FIGURE 6-1. Diagram of variables that influence the impact of a major pollution disaster. The impacts will depend in part on the cultural and community context.

threat appraisal and risk perception; coping strategies; and perceptions of control, cause, and blame.[62] These three aspects of stress and coping are intimately related to each other, the severity of the event, and the severity of the event's impact on a person. In Figure 6.1 I have outlined some aspects of stress and coping.

Risk Perception or Threat Appraisal If you do not regard a situation as threatening, then until you realize you have been harmed, you will not be stressed. On the other hand, if you think something will harm you, then you will likely be fearful, but that risk perception may also lead you to take steps to avoid the hazard if you can.

At TMI risk perceptions were relatively stable for 3½ years.[63] Six months after the accident, those who perceived higher danger also showed higher psychological distress. Three and a half years after the accident, perceived harm to health was still significantly related to psychological distress.[64] The TMI mothers who believed the power plant was dangerous were more likely to show depression or anxiety 3½ years after the accident than those who did not perceive the plant to be dangerous.[65]

Two Coping Styles Perception of a threat calls for coping. Problem-focused coping centers on action. Pack your things and get ready to evacuate your family. Plan what highway to take, make arrangements to stay with friends in another city, start a citizen advocacy group, and so on. In contrast, emotion-focused coping is oriented toward fixing our feelings, not the world around us. Focus on positive aspects of the situation. Make jokes. Seek comfort by talking with someone you feel close to. Drink or take drugs. Exercise or play games. Participate in religious services or rituals, and so on. Some coping actions are better than others. Good coping involves both problem- and emotion-focused strategies.[66]

A research team from the Uniformed Services University of the Health Sciences (a part of the U.S. military), studied coping at TMI. Residents who did the least emotion-oriented coping were also likely to report more depression and more symptoms of psychological distress than those who were high in emotion-oriented coping. Those highest in problem-oriented coping, however, reported the highest symptoms of distress and depression. Stress-related hormones in urine (norepinephrine) were also higher in the high problem-oriented copers.[67]

Why didn't problem-oriented coping reduce stress at TMI? Perhaps because people were not able to alter the situation. This would be true for many kinds of pollution. For example, calling to complain about a smelly factory[68] or about a noisy airport is problem-oriented coping. But such actions are very unlikely to reduce the odor or noise in the short-run. Another interpretation is that people who report high problem-oriented coping might actually be experiencing more negative effects than others. A more severe pollution event is likely to require more action, and therefore more problem-oriented coping. Suppose you or one of your relatives were accidentally exposed to enough radiation to turn your skin red, or make your hair fall out, as some people at TMI testified.[69] You might adopt a predominantly problem-oriented coping style. Write a letter to the company requesting medical expenses, talk to an attorney, discuss options for action with neighbors who had similar experiences, and so on. While you are doing all this, you could also adopt emotion-focused coping strategies of various sorts. Also, taking action about a pollution problem can solidify a person's belief that the problem is a health risk. Then, because the risk is not only actually more serious but also is viewed as more serious, a person may feel even more psychological distress from exposure. Notice the catch-22 here. Doing something about a problem might make it worse. Regardless of why, at TMI problem-oriented coping was associated with higher psychological distress.[70]

Perceived Controllability and Blame When people *think* that they can't control important events, it can decrease motivation and increase negative emotions.[71] The Uniformed University research team at TMI also assessed people's beliefs about the controllability of life events in general.[72] More than a year after the TMI accident, while the accident cleanup was going on, the TMI residents who felt the least control showed higher somatic symptoms (such as digestive problems, nausea, headaches, and so on), higher anxiety, worse depression, and had higher levels of norepinephrine than a comparison group from Maryland or than TMI residents who felt more control.[73]

The Uniformed University researchers also examined how blaming was related to adjustment after the TMI crisis.[74] TMI residents who said they blamed themselves for their overall life situation showed better adjustment than those who took no personal responsibility. Lower self-blame was associated with more somatic symptoms, more depression, and slightly higher norepinephrine. The Uniformed University authors said, "Some assumption of personal responsibility for problems created by a technological accident or mishap is associated with resistance to stress-related difficulties."[75]

These findings on self-blame are a bit surprising. The TMI accident was clearly the responsibility of the utility company, not the public. But blaming others often brings anger. Chronic anger is related to higher chronic stress. Blame is also a term with moral connotations. Blame calls for censure, punishment, and reparation. Injustices provoke anger in victims, producing a double injustice. The unjust situation itself is compounded by the anger and stress of having been wronged. Crime victims are painfully aware of this.

Summary of Stress Aftereffects of TMI

People near TMI at the time of the accident experienced both acute and chronic stress. Worse psychological adjustment years later was associated with living in the 5-mile evacuation zone, being a mother or teenage sibling of a preschool child, believing the plant was dangerous, being high in problem-oriented coping and low in emotion-oriented coping, feeling the situation is not personally controllable, and blaming others.

Psychological stress is also linked to physical health. Evidence in Chapter 5 on noise showed that living in high noise is related to higher blood pressure. Immune system functioning and inflammatory processes are influenced negatively by psychological stress.[76] Stress increases allergic reactions.[77] Higher stress leads to higher susceptibility to upper respiration viruses.[78] After an acute stressor such as an earthquake there is an increase in heart attacks (myocardial infarct) and pulmonary embolisms.[79] Chronic stress increases the risk of heart attack.[80] Finally, stress is associated with a higher incidence of cancer and a lower survival rate for cancer.[81] When the TMI residents reported higher somatic symptoms on the questionnaires, it is possible that the stress of the accident was a contributor to real physical illnesses.

Nuclear Power: Are Public Risk Perceptions Irrational?

Risk perceptions at TMI were associated with worse psychological well-being. Risk perceptions are nearly always the dividing line between advocates and opponents of nuclear power. One argument is that TMI created long-term stress because some people had incorrect perceptions of the risks of nuclear power and the accident. According to this argument, people would not have been so upset if they had known more about the risks of nuclear power and low-level radiation.

Public understanding of nuclear power and radiation could certainly be improved. However, scientists have disputed the short- and long-term safety of radioactivity since it was discovered. What was initially thought to be safe was later found to be harmful. Up until the 1950s, women were sometimes X-rayed during the last trimester of pregnancy to ascertain the position of the fetus.[82] In 1958 Dr. Alice Stewart and her colleagues in Britain found an increased incidence of childhood cancer after prenatal X-rays.[83] Her research met with controversy.[84] Conflicts among scientists regarding the safety of nuclear power escalated in the late 1960s when John Gofman, Arthur Tamplin, and Ernest Sternglass, scientists who worked in the nuclear industry, published papers reporting that the routine releases of radioactivity from nuclear power plants were hazardous.[85] Gofman was the scientist who isolated enough plutonium for the experiments that led to the development of the first plutonium based atomic bomb. He had no regrets about participation in the development of atomic weapons to stop Hitler. After studying the health effects of ionizing radiation, he became an opponent of nuclear power and helped found the Committee for Nuclear Responsibility.[86]

Many scientists took sides. In 1977, before the TMI accident, one scientist wrote, "Evidence of the escalating conflict over nuclear energy policy is particularly abundant in the scientific community.... A leading journal recently rejected an article by nuclear critics because of its advocacy tone and later accepted one by a proponent of nuclear power, which provoked a stinging rebuttal."[87]

More is known about the risks of exposure to low-level radioactivity now. But the state of knowledge about long-term effects still precludes either side from claiming to have "the answer."[88] The National Academy of Sciences Committee on the Biological Effects of Ionizing Radiations (called BEIR) issues reports on radiation hazards at least every decade. Newer reports usually estimate either a higher probability of damaging effects or a lower dosage for damage than do older reports. For example, BEIR V said, "The frequency of severe mental retardation in Japanese A-bomb survivors exposed at 8–15 weeks of gestational age has been found to increase more steeply with dose than was expected at the time of the BEIR III report."[89]

Would improving public knowledge of nuclear power make public attitudes more positive? In the 1970s, approximately 80,000 people in Sweden participated in a project to increase knowledge about energy options. People met in small groups for a total of at least 10 hours. After the educational program, the participants' attitudes still indicated serious concerns about the safety of nuclear power. The proportion of people who were undecided increased to almost three-quarters. In the following year, the pro-nuclear-power Social Democratic government was defeated in elections. The defeat was regarded as due to disaffection with the party's nuclear program.[90] The outcomes of the Swedish educational program suggest that risk perception may increase with more knowledge.[91]

Scientists still take sides on nuclear power. Some claim that nuclear power does not contribute to global warming. Nuclear power is often compared to coal, as if those were our only options for generating electricity. See Chapter 8, "Protect Your Family, Protect Our Planet" for a bit more on global warming and electricity sources.

The Nuclear Reactor Accident at Chernobyl, Ukraine

What Happened

The Chernobyl[92] nuclear power complex included four reactors located about 80 kilometers (50 miles) from Kiev, Ukraine (former USSR). The reactor complex is located near where the Pripyat River meets the Dnieper River (see Figure 6.2). On the night of April 25–26, 1986, the Number 4 reactor went out of control. Two explosions occurred. They blew the top off the reactor and the roof of the building. The reactor caught fire and continued to smolder until May 6. The accident "spilled radiation over 160,000 square miles in Belarus, the Russian Federation and the Ukraine."[93] The Soviet government issued no information about the accident for 35 hours. In Pripyat, a city of 50,000 nearest the reactor, children played outside, schools stayed open, and people continued their regular activities. Meanwhile, both rumors of the reactor accident and the radiation spread through the region. The first radio announcement in Pripyat said that people would be evacuated for 3 days. They were allowed to take only two bags and light clothes with them. The evacuation turned out to be permanent. Meantime, in Kiev, which also received some fallout, thousands of children marched in the May Day parade.[94]

A 30-km ring (18 miles) around the Chernobyl reactor was evacuated. The United Nations (UN) 15-year report says that approximately 116,000 people evacuated.[95] That does not count children who were eventually evacuated from regions in Belarus, and temporarily from Kiev.[96] David Marples,[97]

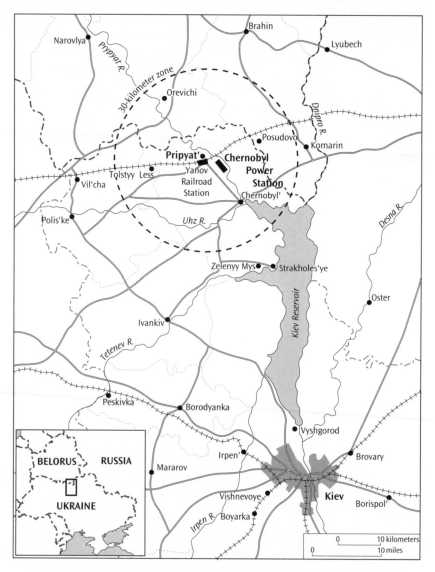

FIGURE 6–2. Map of the location of the Chernobyl nuclear plant disaster

a Canadian scholar of the Soviet Union, estimated that half a million people evacuated at least temporarily,[98] and that almost a quarter million were permanently relocated.[99] Marples's estimate of the number permanently relocated agrees with the UN's 15-year report.

The trauma of the initial evacuation of Pripyat can be appreciated from an eyewitness who worked helping coordinate the evacuation: "The fact is that there was no evacuation scheme, and we did not know in which villages were which Pripyat buildings or microraions...who went where? In Poliske we had a list of children. So I would phone the Village Council and ask: 'Do you have such and such parents? Their children are looking for them.' And they could say to me: 'We have such and such children who are without parents. Generally, we do not know where these children are from.' You sit and phone all the Village councils."[100]

Another evacuee remembered it this way:

Most people did what they were told and didn't even take spare money with them. When the time came, we went straight from the entrance [of the apartment building] and boarded the buses.... We were driven to Ivankov, 37 miles from Pripyat, and then to various villages.... Many of those who were deposited in Ivankov went farther, toward Kiev, on foot; some of them hitchhiked, with no idea of what they expected to find. Some time later, a helicopter pilot I know told me that he had seen, from the air, enormous crowds of lightly clad people, women and children, and old people walking along the road, and on the side of the road, in the direction of Kiev. They had already reached Irpeni and Brovarov. Cars were stuck in the midst of these crowds, as if they were among vast herds of cattle being driven to pasture.... And the crowds of people kept on walking, endlessly.[101]

Those working at the checkpoints to the 30-km zone worked exceedingly long hours. They measured the radiation on people and vehicles to see if they could be allowed out of the zone without being washed down. These workers received radiation exposure themselves. Here is the report of a medical student who worked both at a checkpoint and in the hospital treating radiation victims:

[At the checkpoint] the people got out of the bus, stood in a line.... There was one case where one granddad's boots were "radiating" a great deal. "But I've washed my boots, lads," he said. "Off you go, granddad, you've got to shake some more off." ... We sent him to wash three or four times.... We caught a lot of really dirty [radioactive] trailers, with dust-covered things. We sent them off to be washed.... I remembered a Belarus tractor. In the cabin next to the driver was an old man, his father perhaps. The old man was carrying a hen and a dog. And he said, "Measure my dog." I said: "Granddad, shake your dog's hair well when you get to your destination." ... Around 11–12 May I noticed that I was sleeping a great deal but not feeling refreshed.... A blood analysis was done and I was put on the eighth floor in our department [where those with radiation sickness were being treated].[102]

The evacuation spawned its own controversies. First, the evacuation was delayed, and in the interim no information to protect people was given.[103] Some nurses voluntarily went door to door handing out potassium iodide to protect people from radioactive iodine. Second, the upper-echelon party members and skilled nuclear workers were evacuated first. Third, many people from Pripyat were evacuated to Poliske. Poliske itself had to be evacuated later. Fourth, children were not evacuated from the 30-km zone, except Pripyat, until May 21. That was nearly a month after the accident.[104] Finally, was it really necessary to evacuate people from certain areas? Some Russian scholars believe radiation in some evacuated areas was lower than "background" radiation in non-evacuated areas.[105] Radiation hot spots continued to be discovered. As late as 1994, another 1,000 square kilometers (about 370 square miles) was declared contaminated. People in those contaminated areas did not necessarily relocate. They were reluctant to move unless they could get jobs elsewhere.[106] And who wants to leave home?

At the time of the evacuations no one knew exactly how serious the accident was. The government was undoubtedly hoping to reopen the evacuated areas. But not many evacuated villages have yet been declared habitable. Approximately 20% of land in Belarus was removed from cultivation. The UN estimated that to decrease the radioactivity by 50% would take until approximately 2020. Then it will take another 300 years for a 70% reduction. Radioactivity will continue for approximately 100,000 years for the very long-lived isotopes, such as plutonium.[107]

Radiation Casualties, "Liquidators," and the Environment

How many casualties were there? Official Soviet sources said that 2 workers died in the reactor explosion and 29 others died from radiation sickness. The official deaths were mainly firefighters who are lauded as national heroes. The Soviets said that 299 people were treated for radiation sickness. But a U.S. physician who helped with bone marrow transplants estimated that 500 were treated for radiation sickness. A former Soviet engineer and human rights activist said that 15,000 people died in hospitals in Kiev in the 5 months following the accident and that other diagnoses were used to mask radiation sickness. The Soviet government claimed that "not a single case of radiation sickness had occurred among the population . . . around the nuclear plant."[108]

Hundreds of thousands of Soviet soldiers and citizens worked in the cleanup. Helicopter pilots dropped lead and boron into the burning reactor to put out the fire. They flew through the plume of radioactive smoke. Some worked frantically to construct a cement containment building around the burned-out reactor. They tunneled under and used huge amounts of concrete to keep the nuclear fuel from burning down into the earth where it would contaminate groundwater. Others helped relocate the public, and clean up radioactive debris inside the power station and the surrounding area. The exact number of these "liquidators" is not known. The UN says there might have been 600,000 workers.[109] Their radiation exposure is also mostly unknown. Marples[110] wrote, "Any estimate of direct casualties involves supposition and guesswork....Even 'official' sources are wildly inconsistent—how can one reconcile statements from Ukraine's health ministry and the Chernobyl Union that thousands of union members have died—with an official report from Belarus that only 150 of the 66,000 decontamination workers from that republic have died?"[111] Marples[112] concluded that approximately 6,000 deaths in the *immediate* aftermath of the accident "represents the minimum possible number."

Effects on Children's Psychological Development

There are at least two reasons to expect that prenatal exposure to radiation would affect children's intellectual development. First, if the pregnant mother and fetus are exposed to enough radioactive iodine it will affect thyroid function. Lowered thyroid function can affect the cognitive development of the child. Second, prenatal exposure to ionizing radiation is one known cause of mental retardation. Among the Japanese atom bomb survivors, as the estimated fetal radiation increased, the likelihood of children being born with severe mental retardation increased, especially for radiation between the 8th and 25th weeks of gestation.[113] Even without mental retardation, as estimated fetal radiation increased, the children's IQ scores and school performance decreased on average.[114]

The Minsk Study   Researchers in Minsk, Belarus, gave IQ tests to children whose mothers were pregnant at the time of the accident and living in an area of Belarus that received fallout from Chernobyl.[115] For comparison, the research team also tested children comparable in socioeconomic background but whose mothers were living in a part of Belarus that did not receive fallout. The exposed children and their families had been relocated to the Minsk area when they were about 5 years old. The sample size was 138 exposed and 122 unexposed. Fewer fallout exposed children scored average or higher on the tests compared to the unexposed children. Also, more exposed children scored in the range of borderline mental retardation (IQ score 70 to 79). At age 10, the differences between the groups were smaller, but the average IQ score of the radiation-exposed group (93.7) was still significantly lower than the unexposed group (96.1). Estimated thyroid radiation dose showed a weak relationship to IQ test scores at both ages. At age 6–7 the correlation just missed the cutoff of $p = .05$, but at age 10–11, $p < .05$.[116]

The Minsk team assessed the children's psychological well-being. Each child also took a psychiatric interview and a neurological exam. The parents' answered an anxiety questionnaire. The exposed children had a higher frequency of speech, language, and emotional disorders. This was true both at 6–7 and 10–11 years of age. The emotional disorders in the exposed group were mostly phobias. Some of the exposed children imagined "the Radiation as a cruel monster that could kill them or their parents."[117] The parents of the exposed children had higher trait anxiety. Parent anxiety was correlated with the presence of emotional disorders in the children. Father's anxiety accounted for approximately 25% of the differences among children in emotional disorders. These differences in emotional adjustment of the *parents* years after the accident are reminiscent of the mothers of preschool children at Three Mile Island.

The researchers concluded that the stress of the Chernobyl accident and the social disruption of relocation were important aspects of the negative effects on the children. The research team speculated that high anxiety in the parents can have repercussions on family relationships, which in turn can lead to emotional disorders in the children. The Minsk team did not draw conclusions about radiation and IQ scores. They estimated thyroid doses using as much information as possible about the mothers' diets and locations during the accident. I regard this study as evidence that radioactive iodine from Chernobyl had at least a very small effect on intellectual development.

The Ukrainian Study A similar study was carried out with 6- to 8-year-old Ukrainian children born between April 26, 1986, and February 26, 1987. The sample sizes were 544 prenatally exposed children, and 759 children from a "clean" region of the Ukraine.[118] Some of the exposed children were evacuated from the 30-km zone ($n = 115$), but the rest had been living in areas contaminated to varying degrees. As in the Minsk study, there were more exposed children in the lower IQ groups, and fewer exposed children in the top IQ score groups compared to the unexposed children. When the children were 9–10 years old, some of them (50 exposed and 50 unexposed) were tested again. This time the children took a psychiatric interview, and an EEG (brain wave) test. The exposed children had a higher likelihood of speech and language disorders, motor disorders, emotional disorders, and hyperactivity than the unexposed children. The EEG tests also showed some differences between groups: α-power and β-power were higher, and theta-power was decreased in the left hemisphere of the radiation victims compared to the control children. The Chernobyl children also showed greater lateralization of β-power.[119]

The authors of the Ukrainian study said that all research that has examined the "mental health of the prenatally irradiated children as a result of the Chernobyl disaster...came to the conclusion that the prevalence rate of disorders of psychological development, emotional and behavior disorders, as well as mental retardation, is higher in children irradiated in utero as compared to the non-exposed children."[120] Because of the EEG differences between the exposed and unexposed children, the Ukrainian research team concluded that radiation altered the pituitary-thyroid functioning of exposed children. Altered pituitary thyroid function in turn, altered neurological development and function.

Similar Results, Different Conclusions   The Minsk and Ukrainian research teams found similar results but drew different conclusions about the causes. The Ukrainian scientists leaned toward radiation as the cause of the children's difficulties. They acknowledge a confounding in their work. The mothers' verbal IQ scores were slightly lower in the exposed sample than in the nonexposed sample. The exposed parents' overall mental health was slightly worse. In contrast, the researchers from Belarus concluded that the stress and the social disruption could account for the results. As at TMI, most estimates of radiation exposure were made long after the event.

U.S.-Ukrainian Cooperative Study   A research team from the State University of New York at Stony Brook collaborated with a team of scientists in the Ukraine.[121] One of the investigators, Dr. Evelyn Bromet, had studied the mothers and children at TMI. The Chernobyl exposed sample of 300 10- to 12-year-olds was all children who had been evacuated to Kyiv (Kiev), Ukraine. They were either in utero at the time of the accident or less than 15 months old. Comparison children were drawn from the same classrooms as the exposed children.

The Chernobyl evacuee mothers reported that their children had more somatic symptoms, thought problems, and delinquent behaviors than the comparison sample. The evacuee children rated their own scholastic competence as worse and had more anxiety focused on Chernobyl. Almost half of the evacuees were diagnosed with "vascular dystonia,"[122] compared to about one-seventh of the comparison children.[123] Mothers of evacuees also reported more memory problems in their children even though the actual memory tests did not differ.[124] The children with higher Chernobyl-focused anxiety also scored lower on three attention measures (Trails test, a word-finding test, and teacher rating of attention). Children with higher overall anxiety were rated by their mothers as having more problems than children with lower anxiety.[125] But, there were no significant differences between exposed and nonexposed children on a nonverbal IQ test or other dimensions of the teacher rating scale.

Just as in the TMI study, the mothers of the Chernobyl evacuees showed higher somatization symptoms, expressed higher stress about Chernobyl and its potential health effects, and were more likely than the mothers of the comparison children to have had a depressive episode. Maternal health stress and rating of trauma due to the Chernobyl accident were correlated with the mother's ratings of the child's somatic symptoms.[126]

The authors concluded that "the present results provide no support for the presumption of cognitive or neuropsychological differences between the two groups of children."[127] For the children's psychological functioning, the investigators concluded, "Although radiation and nuclear power evoke deeply rooted fear and anxiety in adults, our study found that 11 years after the explosion, the trauma was not transmitted to children who were unborn or infants when their families were resettled in Kyiv."[128] These conclusions were drawn even though the researchers had found a significant relationship between children's Chernobyl anxiety and performance on three measures of attention.[129]

Studies of Adult Mental Health   A team of scientists from the Netherlands, Russia, and Belarus compared the mental health of people living in the relatively heavily polluted Gomel region of Belarus with people in Tver, Russia. Tver was not affected by Chernobyl fallout.[130] Many evacuees and former liquidators live in the Gomel region. Gomel also received fallout, and some of the villages were evacuated, mostly involuntarily. About 400,000 acres (625 square miles) of farmland was abandoned in Gomel. This had a severe economic impact on the area. Overall, residents of the Gomel region were under a variety of stressors.[131] Four years after the accident people in the Gomel area rated their health to be worse, showed more psychological distress, and were more likely to have recently visited a doctor and taken medications.[132] Six years after the accident, the same research team found that people in the Gomel region who were evacuated or who were mothers with children under 18 years of age were more likely to show psychological distress than other people in Gomel. This was true regardless of the radioactive contamination in the area in which they were currently living.[133] The psychological distress in the mothers echoes the TMI results.

After the break-up of the Soviet Union, people of Jewish ancestry were free to leave. Many went to Israel. Israeli researchers compared the Chernobyl victims with immigrants to Israel from other areas of the former Soviet Union.[134] They stratified the sample into high- and low-radiation-exposure groups. The liquidators and those who had been exposed to the highest radiation had a higher likelihood of posttraumatic stress symptoms a year after emigrating to Israel. After another year, their symptoms had abated considerably. The same pattern held for depressive symptoms. The most exposed group had more somatic symptoms than the unexposed group even 2 years later. The exposed groups differed from the comparison group in systolic blood pressure both 1 year and 2 years after emigrating. The Israeli team concluded that exposure to the combination of stress and radiation at Chernobyl was accompanied by psychological, physiological, and physical symptoms. The good news was that those symptoms tended to abate over the 2-year follow-up. The bad news is that other studies have found a higher suicide rate among liquidators from Estonia as long as 15 years after the accident.[135]

A French and Latvian research team found a higher risk for mental and psychosomatic distress in liquidators who worked on the cleanup for 4 weeks or longer, cleared contaminated forest, or consumed locally grown fresh fruit, vegetables, or meat.[136] These scientists concluded that working as a liquidator could increase psychosomatic disorders and psychological distress by any of three possible pathways: (1) anxiety about radiation exposure could lead to psychological problems; (2) radiation exposure could cause physical disorders, which in turn causes psychological problems; and (3) radiation could induce psychiatric problems directly. Finally, they note that "whether stress-related or radiation-induced, mental distress reflects a genuine human suffering to be taken account of and appears to be an important health consequence of the Chernobyl nuclear accident."[137]

Summary of Chernobyl Studies

The studies of children who were in utero or less than 15 months of age at the time of the Chernobyl accident show differences from comparison groups in psychological symptoms, and sometimes in IQ test scores. In one study the IQ test scores were weakly, but significantly, related to estimated radiation dose. Like the TMI mothers, the mothers of the Chernobyl children have higher rates of psychological adjustment problems, including somatic symptoms and depressive episodes.

The U.S.-Ukrainian researchers commented that the Chernobyl victims underwent "harrowing experiences during the evacuation, arduous battles for residency permits in Kyiv and for government benefits, social stigma, and an irreversible loss of home, belongings, and lifestyle."[138] Eyewitnesses reported that family members were often separated for some time during the evacuation.[139] People faced discrimination in the communities in which they were resettled.[140]

Are the effects psychological or due to radiation? "Although some researchers believe that somatic and neurologic symptoms are psychogenic (psychologic) in origin, others claim that symptoms such as nervous system dysfunction, cognitive disorders, and pain may be the effect of low doses of radiation on the nervous system or the beginning stages of organic diseases."[141] Whether radiation exposure could have direct effects on psychological well-being and neurobehavioral functioning has not been well studied even in animal experiments. Studies of the effects of low-level radiation on animals have emphasized whether prenatal exposure causes cancer or mutations. Behavioral effects have been neglected. There is one study of the neurobehavioral functioning of U.S. Gulf War veterans who have small pieces of shrapnel from depleted uranium bombs[142] embedded in them. The concentration of uranium excreted in urine was the best predictor of performance on a battery of computerized neurocognitive tests.[143] The dose-response relationship adds credence to the idea that exposure to uranium may affect neurocognitive functioning. Whether the results are due to the toxic biological properties of uranium as a metal, the small continuous doses of radiation, or the stress of recovering from injuries is not known. The results suggest the possibility that low-level radioactivity might directly affect behavioral functioning—a possibility that merits further research.

Other Health Effects of the Chernobyl Accident

Summaries of the health effects of Chernobyl invariably include an increase in the incidence of thyroid cancer among those exposed during childhood.[144] There is also an increase in breast cancer, cataracts, and cardiovascular diseases.[145] Some claim that the leukemia rate is not elevated, even for the liquidators who were exposed to relatively high amounts of radiation.[146] There are ongoing studies of leukemia rates in the liquidators, but the radiation doses they received were not measured well.[147] Some studies of workers in nuclear industries have shown higher rates of leukemia.[148] Japanese atom bomb survivors showed increases in leukemia within about 10 years of the bombing.[149,150] Two research groups found that hypothyroidism (underactive thyroid gland) is more frequent among people who were exposed to Chernobyl fallout than among people from adjoining areas.[151] Hypothyroidism affects physical and intellectual development in children.

Some studies of Chernobyl evacuees have reported increases in congenital malformations in embryos and fetuses. However, the results may be due to a decline in nutrition and health care with the break-up of the Soviet Union. Details of methods are not always given in the publications, so reviewers in Western countries tend to be skeptical.[152] Research with British nuclear workers found a higher rate of miscarriages and stillbirths for offspring of male workers, although the results are controversial.[153] Among the Japanese atom bomb survivors, a relationship between the radiation dose to both parents and the likelihood of any untoward pregnancy outcome (malformation, stillbirth, and early mortality) was not quite statistically significant. The authors of the Japanese study regarded their results as an underestimate of the effect of radiation on fetal loss and malformation. They noted that "radiation has caused genetic damage in every species properly studied in an experimental setting."[154] A genetic study of parents and children in the Mogilev area of Belarus found a higher rate of germline mutation compared to a sample in Britain.[155] The summary of an international conference on the health effects of the Chernobyl accident reported, "Stable changes in chromosomes of somatic cells have been identified.... Research is required to determine whether similar changes may lead to increased incidence of disease in offspring."[156]

The international report also concluded that the primary effects of the Chernobyl accident on the public appear to be cardiovascular and neuropsychological.[157] Studies of the Japanese atom bomb survivors also showed increased cardiovascular disease, especially atherosclerosis.[158] The Chernobyl report listed these other health effects: decreased birth rate, worse health of newborns, increased pregnancy complications, and worse child health. That report and a more recent one concluded that health effects were likely exacerbated by declining economic conditions in the area, poor nutrition and food supply, the psychological stress of relocation, and continued residence in contaminated areas.[159]

Wildlife and Natural Resources

A 1995 news article in *Science* reported that the wildlife was flourishing within the evacuated 30-km contamination zone. This was the case even though radiation was still quite high, up to 300 millirads per hour, about 3,000 times the radiation near the U.S. Savannah River nuclear plant. Because the area is fenced, there is very little human disturbance. Researchers found a higher rate of mutation in animals living in the evacuation zone compared to the same species of animals from other areas.[160] For example, Chernobyl barn swallows have a higher rate of mutations that produce partial albinism than either barn swallows from Italy, from an uncontaminated area of Ukraine, or from museums in Ukraine.[161] Wheat grown in a contaminated soil near Chernobyl also showed a higher mutation rate compared to genetically identical wheat grown in an uncontaminated area.[162]

What about water pollution? When the area floods in the spring, the radioactive isotopes from the contaminated areas wash into the rivers. Strontium-90 and cesium-137 are both moving gradually down the Dnieper River to the Black Sea.[163] Radioactive cesium and strontium can enter the food chain via either freshwater or saltwater fish.[164] Strontium-90 and cesium-137 have half-lives of approximately 28 years and 30 years, respectively. Therefore, they will be contaminating the ecosystems for a long time (see Table 6.1 for half-lives of some isotopes). Migration of radioactive isotopes into groundwater is also being studied.[165]

Social Justice Issues with Radiation

The development of nuclear weapons exposed many people, including children, to potentially harmful amounts of radiation. Exposure occurred without the consent of those who were exposed and without sufficient information about the effects of exposure.

Marshall Islands and U.S. Bomb Testing

About halfway between Hawaii and Papua, New Guinea, this tropical paradise of islands and atolls barely rises above sea level. The native people cultivated breadfruit and coconuts, fished in the lovely lagoons of the atolls, and traded with others in the area. During World War II the United States and Japan fought bloody battles in the Marshall Islands. After the war the United Nations designated the Marshalls a United States Trust Territory. The agreement stipulated that the United States shall "promote the economic advancement and self-sufficiency of the inhabitants; . . . protect the inhabitants against the loss of their lands and resources; protect the health of the inhabitants."[166]

The United States exploded 67 atom and hydrogen bombs in the Marshall Islands. The first bombs were in the Bikini atoll. One hundred forty-seven Bikinians were removed in early March, 1946, in preparation for the bombs. They agreed to be relocated. But most authors say the Bikinians thought they had no choice.[167] They had been told that the relocation would be temporary and was "for the good of all mankind."[168] After the first bombs were tested in July 1946, the Bikini leader asked that his people be allowed to go home. There was little food or water on the atoll to which they had been relocated. The fishing was poor, and after several months there the people were starving. The Bikinians were relocated at least three more times by the U.S. government. Some of them were allowed to return home in 1975. But they were evacuated again because of radioactivity. At the present time they still have not returned to their home.[169] The International Atomic Energy Agency Commission issued a report on Bikini radiation in 1998. The report concluded that food grown on the island would be a radiation risk but that visiting or walking on Bikini Island would not.[170]

Other Marshallese fared as badly as the Bikinians. The people of Rongelap atoll received heavy fallout from the most powerful thermonuclear bomb ever detonated, "Bravo," on March 1, 1954. The evacuation of the Rongelapee by the U.S. military was delayed by 2 days. By that time many had radiation burns—on their heads but especially on their feet. They had been walking barefoot in the fallout. Three years later the U.S. military returned the Rongelapee to their atoll, declaring it safe. Medical assessments showed that their body burdens of radiation increased dramatically after they went home. Trust between the Rongelapee and the U.S. doctors deteriorated. In 1972 the Rongelapee asked that Japanese doctors be included in their medical exams. The United States initially refused entry to some of the Japanese doctors.[171] Medical exams documented thyroid cancer, especially among those who were children at the time of the fallout. There was also hypothyroidism and cretinism (severe mental retardation of children due to lack of thyroid hormone).[172] In 1985 Rongelap was evacuated again because of the radioactivity. The Rongelapee still have not returned to their home.[173]

The people (the Enjebi) of Enewetak atoll were removed from their homes in December 1947. Their atoll was the location for the majority of the bomb tests. The lagoon was a missile target. The Enjebi began petitioning the UN for return of their home atoll in 1968. In 1977 the United States finally started cleanup. One island in the atoll, Runit, was the burial site for plutonium-contaminated soils and other nuclear waste. The nuclear waste was put in a bomb crater and covered with a cement dome wider than a football field and 30 feet high. The cement is 18 inches thick, but it is cracked. Birds now lay eggs in the cracks. Some holes in it are 6 inches deep.[174] Contaminated items that were not buried were simply dumped in the ocean. There is concern over the long-term safety of this disposal method.[175] Anything within 3 miles of Runit is declared too radioactive. If ocean levels rise, there is a risk that the plutonium dome on Runit will be submerged. At the present time, the Enjebi people live primarily on another atoll.[176]

The citizens of the Marshall Islands have made claims against the United States for damages to their property, land, and health. Even people who live on atolls that were supposedly uncontaminated have concern about the effects of radiation on themselves and future generations. The Marshall Islanders feel that the incidence of birth defects has risen dramatically since the bomb testing.[177] The Nuclear Claims Tribunal hears claims of personal injury and property damage.[178]

The physical health of Marshall Islanders has been the subject of many studies, but their mental health has been neglected. The disruption of their communities by the U.S. bomb testing was profound. Anthropologists studied the Bikinians during their many relocations. The forced relocation and their need to negotiate with the American government changed the entire social structure of these Polynesian people. One anthropologist, Robert Kiste,[179] commented that "given the competitive nature of the island society when matters of power and land are at stake, relocation is tantamount to the opening of Pandora's box."[180]

The Marshall Island bomb tests also exposed many U.S. military personnel to untoward dosages of radiation. See the section below on atomic veterans.

Other Radioactive Pollution Sites

This section calls attention to some of the problems created by past misuses of radioactive materials. The list is not intended to be comprehensive, only to provide examples of social injustices.

The Nevada Test Site July 1945 near Alamogordo, New Mexico, was the time and place of the first atom bomb explosion on earth. Cattle in the path of the fallout suffered radiation burns and lost the hair on their backs.[181] Starting in 1951 bomb tests in Nevada created fallout across the entire United States and the rest of the globe.[182] Estimates are that in some years the public was exposed to radiation from fallout that doubled the natural background radioactivity from the earth.[183] Fallout was generated not just by atmospheric tests but also by many of the underground tests.[184] Near the Nevada Test Site sheep were killed by fallout. Some people living in remote areas close to the boundary were evacuated.[185] The interested reader can trace some of the fallout controversy by searching back issues of the journal *Science* online.[186]

At the time of the bomb tests, many articles were published in *Science* and other journals documenting the fallout. Articles also tracked the presence of radioactive iodine and strontium in milk and other foods. These same articles usually ended by saying the fallout was safe. But there was dissent and concern for the public. Nobel Prize winner Linus Pauling[187] wrote about the hazards of carbon-14 created by atmospheric tests. Norman Bauer from Utah State University[188] called for information on public exposures to radioactivity in Utah. He pointed out that because of government secrecy the levels of radioactivity in foods in Utah could not be obtained. Bauer also pointed out that the public was being exposed to high levels of radiation involuntarily without information or consent.

Only many years later did epidemiological studies analyze the possible link between bomb fallout and illnesses. In 1987 Dr. Victor Archer showed that the peak incidences of leukemia in the United States were offset from the peak bomb fallout by about 5 years. The graph in his paper is striking. The leukemia rates were also higher in the states with the highest total accumulation of fallout than in states with the lowest fallout. Archer pointed out that because the fallout affected many areas of the earth, it is virtually impossible to find a suitable control group: "The only minimally exposed populations would be in the southern hemisphere, such as Australia or South Africa."[189] In spite of Archer's findings and the conclusions of many expert panels, researchers are still arguing about whether radioactive fallout created a statistically significant increase in cancers in southwestern Utah and the adjoining area of Arizona close to the Nevada Test Site.[190]

Financial compensation for a specific list of cancers became available for people who lived close to the Nevada Test Site in Utah, Arizona, and Nevada.[191] The research on radiation emphasizes cancer and death due to cancer. I have not found epidemiological studies of topics such as early cataracts, the incidence of hypothyroidism, mental retardation, birth defects, special education needs, or intellectual functioning among Nevada Test Site downwinders. These latter conditions were not part of the compensation program as initially enacted. Interviews with Utah residents have shown concern about a spike in children needing special education during atmospheric nuclear weapons testing.[192]

Atomic Veterans Hundreds of thousands of U.S. military personnel were exposed to radioactivity during nuclear bomb tests. Because some of the records are still secret, the exact number is not known. Troops were sometimes in trenches as close as within about a mile of ground zero at the Nevada Test Site. In the Marshall Islands they were on ships at various distances from the blasts or monitoring the blasts from other atolls. After the test blast, soldiers were sometimes ordered to march toward the epicenter of the blast. In the Marshalls, the navy men boarded target ships near the blast zone. They were ordered to scrub them down to test decontamination methods. Pilots flew planes through the bomb clouds to sample the radioactivity. Drones could have been used.[193] Some men in the bombing crews or on maneuvers said they received enough radiation to induce vomiting, hair loss, and sudden loss of their teeth a few months later.[194] Others report fertility problems.[195] Fertility problems and the health of the children of atomic veterans have not been systematically investigated. Approximately 40,000 men were involved in one series of tests in 1946 in the Marshall Islands.[196]

The atomic soldiers said that not all of them were issued radiation badges. Some officers said that the radiation monitoring equipment went completely off-scale and that afterward the allowable radiation limits were raised.[197] Some soldiers saw that others were given special goggles or protective clothing, while they wore only their regular fatigues.[198] The dosages these men received are usually not known. In many cases the veterans have had trouble even documenting their presence at the atomic tests. Some records were destroyed in a fire,[199] others inexplicably misplaced,[200] withheld from veterans,[201] or deliberately destroyed. In 1987 U.S. District Court Judge Marilyn Patel fined and censured the U.S. Veterans Administration (VA) for destroying documents relevant to radiation benefits claims and for failing to establish a system to meet requests for records. In addition, Judge Patel found that VA officials attempted "to stifle full compliance with discovery requests." She found that the VA had threatened retaliation against employees if they revealed that documents being destroyed were relevant to an ongoing lawsuit regarding radiation exposure.[202] At the time of the tests, the soldiers were told never to discuss them outside of the military. This order was given even though many tests were covered by the press.[203]

Early studies suggested a higher rate of leukemia in the atomic veterans who were present at a very "dirty"[204] nuclear explosion at the Nevada Test Site.[205] There are still controversies about whether the atomic soldiers are subject to more cancer or premature death than other veterans of the same era.[206] As with TMI, there has been controversy over whether the radiation doses were high enough to be harmful. William Brady, the principal health physicist at the Nevada Test Site for 40 years, testified in Senate hearings that the dosage reconstructions of the Defense Special Weapons Agency underestimate internal doses by a factor of more than 1,000.[207]

As of 1998 approximately 97% of U.S. veterans' claims for service-related illnesses as a result of radiation exposure were denied.[208] The Radiation Exposure Compensation Act has been amended many times by the Congress in attempts to help the victims. Atomic veterans and their survivors can find helpful information for filing compensation claims in the Web sites of veterans organizations such as the Disabled American Veterans or other specialized organizations such as the National Association of Radiation Survivors and the National Association of Atomic Veterans. One veteran expressed one important reason for bringing to light the ways in which nuclear testing affected those involved. One veteran put it this way: "People don't really care [about the environment] until it really scalds them or burns them or kills them. For the sake of my children, for the sake of you and the people who have to occupy this earth . . . I think I'll tell you what I know."[209]

Other Radiation Release Sites   Locations in the United States and other parts of the world that were part of nuclear weapons development or the early development of nuclear power have varying amounts of contamination with radionuclides. *All* nuclear power plants release some radioactivity into the environment. Nuclear plants are required to file an annual report of their releases.[210] Plutonium in soils downwind of the Rocky Flats weapons center near Denver, Colorado, is 3,000 times the concentration found in other parts of the United States.[211] Radioactive cesium is found in fish near the Savannah River nuclear plant in South Carolina. Radioactive cesium is present in the Susquehanna River downstream from both the TMI and Peach Bottom nuclear plants.[212] Strontium-90 is found in fish near Brookhaven Laboratory on Long Island.[213] Radioactive strontium found in children's baby teeth in Suffolk County, New York (an area near several nuclear power plants and Brookhaven National Laboratory on Long Island), is approximately as high as it was during the peak of atmospheric bomb testing.[214]

The Hanford nuclear facility in eastern Washington State has leaked radioactivity into the air, the Columbia River, and groundwater. Some people in the area felt that their own physicians and the U.S. government were not responsive to their health concerns. They asked individual scientists and the Oregon chapter of Physicians for Social Responsibility to form a research team. The researchers studied the incidence of hypothyroidism by using questionnaires that asked people about their diagnoses and also about their children's learning disabilities. The research team concluded: "Although the self-selected nature of the Hanford sample precludes any effort to estimate the rate of incidence of hypothyroidism in the population . . . the large number of cases, their apparent congruence with estimated but uncertain average doses to the thyroid for the counties of residence, and their onset shortly after massive emissions of 131-I suggest their description as an epidemic phenomenon related to emissions of 131-I."[215]

Uranium Miners and Nuclear Plant Workers   Uranium miners have an elevated risk of lung cancer. Some mines are located on Native American reservations, especially the Navajo reservation. Mine tailings were simply left or were sometimes used as fill in the construction of housing and roads and as a soil conditioner.[216] Children played on piles of tailings, and miners were not warned about changing clothes before going home.[217] In 1979 a uranium mill in Church Rock, New Mexico, spilled radioactive mine waste. The dam that held the waste in settling ponds burst. Ninety-five million gallons of contaminated water polluted the river used by area Navajos. Groundwater is contaminated.[218] In 1986 a uranium hexafluoride cylinder ruptured at a nuclear fuel plant in Oklahoma run by Kerr McGee Corporation. The plant is located in the Cherokee Nation. One worker was killed. Thirty-seven of 42 other workers were hospitalized. Over 27,000 pounds of the radioactive uranium compound was released.[219]

Compensation to miners and ore processors who worked between 1942 and 1971 and contracted cancer is available from the U.S. government under certain circumstances.[220] Compensation is also available under the Energy Employees Occupational Illness Compensation Program Act. This act allows adult children of workers to claim compensation for deceased parents who worked in nuclear weapons plants or experimental reactors. The act provides medical expenses and a lump-sum death benefit for workers who contracted certain types of cancer or who were made seriously ill by radiation in facilities run by the Department of Energy or its predecessors.[221] Contact the Department of Justice Web site. A synopsis of the repeated failures of the U.S. Department of Energy to enforce workplace safety regulations and accompanying ethical issues can be found in a paper by Shrader-Frechette.[222]

Radioactive Pollution in the Former Soviet Union   Some areas in the former Soviet Union have been polluted by weapons testing and nuclear fuel processing, just as they have in the United States. There was an accident and explosion at a nuclear facility in the Chelayabinsk region of the Soviet Union in 1957. Approximately 25,000 square kilometers (approximately 9,000 square miles) were contaminated. This area is near the southern part of the Ural Mountains, drained by the Techa River, which eventually flows into the Arctic Ocean. Radioactive waste was chronically discharged into the river. Radiation has reached all the way to the Arctic Ocean.[223]

The Soviet bomb tests were conducted in the Semipalatinsk area in eastern Kazakhstan starting in 1949.[224] Estimates are that 25,000 people were exposed to highly radioactive fallout from the very first test. Other nuclear facilities were located in Siberia (called Tomsk-7 and Krasnoyarsk-26). An accident at Tomsk-7 in 1993 released plutonium. Not as much is known about pollution at Krasnoyarsk, but radiation above background levels has been found in the Yenisey River.[225]

Radioactive Pollution in Great Britain   One of the first serious reactor accidents occurred in 1957 at the Windscale reactor. The reactor was originally built to produce plutonium for nuclear weapons. It had just started generating electricity a year before it caught on fire and burned for 2 to 3 days. Milk had to be dumped because of contamination with radioactive iodine.[226] There is now a nuclear fuel reprocessing plant there. In 1981, the complex was renamed Sellafield. Pollution from this plant is measurable in nearby forest biot,[227] and in salmon farmed nearby.[228] Controversy centers on the risk of an accident or terrorist act that could release a large quantity of radiation in a heavily populated area.

The United Kingdom conducted nuclear bomb tests at Christmas Island (also used by the United States), southern Australia (Maralinga and Emu), and the Monte Bello Islands (off the coast of western Australia).[229]

Other Nuclear Bomb Testing Sites France tested nuclear weapons in Algeria and in the Pacific at Mururoa and Fangataufa, part of Polynesia. The French government began studying cancer incidence in French Polynesia in 1996. China exploded a total of 44 bombs, compared to 1,127 by the United States, 969 by the USSR, 210 by France, 57 by the UK, and 6 each by India and Pakistan.[230]

Ecosystem Effects of Radiation

The prevailing viewpoint is that if radiation levels are low enough to protect humans from cancer, they are low enough to protect wildlife. This viewpoint is untested.[231] The U.S. EPA considers that radioactivity is safe for the human population if it: (1) does not have more than a 1 in 10,000 chance of inducing cancer in any individual, (2) does not raise cancer risk by more than 1 in 1 million for the population living within 80 kilometers (50 miles) of a facility, and (3) causes only a small number of cases of death and disease.[232] Few studies of ecosystem effects of radiation have been done, and each uses its own criterion for harm.[233] In the former Soviet bomb testing area, researchers have found "evidence that radioactive pollution of the regions has induced lower pollen viability and fertility, rendering more seeds unviable, giving rise to meiotic anomalies and raising concentrations of rare and null variants of storage protein in the seeds of wild populations of Scotch pine."[234] In the section on Chernobyl, I mentioned high mutation rates in humans, wheat, and swallows.

Radiation Research Controversies

The research controversies on ionizing radiation echo the themes of this book. Are statistically significant effects due to radiation exposure or some unidentified confounding variable? What comparison groups are most appropriate? Is there a threshold for safe exposure? These issues are not easily solved, especially when there are virtually no measures of internal radiation exposure used in research on people exposed to radioactive pollutants. Radiation doses are often reconstructed retrospectively. The first research on the Marshall Islanders compared the incidence of thyroid cancer among the Rongelapee with other Marshallese. But the other Marshallese were also radiation-exposed to varying degrees. This is reminiscent of Robert Kehoe's work on blood lead. He compared tetraethyl lead workers with maintenance workers exposed to fumes in the same building, or to people who used lead-glazed cookware, and concluded that relatively high lead exposure was "natural" (see Chapter 1). "Status quo science" also drives the interpretation of radiation research. In the TMI cancer studies, one dispute is over whether the retrospectively reconstructed doses of radiation are theoretically high enough to produce the results. Similarly, for the Chernobyl disaster, the possibility that somatic symptoms could be due to direct effects of radiation is often discounted.

Part II: Chemical Waste Disaster at Love Canal, New York

Leakage from a toxic waste dump in the Love Canal neighborhood of Niagara Falls, New York, became a landmark event in environmental policy in the United States. The worried residents and government officials became embroiled in one of the most controversial environmental events in U.S. history. The events at Love Canal led to the "Superfund" law, signed by President Carter in 1980. The law provided new regulations for toxic chemical waste disposal, new taxes on corporations, and funds for cleaning up abandoned dumps. Love Canal also became a spearhead for citizen activism on environmental issues.[235]

The Love Canal disaster appears at first glance to pit unjust individual suffering against the interests of society as a whole. Should taxpayers bail out victims of pollution? I examine three aspects of the events at Love Canal: the limited research findings, the scientific controversies, and the moral issues raised by the events.

What Happened

The events centered on an aborted canal project that was used for municipal, military, and chemical waste starting in the 1920s. The canal was a clay-lined ditch 100 feet wide and almost 2 miles long. It was constructed in the 1890s as part of a failed project to divert water around the Niagara Falls to generate electricity. Because clay is relatively impermeable, the canal made a pretty good waste disposal site. Hooker Chemical Company (now part of Occidental Chemical Company) discarded chemical waste there between 1942 and 1952. At that time there were very few regulations on waste disposal. After the canal was full, Hooker covered it and deeded it to the Niagara Falls school district. Hooker explicitly declared that the area contained chemical waste.

The 99th Street Elementary School was constructed precisely on top of the canal in 1954. In the postwar baby boom, new homes sprang up adjoining the waste area. Low-income apartments were built on the west side of the canal. Eventually the dump began to leak. It leaked into people's basements and into streams. Wastes came up through the surface during wet years when the water table was high. Testing for toxic substances began in 1976, but it was not until 1978 that the first residents were evacuated.

The pollutants found at Love Canal included dioxins, PCBs, chemical precursors of various persistent chlorinated insecticides, lindane, mirex, toluenes, benzenes, phenols, and other chemicals.[236] The soil on the 99th Street school playground had very high concentrations of lindane, and chemical precursors of two other chlorinated pesticides.[237] Dioxin in high concentrations was found in some basement sumps, sediment in two creeks, and the storm sewers.[238]

At the time decisions were being made at Love Canal, there were no funds available specifically for remediation of waste sites. The lack of funds set a sharp boundary on the actions that government officials felt they could take. Love Canal residents "raised hell" until public money was allocated to purchase their houses. The activism took a huge toll on the psychological well-being of the people at Love Canal and created conflicts among the residents.[239] Conflicts developed between the homeowners association and advocates of the renters in the low-income housing. The renters were relocated later than most of the homeowners.[240]

Eventually the residents relocated voluntarily, but not until after they had held protests in the state capitol, threatened to burn their mortgages, and used every opportunity to publicize their plight in the media.[241] Table 6.3 gives a brief chronology of some key events at Love Canal.[242]

Today the Love Canal neighborhood is called Black Creek Village. The EPA supervised the cleanup and has declared some areas to be habitable. Other areas are usable only as commercial real estate. Homes have been sold at 15%–20% below market value. A fence surrounds the 70-acre area around the former canal. The cap on the dumpsite had to be repaired again in 1993 to prevent further leaking. EPA declared the cleanup of the 102nd Street dumpsite, just a quarter mile from Love Canal, to be complete in 1999.[243]

Very Few Research Findings

Children's Health Effects: Miscarriage, Low Birth Weight, Shorter Stature Very few scientific studies were done on the people living at Love Canal. The New York Department of Health's first recommendation for evacuation was for pregnant women and small children living directly adjacent to

TABLE 6–3. Chronology of key events at Love Canal (Niagara Falls, New York)

DATE	EVENT
Circa 1958	Three or four children burned by chemicals when playing in the area.
1969	City Building Inspectors report that conditions on the old dump site are hazardous, including surface holes, rusting barrels, and chemical residue.
1976	International Joint Commission discovers Mirex in Lake Ontario that is traced to a storm sewer near Love Canal.
	David Russell and David Pollak, reporters at the *Niagara Gazette*, write several stories on Love Canal.
	Children told not to play on open areas in the old dump site.
1977	*April*: City of Niagara Falls hires Calspan Corp. to make a plan for abatement of groundwater pollution at Love Canal.
	August: Calspan reports that storm sewers west of the canal contain PCBs and that organic chemical compounds are visible and malodorous at the top of the canal at all times.
	September: Congressman LaFalce tours Love Canal and urges the EPA to investigate.
	October: EPA begins to sample air in some basements.
	December: Conestoga-Rovers Company is hired to devise a plan to prevent the off-site migration of wastes.
1978	*April*: N.Y. Commissioner of Health Robert Whalen and Commissioner of Environmental Conservation Peter Berle visit Love Canal.
	May: EPA concludes that toxic vapors in homes are a serious health hazard. Karen Schroeder starts to organize residents to get an attorney. State DOH officials meet with residents twice.
	June: Lois Gibbs starts going door to door organizing residents. DOH begins taking blood samples and giving questionnaires to residents who live directly adjoining the dump.
	August 2: Health Commissioner Whalen declares the Love Canal dump a serious threat and recommends that pregnant women and children under the age of 2 years who live directly adjoining the canal be evacuated as soon as possible. Whalen's order is later modified to include closure of the 99th Street school, and evacuation of families that live across the street from the canal. DOH expands blood testing and health surveys to more residents.
	August 4: Love Canal Homeowner's Association is formed.
	August 7: N.Y. Governor Hugh Carey promises state money to purchase the homes of those who are eligible to be evacuated (239 houses). President Carter declares an emergency at Love Canal, allowing some federal funds to be used.
	August 15: Governor Carey meets with residents.
	September: Homeowners group begins its own health survey, with help from Dr. Beverly Paigen.
	October 10: Remediation work begins despite residents' concerns about the safety of digging in the area with residents present.

(Continued)

TABLE 6-3. (Continued)

DATE	EVENT
	November: Dioxin found in sewers, sump pumps, and leaching from the dump.
1979	*February 8*: David Axelrod, new commissioner of health, recommends relocating pregnant women and children under the age of 2 who live within 3 blocks of Love Canal (97th to 103rd St).
	April: Property tax reductions granted by the state legislature.
	Late May: Remedial construction begins on the northern part of the canal.
	August: Approximately 20 residents complain of fumes and illness from the remedial construction; state relocates them temporarily
	September: 425 more residents temporarily relocated to hotels. Cost per day approximately $7,500.
	October: Remedial construction begins at southern end of the canal.
	November: Evacuated residents are notified the construction is done, so state will no longer pay for hotels. Dioxins found in crayfish in nearby Black Creek. Governor's representatives meet with low-income renters to discuss relocation.
	December: Leachate treatment system at Love Canal begins operation.
1980	*May*: EPA pilot study finds evidence of chromosome damage in residents; later results are called inconclusive.
	May 22: President Carter again declares federal emergency at Love Canal, and temporary relocation begins; some residents press for permanent relocation.
	June: Love Canal Area Revitalization Agency created.
	December: President Carter signs the Superfund law.
1981	EPA lists Love Canal as a proposed priority toxic waste site under Superfund.
	EPA fences Black Creek and begins environmental testing.
1982	*May*. EPA declares Love Canal neighborhood safe for habitation.
1985	A new cover is constructed over the canal, and a new waste treatment plant is installed.
1987	EPA issues decision requiring cleaning of streams and sewers. Almost 13 miles of sewers are cleaned in the next 2 years. Sediments are stored on site.
1989	Black and Bergholz Creeks are dredged to remove contaminated silt.
1990	Love Canal Area Revitalization Agency begins offering homes north of the canal for sale at 15-20% below market value.
1992	Contaminated soil around 93rd Street School removed and replaced. School is later demolished to allow other development.
1993	Emergency repairs of the synthetic cap on the canal are done. Hot spots of soil pesticide contamination are found, excavated, and replaced.

TABLE 6–3. (Continued)

DATE	EVENT
1995	EPA reaches agreement with the responsible parties, Occidental Chemical (OC) and the U.S. Army. OC will pay $129 million to the U.S. government, $3 million to be used for health studies. The U.S. Army will pay $8 million to the U.S. government. The funds will go to the EPA and Federal Emergency Management Agency to reimburse previous costs of the cleanup.
1998	Occidental Chemical settles a lawsuit with approximately 2,300 families who formerly lived at Love Canal; each receives payments between $83,000 and $400,000 (total cost to OC between $191 million and $920 million).
1999	EPA completes cleanup of 102nd Street dump, just one-quarter-mile south of Love Canal.
2000	*March*: EPA declares that disposal of contaminated soil and sediment is complete. Approximately 260 homes north of the canal have been rehabilitated and sold (area formerly thought to be uncontaminated) at 15-20% below market value. The area is called Black Creek Village. New senior citizen housing has been constructed.
2002	Occidental Chemical Company remains responsible for ground-water monitoring, maintenance, and operation of the leachate treatment facility.

Sources: Compiled from Center for Online Ethics, 2002; EPA, 2002; Levine, 1982; Mazur, 1998; NYDOH, 1981

the canal.[244] In September 1978 the New York Department of Health issued a report titled "Love Canal: Public Health Time Bomb." It presented preliminary data showing a higher rate of miscarriages and birth defects compared to the rest of the state.[245] This conclusion has been repeated in many other sources.[246] But that study has not appeared in a scientific journal.

In 1984 researchers from the New York Department of Health published a paper on low birth at Love Canal.[247] The data are startling. In 1950, when the dump was still open, 44% (8 of 18) of the babies born to women who lived in homes prone to seepage from the canal were low-birth-weight (less than 5.5 lb or 2500 g). The rate for the rest of upstate New York was 7%. For the time span examined, 1940 to 1978, low-birth-weight births in the "swale" areas (prone to seepage) were significantly higher than the rest of upstate New York and than the rest of the Love Canal area. This study included only people in the single-family homes, not the renters in the low-income housing.

In 1980, biologist Beverly Paigen and her colleagues measured children's growth. They compared more than 400 children who grew up at Love Canal to children from two other neighborhoods matched on socioeconomic status and race.[248] Their analyses also included important confounding variables such as age, family size, income, education, chronic illness, birth weight, maternal smoking, and parent stature. The results showed that the children of Love Canal were significantly shorter on average, by approximately 1 inch. This result held for both African American and white children. Paigen and her colleagues also published data on low birth weight and health conditions of Love Canal children.[249] These studies included both the homeowners and the renters. Love Canal children had a higher rate of seizures, eye irritations, abdominal pain, skin rashes, learning problems, and incontinence. Love Canal families also had 3 times the chance of the comparison population of having a low-birth-weight child.

Why Not More Research on Love Canal? I found only two other published studies. One on chromosomal damage,[250] and the other looked at cancer rates.[251] There were no studies of the psychological well-being or cognitive development of the children of Love Canal, even though news items in *Science* and elsewhere mentioned the high psychological stress of the situation. A major qualitative sociological study was published as a book, but it was not funded by government agencies.[252]

This lack of research at Love Canal contrasts sharply with Three Mile Island. At TMI the long-term psychological studies were funded by several federal agencies—the National Institute of Mental Health, the Nuclear Regulatory Commission, and the Uniformed Services University of the Health Sciences. The Love Canal crisis and the Three Mile Island accident occurred at virtually the same point in time, 1978 and 1979. Although the Three Mile Island crisis involved a larger population, the sample sizes in Dr. Paigen's Love Canal studies were larger than the samples in many of the TMI studies. It was feasible to do research at Love Canal similar to what was done at TMI. It just was not funded. Since that time the United States also has *not* undertaken a major epidemiological study of people living near waste sites. But such studies are done in Europe.[253]

The heated battles between the residents and government agencies at Love Canal brought out controversies among the scientists. At one of the early public meetings, in May 1978, the county health commissioner declared that the waste did not pose much of a hazard. A state toxicologist disagreed with him vehemently in front of the press and the residents.[254] Another scientific controversy centered on one researcher, Dr. Paigen, who questioned the "status quo." One issue was what measure of exposure to use. Then scientists and government officials also battled over the control of data and keeping secret the results of the health assessments.

What Index of Exposure? One of the astonishing facts about Love Canal is that although blood samples from approximately 3,000 people were taken, they were never used to determine internal chemical exposure! This was partly because technologies were not as readily available then.[255] Because toxic chemicals were not measured internally, a proxy or external index of exposure was needed. Two ideas clashed: simple distance from the canal measured by the rings of streets versus historically wet drainage pathways (swales) in the neighborhood. In the end the drainage-ways won out because the chemicals were seeping along them. But before it was settled, the clash over streets versus drainage-ways was costly to Paigen, a genetics researcher then at the Roswell Park Memorial Institute.[256]

Paigen thought that the Love Canal situation was an ideal setting to explore genetic differences in susceptibility to pollutants.[257] The homeowners had done their own health survey. It seemed to show a relationship between living in wet versus dry areas of the neighborhood and illnesses or miscarriages. Based on the homeowner health survey, Paigen sided with the Love Canal residents in promoting the drainage paths as the index of external exposure. But Paigen's position conflicted with the top brass in the Department of Health. As she exercised the academic freedom of a scientist by testifying before Congress and speaking to the Love Canal residents and the press, her estrangement from management grew. She was denied the opportunity to apply for grants through the normal procedures. She was refused permission to use her grant funds for travel if the trip related to Love Canal. She was denied a subcontract of a grant funded by the EPA. Some of her lab specimens disappeared. Mail she received from the EPA or Love Canal Homeowner's Association was sometimes opened and resealed with tape. And when her state income tax return was audited, the tax auditor's file had news clippings about Paigen's activities with the Love Canal residents.[258] The Council of the Association of Scientists of Roswell Park protested the restriction of Paigen's grant applications as scientific censorship.[259]

Control of Research Findings and Secrecy   Because of the controversies and publicity at Love Canal, the evidence regarding the health of the residents was reviewed by at least four panels of experts in 4 years. Some of the panel proceedings and original data were kept secret. New York Health Commissioner David Axelrod even refused to submit data to Congress, citing confidentiality requirements under New York law. But he also applied the confidentiality idea to his deliberations with one expert panel.[260,261] The Department of Health also refused to share data or even technical summaries of results with the Love Canal Homeowner's Association or its consultants, including Paigen. Even when residents sent notarized forms authorizing the release of their health surveys and blood tests, the data were not released.[262]

One expert panel of five men appointed by New York governor Carey (known as the Thomas Panel) recommended that all studies on Love Canal be administered and controlled by the state government under the oversight of a panel appointed by the governor.[263] This suggestion flies in the face of science as an open enterprise. In science the sifting and winnowing of evidence needs to be done from many different points of view.

The Thomas Panel's final report became instant news. The conclusions are still sometimes quoted as showing that there was no evidence of harm at Love Canal. The Thomas Panel concluded that there was no demonstration of acute health effects and that "chronic effects of hazardous waste exposure at Love canal have neither been established nor ruled out as yet."[264]

Although the Thomas report was widely quoted, it lacked the kind of detailed references that a scientific review must include. Professor Levine had to obtain documents about its proceedings under Freedom of Information requests because the panel would not send them to her directly. Those papers, plus evidence from correspondence, indicate that the panel used verbal summaries rather than looking at the statistics directly.[265] Dr. Irwin Bross, then director of biostatistics at Roswell Park Memorial Institute, criticized the Thomas report for failing to faithfully evaluate the data on miscarriages and birth defects.[266] Bross concluded that the miscarriage and birth defect data from Love Canal more than met the $p = .05$ cutoff. He also chided the panel for presenting opinions instead of the facts about reproductive hazards to the residents.

Is Hazardous Waste a Developmental Health Hazard? Whether residing near a hazardous waste site is a health hazard is still controversial in the U.S. One problem is that the exact chemicals, timing, and exposure levels of the people near toxic dumps are usually unknown. Exposure at one point in time does not necessarily reflect exposures years earlier or during certain sensitive periods of development. A multinational study in Europe examined the frequency of birth defects in proximity to (i.e., within 7 kilometers, or 4.24 miles of) hazardous waste sites, compared with births from comparable neighborhoods in a matched control design. The results showed a dose-response relationship between the frequency of excess birth defects and proximity to hazardous landfill sites. Neural tube defects and hypospadias[267] were most affected by proximity to waste sites.[268]

Conflicting Moral Concerns of Residents and Government

Two conflicting moral concerns formed the center of the battle at Love Canal. The first is injustice to the residents. The residents were innocent victims of chemical waste deposited years before. The second is the possibility of injustice to society (taxpayers at large). Have the victims been affected seriously enough to justify opening the government's wallet to clean up the mess? How far should the government open its wallet? These two concerns—individual injustice versus injustice to society—seem to conflict in most policy situations.[269] Different viewpoints on what is moral may be behind many environmental disputes. In this section I try to illuminate some the moral viewpoints of the Love Canal residents and the government agencies. Most situations involving moral judgments of injustice involve strong emotions, as at Love Canal.

The Moral Outrage of the Residents   The life savings of the Love Canal homeowners were tied up in their homes, as is the case for many middle-class Americans. As soon as the waste was publicized, the neighborhood was stigmatized and homes were not marketable. The first neighborhood activism asked for reduction in property taxes and mortgages. The renters in the subsidized housing were also in a bind. They knew that finding subsidized housing in a neighborhood with the same quality of schools, relatively low crime, and a suburban setting would be very difficult.[270] They were upset about the proximity of the Head Start program to the waste.[271]

New York Health Commissioner Robert Whalen did not anticipate the outcry that would follow his announcement that the government would subsidize a selective evacuation. The evacuation was for pregnant women and children under age 2 from the inner ring of single-family homes abutting the canal. After the announcement, meetings between government officials and the residents became more emotional and even disorderly. Below I present some quotes from the residents drawn from Levine's[272] sociological study. I selected these quotes to illustrate the moral outrage the residents expressed. I see two main themes. First, they express a deep sense of the *unfairness* of arbitrary cutoffs for eligibility for evacuation, either by age, pregnancy status, gender, or location of residence. Second, the quotes call for a consideration of *individual undeserved suffering* in the decision making. They express moral outrage at official failure to recognize and recompense that suffering.

- "What about my two-and-a-half-year old; she's out of luck, right?"[273]
- "I threw it [the health survey] on the floor because I was so mad. There was just no place to write about the kids. So they told me that only information about adults was important at this time. And that I could write stuff about the kids on the back of the pages. Is that all they cared about sick kids?"[274]
- People began to shout, many with tears streaming down their faces. One resident cried, "My child was under two last August. It's your fault we didn't leave then." A man whose wife had suffered several miscarriages shouted, "You aren't even human! Humans couldn't do this to each other! You're just trying to pacify us."[275]
- The risk, the [health] commissioner admitted in answer to a barrage of questions, was two or three times greater than the risks present

in a normal population, and he responded affirmatively when asked whether this meant that a pregnant Love Canal woman had a 30 to 45% chance of miscarrying or of bearing a defective child. "What does he mean 'risk assessment,'" commented one resident privately. "It's more like human sacrifice."[276]

- "You're murdering us!"[277]
- "The damage is done! My wife is 8 months pregnant. What are you going to do for my baby? It's too late for my child."[278]
- "You're treating us like the *Titanic*! Women and children first!"[279]
- "Is the school safe? Do you think my children should attend?...it's contaminated! What if the barrels come up out of the ground?"[280]
- "Would you bring your family here?" "Would you trade houses with me?"[281]
- "What does it take," cried one woman, "asthma, three miscarriages, a birth defect, a man with a damaged liver! You're a doctor! You should care about us!"[282]
- "Look me in the face! Are you going to tell me I've got to stay another night with my three children in that house in the light of what is being said here today?" With that, this normally respectful woman shredded the card bearing his name and his title and threw the pieces down, crying out, "You are not a doctor! Every time I think of you, I'll think with disgust!"[283]
- Some of the women tried to reach Dr. Axelrod in a personal way as well. One woman sent him a Father's Day card bearing the names of women who suffered miscarriages. Letters were sent to him listing the deaths in the area. In these and other ways, the residents tried to make the commissioner truly feel their plight, feel the hurt as a fellow human being.[284]
- Someone shouted, "Governor, I live right across the street from the homes next to the canal. My house is contaminated, too," and the governor said that the state would buy those too. Then someone from 102nd Street spoke up and he [the governor] said, "No, that's it. There is going to be a line. We'll put a fence up."[285]

The Apparent Moral Stance of the Health Department The Health Department officials seemed to want smoking-gun type of evidence in order to provide government assistance for evacuation. To the Department of Health, fairness seemed to mean applying the same rules or criteria to everyone objectively. Both principles try to emphasize *procedural* justice.[286] But the Department of Health initially excluded the renters in the low-income area. The Health Department's emphasis on procedural justice showed when Commissioner Axelrod, in August 1979, wrote, "Women who became pregnant prior to February 8 obviously were unaware of the increased risks.... These women were placed at involuntary risk because they lacked the knowledge of the relevant health data when they became pregnant. Since the availability of the information…those women wishing to become pregnant are now in a position to make a choice with full information concerning the risks.…We have reviewed the requests of the Homeowners Association to relocate women contemplating pregnancy prior to their conception, and *can find no fair and equitable reason* for distinguishing between those contemplating pregnancy and other women in the canal area."[287]

Then construction for the cleanup began. Residents who could verify that the odors from the canal made them ill could be moved out temporarily at state expense. One doctor wrote statements for 112 residents certifying that the chemicals had caused their acute depression. Axelrod rejected their requests: "The simultaneous certification by a single physician of 112 consecutive cases of acute depression among residents…and its causal link to remedial construction at the landfill, challenges our credulity. We reject his certification of a virtual epidemic of acute depression, pending individual clinical evaluations to be arranged by qualified physicians of the State Office of Mental Health."[288] The New York Supreme Court ordered that the people claiming depression be moved while they were given time to get the necessary examinations. Axelrod's rejection of the residents' mass claim of acute depression was apparently viewed as procedurally unfair by the New York Supreme Court.[289]

The Health Department tried to rely on the hard data of research findings in deciding whether to evacuate the residents. To the Health Department, procedural justice in research meant sticking to 5 chances in 100 or less ($p = .05$) as an "objective" cutoff for interpreting results as significant. The standard statistical cutoff was perceived as procedurally fair by the Health Department because it was incorrectly believed to be "objective." Many times the residents were told that there was no evidence that their health had been harmed. Within the health department, no one seemed to question whether $p = .05$ was the right chance of a false positive error for the situation of public exposure to toxic chemical waste. There seemed to be no discussion of whether the false negative error of *failing* to relocate people in the face of uncertain harm might be more serious than relocating people when it might not be needed. Also, the Health Department wanted to be sure that the high rate of miscarriages couldn't be explained by other variables, such as employment in the chemical industry, smoking, alcohol use, and socioeconomic status.[290]

But the Health Department also neglected to consider the procedural fairness of other aspects of their investigation. The residents' perceptions of the fairness were damaged by the fact that the health surveys and blood samples were initially collected only from the residents of the inner ring of homes. Under an outcry of protest from the neighborhood, the health screening was expanded. The decision to relocate those in the inner ring of homes was based on the initial health screenings. The residents were told that there was not enough evidence to relocate others. But the health surveys and blood samples from the rest of the neighborhood had not been examined yet![291] This obviously violated procedural fairness and the residents knew it.

Fighting Dirty for the Moral High Ground   The two sides in the Love Canal crisis began to derogate each other's motives. Some government officials came to perceive the residents as "like spoiled children."[292] Such a perception is not only derogatory but implicitly undercuts their moral claim as victims of injustice. When the residents asked, "Would you trade houses with me?" and "Would you bring your children here?" the intent was to show that any rational person would respond similarly. Of course, not all government officials felt the residents were spoiled children. One health department official stated publicly in May 1978 that if he lived there and could afford to move out, he would do so.[293]

Statements from the residents such as "You're just trying to pacify us" accuse the agency representatives of failing to carry out their obligation to protect public health. Some residents also personally derogated Health Commissioner Axelrod with statements such as "You're not human" and "You're not a doctor." To the residents, their unjust suffering constituted the moral high ground. Sticking to established procedures and "objectivity" was the moral high ground for the Health Department. Both are legitimate moral principles, but they can conflict in policy decisions. Even if an agency acts with impeccable procedures, the procedures may fail to bring about a fair outcome for victims.[294] At the same time, it may be unreasonable for public policy to remedy the suffering of every individual who is negatively affected by pollution.

Policy decisions in a pollution crisis are usually an attempt to balance the interest of the society (the taxpayers' interest in avoiding unnecessary expenditures of public money) or the bureaucracy (preserving the institution and the desire of its officials to save face) with the unjust suffering of the victims. In an unfolding pollution crisis the uncertainties make decisions exceedingly difficult. But as shown at Chernobyl and TMI, delaying a decision while gathering information can also increase damage to public health.

What about Hooker Chemical Company?   Why didn't the Love Canal residents focus their outrage on Hooker Chemical Company? Some did. But the Hooker representatives who attended public meetings and press conferences steadfastly maintained that the company was not legally responsible. It had included a disclaimer when it deeded the Love Canal property to the school district. Legal actions were eventually filed against Hooker (and parent company Occidental Chemical Company) by the EPA, the State of New York, the City of Niagara Falls and its board of education, the Niagara County Health

Department, and families residing at Love Canal. Hooker published full-page newspaper ads in which it repeated the conclusions of the Thomas Panel report. The company distributed pamphlets emphasizing the company's importance as a local employer, its concern about people in Niagara Falls, and its use of state-of-the-art waste disposal techniques.[295] Most of the legal actions were settled only in 1995 and 1998, almost 20 years after the crisis.[296]

Other Chemical Pollution Crises and Disasters

Here are a few examples of other crises and disasters that involved public exposure to toxic chemicals.

1. *Bhopal, India, 1984.* On December 2 and the early hours of December 3, 1984, an explosion at the Union Carbide chemical plant released approximately 45 tons of methyl isocyanate. Between 3,000 and 10,000 people were killed, and up to 500,000 were injured. People woke in the night choking, with burning eyes, and ran for their lives. The victims who did not die continue to suffer from many illnesses. The victims are still seeking compensation. Many of the victims were very poor people living in a shantytown that abuts the property of the Union Carbide plant. Union Carbide claimed that the accident occurred because of sabotage by a terrorist. But there is no solid evidence of sabotage. Indian sources blame the many blatant safety violations at the plant, combined with questionable management policies.[297]

2. *Meda (Seveso), Italy, 1976.* On July 10, a pesticide plant accident during the manufacture of the herbicide 2,4,5-T released dioxin into the atmosphere. The other name for 2,4,5-T is Agent Orange, the code name used by the United States during the Vietnam War. The contaminants drifted into the nearby community of Seveso. Approximately 47,000 people were exposed, 700 of them to very high levels of dioxin. The 700 in the high-contamination zone were evacuated. Their serum dioxin levels are among the highest ever recorded in people. Thirty years later the dioxin levels are still elevated compared to those from surrounding communities.[298] Health studies of the victims showed initial chloracne (skin rash). Recent follow-ups show an increase in cardiovascular diseases and cancers, and research continues.[299] Children born to men exposed in the accident have a higher-than-normal likelihood of being female, a result that suggests altered endocrine function.[300] Infants born to mothers from the highly contaminated areas have higher levels of thyroid-stimulating hormone (b-TSH) than a comparison group.[301] High b-TSH is a risk for thyroid insufficiency.

3. *Times Beach (and other areas), Missouri, 1970s.* In 1985, approximately 600–700 homes were evacuated in Times Beach, Missouri.[302] Dioxin-contaminated waste oil was sprayed in the early 1970s to control dust on roads in Times Beach, as well as on some horse arenas. Three days after one horse arena was sprayed, birds that lived in the rafters were found dead on the floor. Over the next few weeks dogs, cats, and 48 horses died. Even though some of the soil in the arena was removed and replaced, horses continued to die almost 3 years after the initial spraying. Chickens exposed to a sprayed road died. Four children who played in the horse arena developed chloracne, and one child was hospitalized with hemorrhagic cystitis. Tests by scientists from the CDC and the Missouri Division of Health showed that the soil in the horse arena contained approximately 32 ppm (32,000 ppb) of dioxin, apparently enough to kill.[303] Cleanup of Times Beach did not begin until 1988. The most highly contaminated soils were excavated and incinerated to a standard of 20 ppb dioxins. Less contaminated areas were covered with fresh soil. The area that was formerly Times Beach is now a state park.[304] The dioxin-contaminated oil was traced to a salvage company that obtained it from a manufacturer of hexachlorophene. Hexachlorophene is an antibacterial substance that used to be added to soaps. Often called "green soap," hexachlorophene was withdrawn from the market partly due to association with birth defects in children born to nurses who washed with it many times a day. The sludge at the manufacturer had more than 300 ppm dioxin.[305]

4. *Herculaneum, Missouri, ongoing lead pollution.* A lead smelter that has operated in Herculaneum for approximately 100 years has polluted the city. Pregnant women and families with small children were evacuated in 2002 while the company, Doe Run, a subsidiary of Renco Group Inc., is ordered to clean up the lead. The Missouri Department of Natural Resources found that 15% of children less than 6 years old who live within a mile of the smelter have elevated lead (>10 µg/dl). The Missouri DNR reported that "the smelter has violated the clean air standard for lead. Over the last twenty years, the U.S. Environmental Protection Agency (EPA) and the Missouri Department of Natural Resources (MoDNR) have taken repeated action to bring the smelter into compliance with the clean air standard."[306] Dust very high in lead was regularly spilled from trucks coming to and from the plant. Furthermore, "MoDNR continued to observe violations of clean air law related to fugitive dust from the facility. On September 17, 2001, EPA and MoDNR issued a letter to Doe Run requiring them to completely characterize residential soil contamination in Herculaneum within 60 days, and to begin residential soil replacement."[307] The Missouri

DNR also issued an "Abatement and Cease and Desist Order" that required the company to take 10 remediation steps to prevent trucks from spreading lead in the city. "Cease and desist" orders are used as a last resort in environmental enforcement. Doe Run replaced soil in the yards of 487 homes by July 2007. In 2007 Doe Run agreed to install truck wash stations, clean the streets where its trucks operate, and assure that trucks do not leak lead during transportation.[308] Missouri DNR previously ordered Doe Run to stop the trucks from leaking in 2002. The Missouri DNR Website recorded that the lead air emissions were in violation in the first quarter of 2008.[309]

Social Justice Issues with Chemical Toxic Waste in America

The problems of chemical waste gave rise to the environmental justice movement in the United States. In 1979 the Urban League and the Sierra Club held a conference on the issue of racial and socioeconomic inequities in pollution in Detroit. The EPA began attempting to address the issue of inequities in the siting of new facilities as early as 1978. Other situations like Love Canal still occur.[310] Unfortunately, the delays in dealing with hazardous waste situations after they are discovered are often longer than the delay faced by the Love Canal neighborhood.

St. Regis Indian Reservation, Mohawk Nation (Akwesasne), New York and Ontario This Indian reservation along the St. Lawrence River is adjacent to a General Motors factory. The company discarded PCB-laden hydraulic oil. The General Motors location was placed on the Superfund list in 1983. EPA and GM began negotiating in 1985. The groundwater is contaminated with PCBs. Two plants now owned by Alcoa are also polluted with PCBs, and there is a paper mill on the Canadian side of the river. After cleanup to reduce the concentration of PCBs migrating off the site, General Motors dredged contaminated sediment out of the St. Lawrence River and adjacent creeks in 1994. EPA regards this project as a model for other river cleanups.[311] The tribal government's Environmental Division meets with the EPA regularly. Partly because of tribal concerns, the PCB-contaminated materials will be discarded in a regulated landfill elsewhere rather than on site.[312] GM started engineering for groundwater remediation in 2007, and monitoring wells were installed.[313] The tribe is still seriously concerned about on-site disposal for materials in one of GM's landfills. The EPA says, "There is strong public opposition to the containment remedy."[314]

Relocation is not an option for an Indian reservation, the tribe's ancestral homeland.

Health surveys have found that the Mohawks are at greater risk from PCB-contaminated fish than the average New York angler. Mohawks consume more fish than the average New Yorker.[315] The most popular food fish is yellow perch. The pattern of PCB congeners found in the Mohawk women is very similar to the congener pattern in the perch.[316] In men who do not eat many fish but live close to a polluted site, congeners in their blood matched the congeners in air.[317] Girls with higher estrogenic PCB congeners reach menarche (first menstruation) earlier than girls with lower estrogenic PCBs.[318]

Agriculture Street Dump, New Orleans The New Orleans metropolitan area has at least 35 hazardous waste sites. The Agriculture Street Dump was closed in 1958. Housing development on top of the dump that started in 1969 includes a school and senior citizen housing. After Hurricane Betsy in 1965, part of the dump was reopened. For 6 months approximately 300 truckloads of debris per day were delivered and burned in open fires. Following that, the homes built by the Housing Authority of New Orleans were offered to low- and middle-income families on a "rent to own" program. The residents are primarily African American. More than 100 toxic chemicals have been found in the soil around the homes. Lead is particularly high and EPA declared it a Superfund site. Residents then could not sell their hard-earned homes, and the area is stigmatized. In 1996 some homeowners posted signs in their yards saying, "Move us out NOW!"[319] In 1998 EPA removed contaminated soil from people's yards, laid down a barrier, and then laid 2 feet of new soil. In 2005 Hurricane Katrina flooded the Agriculture Street site. Some homes and apartments became uninhabitable. Residents were concerned that the pollutants were mobilized by the flood. Other people were illegally dumping in an undeveloped part of the old dump area. The undeveloped section is fenced but gates are commonly broken open. After Katrina the EPA tested soil for lead, arsenic, and some other pollutants. EPA concluded that the cap on the dump was not damaged by Katrina.[320] But the Natural Resources Defense Council reported, "…leachate was visibly leaking from the site and spreading across the street onto the grounds of the local senior citizen's center." Tests showed chemical pollutants (polyaromatic hydrocarbons) exceeded allowable levels by 2–20 times.[321] There is ongoing legal action in the state courts.[322]

EPA's Relocation Policy Relocations for cleanup of pollution are controversial. They disrupt the lives of individuals and families, cause psychological stress, and stigmatize and divide communities. They cost money for the government, the companies or other parties found to be responsible for the dumped waste, and for the relocated citizens. The necessity of relocation, and whether it is carried out with respect for the residents, is always questioned.

To make relocation decisions fairer and more compassionate, the EPA sponsored a 2-day meeting in 2000 on its interim relocation policy.[323] Some representatives of civil rights organizations said they believed the EPA was biased against relocating residents near Superfund sites, especially minority residents. Also, relocating communities of color can force those residents into less desirable neighborhoods. Other participants commented that the EPA needed to consider the psychological impacts of both the toxic waste and the relocation. They also suggested that communities that are relocated should be kept together if possible. The participants in the meeting pointed out that the "health and welfare" of people includes overall quality of life, stress, risk perceptions, security of the new neighborhood, property values, community viability, and the stigma attached to toxic waste.[324]

But the EPA is not the only agency in the United States that moves people from their homes. The Department of Transportation, state, and local governments regularly purchase people's property in order to build highways, expand airports, build convention centers, and install electricity power lines. In these kinds of relocations the "public good" also stands against individual rights. But what constitutes public good is interpreted very differently in toxic waste cleanups versus public works projects. With public works the cost of purchasing homes and relocating people is merely routine. The focus is on accomplishing highway expansion or other projects. With toxic waste cleanups the focus is on whether there is an *imminent* danger to health, as well as on saving money and avoiding relocation. Most people resent having their homes taken from them for public works projects. But protesting the seizure does nothing. In toxic waste situations, the government usually tries to *avoid* purchasing people's property. This is true even when neighborhoods lobby for a buyout, as at Agriculture Street and Love Canal. The fact that the cleanup of a leaky dump is a public good in itself is usually not given consideration.

CHAPTER 7

The Best Science, Values, and the
Precautionary Principle to Protect Children

Environmental regulations have saved many children and their families from much unnecessary suffering. Mercury is no longer discarded from factories in industrialized countries the way it was at Minamata. PCBs have been phased out. Lead is no longer added to U.S. or European gasoline and paint. Better regulations apply to new dumps in which toxic chemicals are discarded. At the same time, the scourge of pollution victimizes almost every child in one way or another. There are serious social inequities regarding whose children are most affected. Children continue to be harmed by lead, mercury, PCBs and pesticides, and noise. Toxic chemical dumps are still leaking into people's homes, long-lived radionuclides contaminate some areas heavily, and the planet's food chain is still bioconcentrating pollutants such as mercury, PCBs, and PBDEs in remote areas such as the Arctic and in industrialized areas.

The best science by itself cannot solve the pollution problems we face today. And science cannot prevent problems in the future. Science by itself does not make environmental policy. Ethical and value considerations are central to policy. The best science, no matter how good it is, will always involve uncertainty, interpretation, and arguments among scientists. Whether the scientific uncertainties and arguments should delay policies to protect the public is a matter of one's moral values.

The Precautionary Principle was proposed in its basic form at least as early as the Surgeon General's 1925 conference on tetraethyl lead. It was reiterated in 1936 in the meeting over PCBs. But policy is still not precautionary. To be precautionary in a rational way requires careful consideration of the value judgments that are embedded in the scientific research, risk assessments, and benefit-cost analyses. Before we consider the Precautionary Principle in detail, let's recap what we have learned about how pollution affects children's behavioral development.

What We Know about How the Pollutants Affect Children

The research shows that virtually all children are exposed to many kinds of pollution that can harm their development and overall quality of life. Many of these pollutants are in the food chain. Research on lead, mercury, PCBs, and noise has produced the most evidence that children's intellectual development and behavioral functioning are being harmed. The effects of pesticides (OP and carbamate) and low-level radioactivity on psychological development have not been well studied. But there is a logical chain of evidence that they can alter early brain development and later functioning too. The psychological stresses of pollution disasters can have long-lasting effects on families and communities. For the latter hazards to children's development, the scientists will continue their arguments about what level of exposure produces what magnitude of negative effects, and whether there are alternative explanations of the results. Scientists will also continue to study the exact developmental processes that are altered by exposure to a toxic substance, whether there is a sensitive period during development when a fetus or child is most susceptible to exposure, how combinations of toxic substances affect development, and what other factors in the child's home and school environment or genetics of the child might exacerbate or ameliorate the effects of pollutants.

How serious are the pollutants discussed in this book? The pollutants that I chose to write about are those for which there is at least some good research on behavioral and psychological effects. That fact does not necessarily imply that they have worse effects on children than other pollutants. I have included physical health effects only in passing. The pollutants described in this book came to the attention of public health officials because of some extraordinary events—children killed by lead poisoning, methylmercury disasters in Minamata and Iraq, PCB poisonings in Japan and Taiwan, public complaints and lawsuits over airport noise, pesticide poisoning incidents, nuclear accidents, and blatantly leaking toxic waste dumps.

These pollutants are a public health concern because virtually all children are exposed to them to some extent, whether they live in industrialized countries or in remote areas. The magnitude of any public health issue depends on the size of the population that is exposed. Even if there is only a 1 in 1,000 chance that a child will be harmed by prenatal exposure to a toxic substance, if 1 million children are exposed, then we would expect 1,000 children to suffer negative effects. Whether it is worthwhile to save those 1,000 children depends on one's moral values. Because children cannot protect themselves, we give them extra legal protection in our society. Children are "captive riders" who must go along with whatever their families and the society around them decide. Protecting children from pollution is plainly a moral choice.

Ethics and Values in Risk Assessment and Benefit-Cost Analysis

The best science by itself cannot determine policy. But, the research on how pollution affects children's development can and should be used to inform environmental policies. As I showed in each chapter, the many layers of decisions between research and policy all contain value judgments and subtle interpretations. Because policies are fundamentally ethical issues, there is no a priori reason to give the ethical viewpoints of scientists more weight than the ethical viewpoints of anyone else. Throughout I have pointed to instances in which particular research decisions made by scientists influenced the conclusions that they drew about the harm caused by a pollutant. For example, Ernhart did not conclude that lead exposure harms children because the results had 7 in 100 chances of being wrong rather than only 5 in 100 (see Chapter 1). Schultz (see Chapter 5) decided to tabulate the percentage of people highly annoyed by aircraft noise, rather than the percentage moderately annoyed. In studying exposure to pesticides, one research team chose to compare the average exposure to the EPA's reference dose, and another used the most highly exposed child for comparison (see Chapter 4).

Research conclusions, with the scientists' decisions and interpretations embedded in them, then become a part of risk assessments. The risk assessor must decide what constitutes an adverse event due to the pollutant, and how much of an increase in adverse events is an "acceptable risk." Is it acceptable to have twice as many developmentally disabled children due to lead, mercury, PCBs, pesticides, noise, or ionizing radiation? That decision made as part of a risk assessment is an ethical decision. But such decisions are not normally reported to the public, the public is not normally consulted about them, and the implications of these value decisions for the results of the risk assessment are not normally discussed. All the rest of the steps in the risk assessment process also involve value judgments (see Chapter 2). Risk assessments should inform environmental policies. But because of the value judgments and scientific interpretations embedded in them, risk assessments should not serve as black-and-white decision criteria or as the final word about the effects of a pollutant. Instead, the risk assessment can be used as a medium for discussion of crucial value questions such as, "How much harm from this pollutant is too much?"

After a formal risk assessment, agencies will normally conduct a benefit-cost analysis based in part on the risk assessment. Our technologies are intended to give us better quality of life. For example, fungus on seed grain is effectively deterred by methylmercury. But is the benefit worth the pollution it leaves behind, and how does methylmercury compare to other options for controlling seed fungus? Our society decided the pollution and poisoning risks of methylmercury on seed grain were too high. We can easily count up the dollar profits as the benefits for the producer of a technology. But all products cause pollution as a side effect (that is, they create what economists call an "externality"). The costs of pollution are difficult to measure in dollars.

Benefit-cost analysis also involves both judgment calls and ethical issues.[1] One fundamental ethical issue is rooted in the fact that the costs and benefits are simply added up. This favors the option for which the gains outweigh the losses by the most. But what is left out? Are gains and losses distributed fairly in society? How large is the worst loss? Is the worst loss a regional catastrophe like Chernobyl? Will compensation be available to those who suffer losses?[2] For example, when an airport is expanded, nearby residents are going to be involuntarily exposed to the added noise. Will they be compensated for their loss in property values? Will they be compensated for the fact that their children may have more trouble learning to read, perform more poorly in school, have high blood pressure, or show high chronic stress? Environmental injustices occur for most pollutants. Because of these inequities, the ethical issues of involuntary exposure are critical.[3]

One of the reasons I wrote this book is that the behavioral effects of pollutants on children are often neglected in the risk assessments and benefit-cost analyses. An important ethical decision in any benefit-cost analysis is what to include in the calculations. The monetary benefits to the airlines of fewer flight delays are easy to estimate. It is possible, but more difficult and uncertain, to estimate the monetary costs of lower property values due to noise. But outcomes such as psychological stress and poor reading do not have dollar values in the marketplace. Dollar values for stress and children's school performance can be estimated, but these more often are just omitted.[4] Even if dollar values for intangibles such as educational attainment and psychological stress are estimated, those dollar values will involve implicit value judgments. Dollar values simply cannot address most ethical issues. Instead, equating one's well-being with financial matters can add insult to injury. Would most people knowingly sell their health?[5] Should people be allowed to sell their children's psychological well-being and future educational attainment? Another key ethical question is whether to place higher importance on the well-being of people in the present than of people in the future, such as our children and grandchildren, and their grandchildren to the nth generation.

Both benefit-cost analysis and risk assessment usually omit other issues of ethical importance.[6] Does the population voluntarily consent to a risk or is exposure involuntary? What kinds of societal benefits can justify involuntary exposures to pollutants? Is information about risks given to the public or deliberately withheld? Can a pollutant be easily cleaned up later if it is discovered to be more harmful than was originally thought? Are effects irreversible? Does the pollutant alter nature, how well do scientists understand the pollutant, and what unanticipated effects might arise in the future?

Risk Perception Includes Factors Excluded
in Risk Assessments

Although many of the ethical concerns listed in the previous section are omitted from formal risk assessments and benefit-cost analyses, many of them do influence how people perceive environmental hazards. The moral implications of such dimensions may be a large part of why they are related to people's risk perceptions.[7] In psychological studies of risk perception, people have been asked to rate different hazards on a variety of dimensions, such as voluntary/involuntary, global catastrophic/local catastrophic, increasing/decreasing, equitable/inequitable, dreaded/not dreaded, controllable/uncontrollable, fatal/not fatal, observable/not observable, and so on. Research shows that these dimensions can influence people's risk perceptions quite dramatically.[8]

These dimensions that are omitted from risk assessment help explain why the public's viewpoints on some technologies, especially nuclear power, differ so dramatically from the numerical risk assessments. For example, Paul Slovic[9] found that college students and members of the League of Women Voters ranked nuclear power as the most dangerous technology in a list of 30 items that included things such as handguns, motor vehicles, contraceptives, X-rays, home appliances, pesticides, and so on. In contrast, a sample of risk assessment experts ranked nuclear power 20th out of 30. Members of the public do think about many aspects of hazards that are excluded in risk assessments.[10] Most of the omissions raise ethical issues that are neglected or swept under the rug in a formal risk assessment.

The Rationality Criterion

Sometimes economists and researchers who study decision making say there are specific criteria by which public policies could be made more rational. One idea is that the government should arrange its policies so that the same amount of money is spent saving lives threatened in different ways.[11] For example, one estimate of the cost of airline security for saving one life from a terrorist hijacking was $6 million (prior to the terrorist aggression against the United States on September 11, 2001). On the other hand, an estimate of the cost of removing asbestos from schools is approximately $250 million per life saved.[12] Do such comparisons show that public policy is irrational? One criterion of rationality would require spending the same amount per life saved, or at least the same amount per year of life expectancy saved.[13]

There is merit in comparing government expenditures in this way, but it leaves out other moral criteria.[14] The same expenditures per life do not necessarily imply equal protection, nor do they imply equal quality of life. The pollutants in this book affect quality of life even when they do not kill. The idea of equal safety expenditures per life saved also does not consider the voluntary versus involuntary nature of a risk. Voluntary exposure with full information is critical where risks are not distributed equitably. And throughout the book I showed that exposures to pollutants that affect child development are not equal-opportunity events.

The Roles of Government in Regulating Risks and Promoting Commerce

One ethical argument in favor of more protective government environmental policies can be based on Jeremy Bentham's utilitarian viewpoint that the function of government is promoting the greatest good for the greatest number. However, Bentham argued that the primary duty of government is to *protect* its citizens against serious harm. Maximizing enjoyment or benefits ought to be left to the citizens themselves.[15] To put this another way, each of us is free to try to get rich in any way we want as long as we don't hurt others. In 1925, Kehoe argued that the benefits of lead in gasoline were certainties that should be given more weight than uncertain harm (see Chapter 1). Therefore, Kehoe said the lead additive should be used until harm was proven. But if the government's primary duty is to *protect* citizens against harm, then a protective policy stance should be taken. On Bentham's view, the government's duty is primarily to protect us, and only *secondarily* to promote commerce and business innovation. The industries and citizens should be left to figure out how to promote their own welfare *without* harming each other and the environment. If the role of government is to protect against harm then agencies like the EPA would place higher weight on risks (especially widespread ones) than on the potential benefits of polluting technologies.

The Precautionary Principle

The record of the past shows that environmental pollutants once thought to be safe turned out to be harmful to children's development. In 1925 at the Surgeon General's conference on lead in gasoline, Alice Hamilton and other public health doctors argued that the proponents of lead should provide evidence of safety. They lost that argument. For PCBs, mercury, high levels of noise, and other toxic chemical wastes, harm to children's psychological development was discovered only after exposures had gone on for a generation or more. Should we continue approving new technologies that pollute unless we have smoking-gun type evidence of harm?

The Precautionary Principle has become a centerpoint of policy debate. As adopted in a declaration at the United Nations conference on environment and development in Rio de Janeiro in 1992, the Precautionary Principle was stated this way: "In order to protect the environment, the precautionary approach shall be widely applied by States according to their capabilities. Where there are threats of serious or irreversible damage, lack of full scientific certainty shall not be used as a reason for postponing cost-effective measures to prevent environmental degradation."[16] Four principles are often given for bringing the Precautionary Principle into practice: (1) preventative action should be taken in the face of uncertainty, as long as there is some evidence of potential harm; (2) the burden of proof should be shifted to the proponents of a potentially harmful activity; (3) a wide range of alternatives to potentially harmful actions should be examined; and (4) public participation in the decision process should be encouraged and increased.[17]

The Precautionary Principle is as controversial now as it was at the 1925 Surgeon General's tetraethyl lead conference. The arguments are similar, too. Echoing Kehoe's arguments in favor of tetraethyl lead, one side claims that applying the Precautionary Principle will squelch all technological innovation. Others claim that the Precautionary Principle provides long-awaited wisdom by correcting the current bias in favor of adopting new technologies before enough is known about them.[18]

But how precautionary do we want to be? Expressions of the Precautionary Principle lack specificity and admit a wide range of interpretations.[19] Here I examine the four principles of precaution and their relation to scientific uncertainty.

Preventative Action in the Face of Uncertainty

How certain should the evidence be in order to warrant protective action? Most government decisions regarding pollution rely on published scientific research. When a regulatory agency decides whether or not to regulate a pollutant there will never be absolute certainty about the correctness of that decision. As one of my colleagues says, "Being in statistics means never having to say you're sure."[20] The issue is not *whether* there is scientific uncertainty but *how much* scientific uncertainty should be tolerated when making policy decisions. The highest scientific uncertainty is not attached to claims that pollutants are harmful but to claims that pollutants are safe.

There are several kinds of scientific uncertainty, and I have given examples of them throughout this book. Each type of scientific uncertainty functions indirectly as a criterion for regulating pollutants. If there is too much uncertainty, the pollutant will not be regulated. These sources of uncertainty are just as important for the conclusion that a pollutant is *not* harmful as they are for the conclusion that a pollutant is harmful. The *first* kind of uncertainty is embedded in the $p = .05$ decision standard used to publish research—the chance of a false positive error. A *second* contributor to scientific uncertainty is found in the fact that it is very rare for *all* individual research studies on the same topic to yield the same results. A *third* category of scientific uncertainty is rooted in different viewpoints about the correct theoretical explanations for a phenomenon. A *fourth* type of uncertainty is the generalization of scientific results from one situation to another.[21]

The $p = .05$ Cutoff

The scientific community has a tradition of using a low probability ($p = .05$ or lower) of false positive error for publication (i.e., for declaring there is an association between a pollutant and an adverse outcome, when there might not be). The Precautionary Principle might suggest relaxing that criterion. Precaution would allow protective action to be based on results that point in the direction of harm, even if they do not quite reach the .05 cutoff. This interpretation implies abandoning Kehoe's criterion that requires smoking-gun type of evidence of harm prior to regulation. This would also shift the burden of proof of safety to the polluter. Right now we don't really require evidence of safety.

In policy decisions the chance of a false positive error might be called "industry risk." It is the chance that a pollutant will be declared harmful when it is not, a situation in which the company will be deprived of profits.[22] The likelihood of a false positive error is inversely related to the chance of a false negative error, the likelihood of *failing* to link a pollutant to harmful effect when the pollutant is actually harmful. The false negative error in policy can be called "public risk." If a pollutant is harmful but agencies fail to regulate that pollutant, the public stands to be harmed. The more certainty we require for declaring that a pollutant is harmful, the more risk we impose on the public (i.e., the higher the chance will be of a false negative error). (See the Appendix for a graph showing the trade-off between false positive and false negative error.) If researchers used $p = .10$ instead of .05, then there would be a better chance of detecting harm from a pollutant. The linkage between false positive and false negative error is especially important if the producer of a pollutant is required to provide evidence of safety. A lack of evidence for harm ("nonsignificant results") is not equivalent to evidence for safety unless the chance of false negative error is small (say, .05 or less). It is rare for the chance of a false negative error to be that low. This means that at present we allow greater scientific uncertainty for claims of safety than for claims of harm.

Not All Studies Find Harm

Individual research projects do not always agree. For example, the effects of mercury exposure in fish on children's intellectual development did not reach the $p = .05$ criterion in the initial studies in Seychelles. In the Faroe Islands the results did meet the cutoff (Chapter 2). Noise exposure did not have an effect on children's reading scores in the LAX study (Chapter 5). One of the studies of children of Chernobyl failed to find an association of IQ test scores and residence near Chernobyl at the time of the accident, whereas two other studies did (Chapter 6). Throughout the book I have emphasized that the sensitivity of a study (also called its statistical power) is very important. Poorly designed research will usually yield a "no evidence" conclusion.[23] But the chance of a false negative error will often be very high.

Taking account of alternative explanations and thinking through details analytically are two essential aspects of science. The failure of all individual studies to point unequivocally to the same conclusion is usually dealt with by analytical comparison of key research details that may account for the different results. I have done this for my readers throughout. When researchers draw a conclusion that there is no evidence linking a pollutant to harm, the details of the study and its assumptions should be critiqued just as seriously as when a study concludes there is a link. This brings us to the third source of scientific uncertainty: how research results are interpreted in the context of other findings and the prevailing theories.

Theoretical Differences Yield Different Conclusions

Different conclusions may be drawn from the same research studies because the researchers differ on key theoretical disputes. Points of theoretical dispute are also points of opportunity for research discoveries if a scientist collects new evidence to refute the claims of others. But in epidemiology research, where variables cannot be controlled by the experimenter, such breakthroughs are unusual. Instead, key research issues in epidemiological research of any sort are (1) what comparison group to test, (2) what confounding variables to measure, and (3) how to include those confounding variables in the analysis of the data. In Chapter 1 I described how Clair Patterson questioned Kehoe's theory of "normal" lead exposure and collected data that refuted Kehoe's ideas. By including omega-3 fatty acids from fish in their recent study of methylmercury from fish, the researchers in the Seychelles resolved the differences between the earlier Faroe Island and Seychelles findings. For noise annoyance (Chapter 5) we saw that separating aircraft noise from road and rail traffic yields different conclusions. For PCBs (Chapter 3) we saw how discussions of the pros and cons of the details of the Michigan study among the scientists were interpreted in the GE Web site to imply that the study was inconclusive. In the aftermath of Chernobyl (Chapter 6), where two research teams both found an association of residence in polluted areas with IQ test scores, one interpreted the results as evidence that radiation had effects on cognitive development, while the other research team demurred.

There are no hard-and-fast rules in science for interpreting results. There are no cut-and-dried standards for comparing the findings to other research or explaining the pattern of data with theoretical ideas. Interpretation involves many nuances of the data and research setting that lend credibility to some explanations of the results in comparison with other explanations. Science is partly art.

Generalizing Scientific Results

The fourth source of scientific uncertainty is whether results and conclusions from one research context can be applied to another situation. How far can we extrapolate the results? This issue is enmeshed in the prevailing theoretical explanations of the phenomenon and the other sources of scientific uncertainties. Where the biological or psychological mechanisms that link a pollutant to the outcome are well understood, then scientists will have more confidence about generalizing from one research setting to another. Conversely, where the theory of a phenomenon is weak, scientists are more reticent to make predictions beyond the bounds of their actual data. For example, a very rich network of data from both laboratory and field studies supports the theory that early childhood auditory word discrimination is a prerequisite for learning to read. This bolsters our confidence in generalizing the finding that high noise exposure impairs children's learning to read (Chapter 5).

Extrapolation to lower levels of exposure of a pollutant requires both theory and assumptions about the shape of the dose-response relationship. In behavioral epidemiology, theories that are detailed enough and firmly enough grounded in data to specify the shape of a dose-response relationship are rare. At present we have the most evidence about the dose-response curves for the behavioral effects of lead, mercury, and noise. Because dose-response curves are not well worked out for many pollutants, the question of whether there is a threshold continues to be a source of uncertainty and dispute. The potential for dispute is best illustrated by the continuing arguments in the scientific literature about low-level chronic exposure to ionizing radiation.

Similarly, there is uncertainty in generalizing from animal research to humans. These generalizations depend on how well we understand the mechanisms that explain the effects in animals, and whether those mechanisms apply to humans. For example, the roles of neurotransmitters in directing early brain development have important commonalities across mammals. Research with rodents has illuminated some key steps in the process by which nicotine and at least one OP insecticide alter brain development. To the extent that the mechanisms are well understood and common across species, an argument can be made to apply those results to human risk assessments. The application to humans gains further credence when epidemiology findings agree with expectations based on the rodent research. Where there is a reasonably well-supported theory of the mechanisms by which a pollutant does its harm, generalization seems a short step rather than a long leap. Without such a well-supported theory, the inference to humans seems more uncertain, whether the conclusion is for harm or safety.

As with the other sources of scientific uncertainty, the generalizability issues apply with equal force to studies that conclude there is no evidence of harm. Generalizing a "no evidence" conclusion should also require examination of: (1) the chance of a false negative error (preferably it should be 5 in 100 or lower, just as we normally demand for a conclusion linking a pollutant to harm), (2) studies that have yielded conflicting conclusions about the same pollutant, (3) plausible alternative interpretations and (4) whether there is a well-supported theory of the mechanisms by which the pollutant *fails* to affect humans or other species.

When confronted with this daunting list of scientific uncertainties, what can we conclude? The research findings for the pollutants in this book are as strong as many research findings on other topics that are presented as "fact." There is *not* more scientific uncertainty in the research on how pollution affects child development than in other research. I have highlighted the uncertainties and arguments in order to bring my readers into the debate over how we *should* make decisions. The conclusions from the "best science" will inevitably contain uncertainty.

The Burden of Proof Is on the Polluter

In the section above I emphasized that scientific uncertainty is a two-edged sword that can cut with equal force both those who would conclude that a pollutant is harmful and those who would conclude that there is no harm. However, since the 1925 Surgeon General's conference on lead, scientific uncertainty has been invoked as a reason *not* to regulate potentially harmful pollutants. If the burden of proof were actually shifted to the proponents of potentially harmful activities, then studies of a new technology would need a false *negative* error of .05 or lower to show "safety." This would often require that the chance of a false *positive* error be allowed to be quite high, perhaps .20 or higher. So this interpretation of the Precautionary Principle would change the way scientific research is interpreted when it is applied in policy. Not many epidemiological studies would reach a conclusion of "no harm" using these criteria. Experiments in which laboratory animals are exposed to toxics would similarly have a difficult time in drawing a conclusion of no harm. The research would also be more expensive, in both dollars and in animal lives. Larger sample sizes would be needed in order for the chance of both a false positive and false negative error to be low.[24]

Requiring that the probability of a false *negative* error be .05 or lower while simultaneously allowing the chance of a false positive error to rise to .20 or even higher would not be acceptable to most scientists. Science uses $p = .05$ cutoff for significance to avoid accepting false information as true.[25] Ironically, often a researcher will discuss a *nonsignificant* result as if the finding has a low chance of being wrong.[26] Shifting the burden of proof of safety to producers of potentially harmful pollutants requires us to look at the probability of false negative errors. At present false negative errors are either neglected or .20 is regarded as good enough. The likelihood of a false negative error is a numerical measure that tells something about the sensitivity of a study for detecting harm due to a pollutant.[27]

Consider a Wide Range of Alternatives

The Precautionary Principle exhorts us to look beyond the status quo for the solution to pollution. Change from the status quo can be difficult. I pointed to instances in which status quo science impeded progress or tipped the interpretation of results in one direction. This aspect of the Precautionary Principle encourages us to exercise the creativity of the human imagination to avoid blind spots induced by the status quo. For example, had the 1925 Surgeon General's committee asked the auto makers and gasoline refiners to consider other ways of enhancing fuel efficiency and engine power, the history of lead exposure in the United States would have been altered. If we ask electricity companies to look beyond the nuclear versus coal dichotomy for the solution to global warming, we may alter the future of the planet.

Failure to consider a wide range of alternatives can cause seemingly protective government policies to fail. One example is the gasoline additive MTBE, which reduces the carbon monoxide produced by engines. The EPA required it in some areas of the United States in order to reduce air pollution. The EPA thought that requiring MTBE would be precautionary. But MTBE easily migrates into groundwater, some individuals feel ill after breathing its fumes, and animal research suggests it is a carcinogen.[28] After these problems with MTBE surfaced, the EPA evaluated the fuel additive using criteria similar to those used to evaluate lead in gasoline. The EPA asked whether it had enough smoking-gun type of evidence against MTBE to mandate its removal. EPA apparently forgot that the original purpose was to protect people from air pollution. Consideration of a wider range of alternatives would have included options for lowering air pollution such as allotting more money for mass transit, requiring engines with higher fuel efficiency, and so on. Worse yet, Goldstein[29] reported that the EPA was planning to replace MTBE with other fuel additives that have not been studied thoroughly enough to assure that they will not create similar problems.

Another example of status quo thinking and failure to consider a wide range of alternatives occurred with boats and manatees in Florida. Manatees are an endangered species. It seemed logical to set a lower speed limit on the boats.

That was done more than 20 years ago. The boats slowed down, but the number of manatee deaths due to collisions actually increased. Wildlife managers assumed that the manatees were too stupid to learn to avoid boats. Many of the animals have multiple propeller scars. But it turns out that manatees cannot hear the engines of slow-moving boats, whereas they can hear revved-up engines. The solution should be to keep the speed limit and put underwater warning beepers on the boats so that the manatees can hear them.[30]

The manatee and MTBE examples illustrate the important way that ongoing scientific research can support precautionary government action. Remember that in 1925 the Surgeon General's tetraethyl lead conference asked for more research. That research was not done. When lead in gasoline finally came under regulatory scrutiny more than 50 years later, the industry resisted efforts to phase it out. Similarly, voluntary limits on lead in paint were not effective. When new technologies start to come into wide use, follow-up research on the effects of exposures is essential. Studies monitoring changes in ecosystems and human health should be designed so that they are sensitive enough to detect important changes. Achieving good sensitivity may require allowing the false positive error to rise considerably. For example, a study of the impact of a dam on Hudson River fisheries concluded that the most sensitive fish counting methods would have a 50% chance of a false negative error (chance of concluding "no harm"). This would be true even if the survival of yearlings were reduced by half for 20 years.[31] Precaution requires us to ask whether science will be able to tell us in a timely manner if a new technology is harming the environment.

Encourage Public Participation in the Decision Process

In 1996 the National Research Council recommended that risk assessments by government agencies be more than number crunching. NRC recommended a deliberative process involving the ethical, social, and economic concerns of affected parties.[32] Partly as a result of these recommendations, government agencies now invite "stakeholders" to participate in the process of characterizing risks and formulating policy options for risks. The goal is to build consensus by including the important concerns of different parties from the outset. Every affected citizen is potentially a stakeholder, but in many cases "stakeholder" is more narrowly construed. In the United States there has been progress in including citizens in the risk assessment process. Including citizens on panels of industry and scientific experts is often merely token representation. The final decisions are still driven largely by highly technical information. Numerical risk assessments, FAA noise contour maps, estimated annoyance, average estimated exposure to pesticides, and so on. The *ethical* issues embedded in virtually every step are rarely discussed. Issues such as inequities in who receives most of the pollution, and the potential for irreversible effects on the earth are not only omitted from the numerical risk assessments but are also issues that reach into many people's strongest moral sentiments.

One proposal for better public input is to select a panel of citizens to negotiate with industry representatives. The negotiations might involve future compensation to a community for the pollution or natural resource loss that will result from a proposed new industrial facility. The idea behind such citizen negotiating panels is that the local people know their own needs best and can negotiate appropriate pollution compensation on behalf of the community.[33] For this to work, the negotiating panel has to represent the views of the community faithfully, take a long-term view of the proposed facility, understand the risk assessments thoroughly, be able to question the assumptions embedded in the risk assessments, and have power and resources equal to the industry's. In principle, such a negotiation could result in a safer and less polluting industrial facility than would be achieved through current standard regulatory means.

In Wisconsin the citizen negotiating panel approach failed spectacularly in the case of the as yet unconstructed Crandon mine. The company that claimed the ore body negotiated a local agreement with the town board (township of Nashville, Wisconsin). The company would have made annual payments as compensation for the noise, dust, water, and air pollution of the mining facility. But people protested because the negotiation meetings were closed to the public. The agreement that came out of the negotiations seemed one-sided. The *entire* town board was voted out of office in the next election. Such an occurrence shows how citizen negotiating panels can also be divisive and stressful to the community.[34,35,36]

Is the Crandon mine in Wisconsin a simple case of obstructionist NIMBY ("Not in my backyard") tactics? Or is it an example of effective public participation in environmental decision making to keep a community safe from pollution? The people of Wisconsin spoke clearly through the voting booth and citizen organizations in favor of a more precautionary stance with respect to mining pollution. In this case, the attempt to use a citizen negotiating panel led to more litigation and conflict.

The Democratic Process and Environmental Protection

In the United States and many other parts of the world we are fortunate to have a democratic society in which each of us is free to express an opinion on any issue. Nonprofit citizen interest groups help inform the public about environmental issues, urging citizens to write to legislators about those issues. Nonprofit environmental groups also provide information to members of the government at all levels, and take legal action when the government fails to enforce environmental policies. For their part, corporations lobby members of Congress, make campaign donations, hire attorneys to take legal action, purchase advertising to influence public perceptions of environmental issues, and tout their environmental records.

In order for us to have a representative democracy, it is necessary for citizens with opinions about environmental issues to make their ethics and values heard at the voting booth and more directly by communications with elected officials. Although the best science is very important in environmental policy, the best science is never beyond uncertainty, and ethical values are the crux of environmental controversies.

I have laid out the chain of evidence showing that many different pollutants harm children's development. If you are on a camping trip and see tracks in the mud and scratches on a tree, you do not need to see the bear before you take steps to protect your food from the hungry bruin. If a more precautionary environmental policy stance is adopted, tragedies such as the pollution disasters recounted in this book will not be repeated with new untested technologies.

Our generation faces the same choice faced by the Surgeon General's committee in 1925. Do we bring aspects of the Precautionary Principle into action in ways that will be genuinely protective of the psychological well-being of future generations? Or do we continue introducing new pollutants unless we have smoking-gun evidence of harm? Where is the line between squelching technological innovation and preventing pollution of the entire planet with the next long-lived untested pollutant?

Our society needs the courage to examine the moral values that are embedded in our risk assessments, regulatory processes, and ways of life. We need to debate those ethical issues together and find the creativity and courage to bring environmental justice and liberty to all.

Chapter 8
Protect Your Family, Protect Our Planet

Government policy and personal action are both needed to effectively protect children from pollutants. And when we protect our families, usually we protect the planet too. This chapter is mainly about what you can do to protect your own family. For each pollutant I give tips for protecting both your family and the planet.

But personal action is not protective without government policy. Air pollution is a good example. In the prologue I mentioned that my personal interest in pollution is because I grew up during the smoggiest time in history in Los Angeles. In high school when I worked out with the swim team, my lungs hurt and I coughed afterward. This was true for the whole swim team. To decrease the smog-producing emissions from your own car, you can drive less, or not at all, on a day that is predicted to be smoggy. If everyone did that, it would work. But everyone won't. This is called a "tragedy of the commons."[1] People opt for the thing that is easiest for them personally in the short term. So even on a smoggy day, most of us hop in our cars and drive as much as we want. But short-term gains often create long-term tragedies.

The L.A. smog problem was not solved by individual people refusing to drive. Instead, smog was reduced dramatically by state regulations on automobile emissions starting in 1966. Between 1970 and 1996 the peak air pollution declined almost 60%. This was even though Californians drove more than twice as many miles in 1996 as in 1970.[2] Government action was the key.

Government regulations for lead and PCBs reduced children's exposure enormously, as we saw in Chapter 1 and Chapter 3. But effective regulation of lead in children's toys and jewelry, mercury emissions, mercury in fish, pesticides, and noise exposures are still a wish rather than a reality.

In this book I have covered just a few pollutants for which there is solid evidence of negative effects on children's behavioral health. There are other pollutants that also affect children's behavioral function, such as hormone disrupters. But the evidence for effects on humans is not yet as clear even though the animal research evidence of harm is accumulating rapidly. Also, any pollutant that affects physical health will affect children's behavioral functioning. Asthma is a good example. High air pollution can trigger asthma attacks. Children with asthma are likely to miss more school. Absence from school is a risk for poor academic performance. So, the range of pollutants that influence children's behavior goes much beyond those in this book.

My aim in this chapter is to help parents put the information in this book into practice without panicking. Those of us who are lackadaisical about protecting ourselves are much more motivated to protect our children. Here's a conversation I had with one of my relatives who is an artist. She knows that her "flake white" paint contains lead, and she paints without gloves. She told me she enjoys painting much more without gloves. And she says the leaded "flake white" works better than other white paints. When I said, "The lead will be stored in your bones," she merely nodded. But when I added, "When you breast feed your first baby, the lead will be mobilized out of your bones," her expression changed. She said, "I think I better find a lead-free substitute for my 'flake white.'" Cutting back your own exposure can reduce child exposure for pollutants in the home or for pollutants that can be transferred prenatally or via breast milk.

Lead

Lead is a useful industrial metal, but it should be kept out of the environment. Less lead exposure is better. The scientific consensus is now that there is no threshold for negative effects of lead. If you wanted to poison people with a toxic substance, what better way than to coat the interiors of their dwellings and distribute the substance broadly in the air.[3] That's what happened with lead. But we also have to watch out for other sources of lead such as from toy and hobbies. Here are some tips for reducing your own family's lead exposure and preventing further dispersal of lead in the environment.

1. *Blood testing.* If you live or have lived in a home built before 1978, *ask your doctor* about testing your children's lead. If your child goes to school or daycare in an older building his or her lead should be tested. Lead testing is especially important if you have done remodeling while you lived in an older home.

2. *Children's toys and jewelry.* In 2007 the news headlines reported the recall of Thomas the Train because of lead-based paint. This toy set was manufactured in China, and there was finger pointing over who to blame.. Were companies in China cheating on the toy standards by using lead paint? Did the U.S. companies fail to specify no lead paint? Was the enforcement unfair? But toy recalls are not a new problem. Crayons made in China were recalled

in 1994. Toys manufactured in other countries have also been recalled. In the last two years many other popular toys have been recalled, including some Disney items, Sesame Street items, and Winnie the Pooh items. Millions of pieces of children's jewelry have been recalled. Remember that a child died in 2006 from swallowing a free piece of jewelry that came with a pair of shoes.

Instead of prevention, the U.S. government is using product recalls. Recalls occur after a toxic toy is in hundreds of thousands of children's hands and mouths. I check the Consumer Product Safety Commission (CPSC) Website for recalls before each speech I give on pollution and children. I am always surprised at the number of recalls. Each month there are at least a few recalls for lead in children's jewelry and/or toys. Most are not covered in the news.

What's a parent to do? If you have Internet access, check the CPSC Website prior to toy shopping. Print the latest recalls so you can remember them. Because of social inequities in Internet access, this isn't going to help most lower income families. If you find recalled toys in your toy chest, toss them out or return them for a refund. Don't donate them to second-hand stores.

If you can afford it, try toys from artisan toy manufacturers in the United States.

Most importantly, write your representatives in Congress and ask for meaningful and preventative monitoring of lead in toys.

3. *Lead paint and dust* in the home. If you live in a home built before 1978, your home likely has at least some lead-based paint inside or outside. You can test the paint in your home with a kit from the hardware store that costs about $6, but the EPA says those kits are not very reliable. If you have lead paint, dust in your home will contain lead. Lead dust is absorbed from the air by breathing it. The EPA says to mop with an all-purpose cleaner or water and rinse the mop very thoroughly. *Vacuum* frequently rather than using a broom or carpet sweeper, and if possible get a HEPA (high-efficiency particle accumulating) vac. If you rent your home, ask your landlord about the age of the building and testing for lead-based paint. Or test it yourself and ask the landlord about doing lead abatement. If you are purchasing a home, ask your seller about the age of the home, lead-based paint, and if possible negotiate with the seller to carry out EPA-approved lead abatement as part of the purchase price.

4. *Painting and remodeling.* When you do remodeling projects, protect your family from the dust of the project. The EPA has an excellent online pamphlet on lead and renovation. If you do it yourself, follow EPA's guidelines. Isolate the work area from the rest of the home. Wear a proper respirator. A respirator for lead dust costs only about $40. Make sure you *wash your work clothes separately* from other clothes, and wash your own hands and face with soap before eating. Don't track dust into the rest of the house. Mop and vacuum the house much more frequently during projects. *Wash children's hands* and faces before snacks and meals. Change your furnace filter frequently. If you

hire a contractor, give the contractor the EPA guide and ask the contractor to follow it.

5. *Lead in soil* near your home. For homes built prior to 1978, the soil nearest the house can be high in lead. Plant ground cover close to the house to prevent bare soil from being accessible to children and wildlife. If you are a vegetable gardener with an older home, have your soil tested for lead. The EPA limit on lead in soil is 400 ppm (parts per million). When I had my own garden soil tested, it was double the limit. Now I garden in containers with topsoil I bought. For housing built before 1978, when you or your landlord paints the exterior, make sure the chips are captured on plastic ground cloths, bagged, and sent to your community's toxic waste collection program. Keep windows closed when lead dust is being produced by exterior painting projects.

6. *Lead in drinking water.* If you live in an older house or older part of a city, run the water for about one to two minutes in the morning before drinking. Many cities use lead water mains, and many older homes have lead pipe connecting the house to the water main. Older solder on copper pipe contains lead. The good news is that the amount of lead (and other pollutants) permitted in municipal water supplies is regulated by the EPA. The regulatory standards for bottled water are lax compared to those for city water. Cities with lead water mains are now in the process of replacing them. There have been cities in which the water utility tried to skew test results to avoid replacing lead water mains.

Running the water to clear lead has an environmental downside. Water conservation is important, even in the water-rich upper Great Lakes area, where I live. Even if the city water supply uses water from wells, as the population and water use grows, wells draw down the water table, and natural springs decrease their flow and can dry up. This affects aquatic ecosystems negatively. If you live in an arid climate, consider a water filter that will remove lead, and change the filter regularly.

7. *Hobbies.* Fishing weights are normally made of lead, and those of you who tie your own fishing flies may have lead in your kit. Lead is illegal in Yellowstone National Park. Canada is phasing out lead weights for fishing. Use alternatives to lead. Ask your fishing shop to sell non-lead fishing weights. If you do use lead when fishing, handle these items carefully, do not lick your fingers, and do not use your teeth as pliers to pinch a weight onto your line. You know it isn't possible to retrieve all lead items to prevent wildlife from swallowing them, but do your best. The EPA had a Web page on lead in fishing weights, but there are no regulations yet.

Handle *gunshot* very carefully if you load your own. Steel shot is required for waterfowl. Can you use it for other types of hunting as well? The U.S. military is developing nonleaded ammunition.

Solders for electronics and for making *stained glass* usually have lead in them. If you make *pottery*, use lead-free glazes. If you *paint*, know your materials and handle them carefully. The label "Conforms to ASTM D-4236" by itself does not mean an art product is lead-free. If a product conforms to ASTM D-4236 and it has lead, the label is supposed to say "Warning. Contains Lead."

8. *Good nutrition.* Maintain a good intake of calcium in your family's diet. Calcium and lead have chemical similarities. Less lead is absorbed if calcium intake is adequate. Because lead interferes with synthesis of heme, iron is also important. Do not use *utensils* made with lead (pewter) or with lead glazing. Check suspicious items with a test kit from the hardware store. Keep your precious antiques but don't eat with them if they contain lead.

9. *Be safe.* If you *work* in an environment with lead, use your personal safety equipment properly to reduce your own exposure. Wash your hands and face with soap before eating. If possible, change clothes before coming home after work, and do not bring lead home on your work clothes.

10. *Dispose* of car and motorcycle *batteries*, computers, cell phones, and electronic appliances in your community's toxic waste disposal and recycling program. Car batteries have lead in them, and the electronics usually have lead shielding and lead solders in them.

11. *Imported* traditional medicines and foods. The CDC finds lead poisoning in both children and adults from imported remedies (pills and potions that contain lead) and foods that are brought in by relatives.[4] Be exceedingly careful about any substance not authorized for import into the United States. In one case, pills imported from Hong Kong were found to contain 1–3 ppm lead, and a woman suffered lead poisoning. In other cases, children who ate candy from Mexico that was in a lead-glazed jar were lead poisoned. Food coloring from Iraq lead-poisoned nine members of one family.[5]

12. *Advocate and educate.* Advocate for strong enforcement of lead paint abatement in both single-family and rental housing in your city, county, and state. Advocate for lead-free children's toys and jewelry. Advocate for a phase-out of lead in fishing weights. Share information with others, especially those with children living in older homes or apartments. Resist efforts to relax current lead regulations. As crude oil supplies dwindle over the next 20 to 30 years, I personally would not be surprised to see gasoline makers propose using tetraethyl lead in gasoline again.

Mercury

1. *Fish is good for you—mercury is not.* Don't stop eating fish. Just be choosy about what fish you eat. Fish contain nutrients that are essential for brain development such as omega-3 fatty acids. Eating fish that are low in mercury is a good thing to do during pregnancy. Big sport fish (northern pike, musky, walleye, bass) are usually higher in mercury than many types of commercially sold ocean fish or sport-caught panfish. If you or a member of your family fishes or hunts for food, get a copy of the fish consumption advisory for

the state in which you are fishing or hunting.[6] The State of Wisconsin has documented four cases of mercury poisoning in adults from eating contaminated fish almost daily. Follow the sport-fish consumption recommendations, especially for children's meals, and especially if you are a woman who is pregnant, nursing, or planning to become pregnant. See Table 2.4 for the levels of mercury and the good omega-3's in some commercially sold fish in the United States. If you are pregnant or nursing, use my "Weekly Methylmercury Counter" in Figure 2.3 to track your weekly methylmercury.

2. *Ask your doctor* about nonmercury vaccines. Vaccination is very important for both you and your child. Nonmercury alternatives are available, even for influenza vaccine.

3. Clean up *mercury spills* according to the recommendations in the Health Care Without Harm Web site.[7]

Never vacuum up any mercury-containing product (including fluorescent light bulbs). Using a vacuum puts the mercury into the air. Eighty percent of mercury that is inhaled is absorbed. *Never* wash mercury down the sink. The mercury will collect in the trap and evaporate into the air in the house.

To clean up a small spill wear rubber gloves. Change into work clothes because you are going to throw the clothes away. Keep everyone else out of the area. Remove your jewelry so mercury doesn't stick to it. If the mercury is on a hard floor, scoop up a small puddle of mercury by pushing it with a piece of cardboard into a disposable dustpan. Then carefully put it into a container with a lid. Find tiny beads of mercury by using a flashlight. Use duct tape to pick up the small beads of mercury. Wipe the area with disposable towels. Put the lid on the container. Put everything you used in the cleanup, including your clothes, into a garbage bag and seal it. Dispose in your community toxic waste collection. Ventilate the area for at least 24 hours.

If you spill mercury on a carpet, carefully cut out a generous section of carpet around the spill. Follow all the other directions above.

4. *Dispose* of mercury-containing products in your community's toxic disposal program. This includes latex paints made before 1991 (which may contain a small amount of mercury as a mildew inhibitor), "silent" light switches, cosmetics containing "calomel" (mercurous chloride), older button batteries, fluorescent bulbs, old thermostats, and mercury thermometers. Proper disposal reduces the amount of mercury that enters the environment.

5. *Reduce* electricity use. Electricity from coal-burning power plants puts mercury into the air. Turn off appliances and lights when they are not in use. One of the best ways to reduce electricity use is to put compact fluorescent bulbs or fluorescent circle lights in most of your lighting fixtures. Dispose of used fluorescent bulbs in your community's toxic disposal program. The bulbs contain small amounts of mercury, but by reducing electricity use you reduce overall mercury emissions. Pay attention to the estimated energy consumption of new appliances, TVs, and computers that you purchase.

6. *Herbal remedies can contain heavy metals.* I scraped my knees and elbows in a bike accident a few years ago. One of my friends generously brought me a tube of ointment called Traumeel (a trademark). I smeared some on my wounds. Then I read the label. It said 'mercurius solubis.' Sounded like mercury to me! I quit using the ointment. Then I wrote a letter to the FDA. The FDA contacted the company. Yes, there is mercury in Traumeel. The company said it is part of the traditional homeopathic formula. So, mercury is allowed in Traumeel, and in other homeopathic preparations.

The 'mercurius solubis' in Traumeel is actually a mercury ammonium nitrate compound. Traumeel is also available in tablets and drops. Websites advertise the drops as a medicine for children. There has been very little research on herbal products containing mercury. How much of the mercury in the drops is absorbed? We don't know.

A team of Boston-area researchers found that 20% of Ayurvedic herbal medicines contained lead, mercury, or arsenic.[8] Taken at the recommended dose, some of these medicines can create flagrant poisoning. Some of the medications list the toxic ingredients on the label, others don't. Read the label before you purchase it. Don't use preparations containing lead, mercury, or arsenic, even though they are "traditional."

Why might mercury be in such preparations? Remember that it is a powerful anti-bacterial and anti-fungal agent. We have safer anti-bacterial and anti-fungals now.

7. *Red tattoo dyes often contain mercury.* Some people develop an allergic reaction to the mercury in the dyes. One Japanese man with tattoos all over his chest had a severe allergic reaction after eating a swordfish dinner.[9] Swordfish is high in mercury.

8. *Cosmetics for skin bleaching* used to contain mercury compounds.[10] The U.S. FDA banned the use of mercury in cosmetics in 2007. The exception is that mercury may be added to eye cosmetics as a preservative, and in other cosmetics as a trace of less than 1 ppm. Minnesota completely banned mercury in cosmetics and fragrances in 2008. State law in both California and Minnesota requires that if mercury is present, it must be listed on the label.

9. *Educate and advocate* for regulations on mercury emissions from coal burning in your state and the federal government. Advocate for stricter energy-efficiency requirements for electrical appliances. Another source of mercury pollution is mining. Advocate for corporate responsibility for environmental pollution by Northern Hemisphere companies operating mining and other industry in developing countries. Advocate for better food testing, sport-fish testing, and reporting of numerical mercury values so that the public will be better informed about this food contaminant. The long-term issue is how to reduce mercury in the food chain for ourselves, future generations, and the earth's ecosystems.

PCBs, OCs, and PBDEs

PCBs were spread around the globe before the chemists even knew how to separate them from DDT in their assays. PCBs were thought to be safe because of their low *acute* toxicity. Now we know that acute toxicity is not the only problem. Developmental and behavioral effects, as well as chronic toxicity, should also be considered. The pollution mistakes of the previous two or three generations rest on us. For chemicals that biomagnify up the food chain, it is important to get them out of global circulation.

1. *Follow Sport-fish consumption advisories.* It isn't just the Great Lakes and the Hudson River that are polluted with PCBs. In 2006 there were over 1,000 fish consumption warnings in the United States for PCBs. Many former and current military bases have PCB pollution. If you or a member of your family fishes, read and follow the sport-fish consumption advisory for your area. Most of the advisories have different recommendations for men versus for children and women of childbearing age. Because numerical values of PCBs are not available for different species of fish, I cannot give you a PCB counting chart analogous to the methylmercury counter. Some anglers summarize the recommendations for the Great Lakes as "Eat only small (less than about 18 inches) fish, and less than once a month." Get the booklet for your location! *Do not make soup* out of PCB-contaminated fish. Soups retain the fat, which is where the PCBs and other lipophilic contaminants are.

Other oily predator *fish purchased commercially* can also contain PCBs. *Trim off the fatty part of fish near the spine and the body flaps* (see trimming guidelines in any Great Lakes State fish consumption advisory). As with mercury, *fish is good for you, but PCBs are not.* Be choosy about what kind of fish you eat, and prepare it to minimize contamination.

2. *Reduce total organochlorine (OC) chemical intake.* A Dutch study estimated that dairy products, processed foods, and meat (in that order) are our major sources of PCBs, OCs, dioxins, and furans.[11] Another Dutch study tested a "dioxin-free" diet in women who were breast-feeding. The low-dioxin diet in Holland consists of low-fat or fat-free dairy products; vegetable oils; any kind of vegetables, fruits, and grains; soy protein; but only pork in the meat category. In The Netherlands pork is the meat lowest in OCs. A diet such as this can reduce dioxin and PCB intake in the long term in order to affect the dioxin content of breast milk (they did not measure PCBs).[12]

In the United States the Institute of Medicine[13] recommended a reduction of contaminated animal fats in animal feed. Over a billion pounds of slaughterhouse fat are added to animal feeds in the United States each year. Organic meat production is an exception. Neither poultry nor mammal slaughterhouse by-products may be fed to poultry or mammals that are raised according to the U.S. organic standards. This implies that certified organic meat and dairy products should be lower in PCBs and OCs. But I haven't yet found research on this topic.

3. Discard *old appliances* (manufactured before about 1980) in your community's toxic waste disposal program. They may have PCBs in their capacitors, caulks, paints, and lubrication oils. Some communities charge a special fee for discarding old appliances. The fee goes to help cover the cost of disposing of these toxic materials in the proper hazardous waste disposal facility. Don't dump these items illegally. Pay the fee and help control these pollutants.

4. Capacitors in *old fluorescent light fixtures* (manufactured before 1979) may also contain PCBs. Take these items to your community's toxic waste program if they are being replaced. If a ballast is leaking, clean it up according to the EPA guidelines.

5. Older floor varnishes can contain both lead and PCBs, as can the glues used with lineoleum. When you have floors sanded or replaced, isolate the work area from the rest of the house.

6. *Advocate and educate.* Share information about the effects of PCB contamination on children's development, especially with women planning to have children. Advocate for better information for sport anglers. In most Great Lakes States in the United States, the fish advisory is published separately from the fishing rules and regulations. Many anglers and members of the public are unaware of the guidelines.

Advocate for better information, including more testing and labeling for this category of contaminants in commercial fish, meats, and dairy products. Advocate ratification of the Persistent Organic Pollutant Treaty by the U.S. Senate. Advocate better control of PCBs, PBDEs, and related chemicals in the food chain, including those from recycled fats.

Pesticides

1. *Reduce* household pesticide use. Consider least-toxic alternatives to grabbing the spray can. Keep your home clean of food residues to reduce habitat for insects and rodents. Garbage containers, both indoors and outdoors, should have tight lids. Seal cracks and crevices where insects enter. New reports indicate that catnip may be a powerful insect repellent for roaches and ants. Boric acid powder is a less toxic insecticide that can be used in cracks and crevices, but boron is also a neurotoxin. Diatomaceous earth is another alternative, but it is an eye and lung irritant. Use all pesticides with great care. Store pesticides out of reach of children. Integrated pest management in New York City reduced cockroaches and pesticide exposure.[14]

2. *Read and follow* instructions on pesticide labels. The pesticide label is the law. Many pesticide poisonings involve misuse. *Wash your hands* thoroughly after handling any pesticide container. Do not let your children apply

pesticides. *Dispose* of empty pesticide containers, or old unused pesticides, according to recommendations from your community's toxic waste disposal program.

3. *Consider purchasing certified organic food* if you can afford it. Certified organic food is grown without OP, carbamate, pyrethroid, organochlorine, or neonicotinoid pesticides. Organically grown food is more expensive but reduces pesticide exposure all along the line—to the wildlife and ecosystems on farms, to the farm workers and their families, to workers involved in food processing, and to the food consumer.

4. *Wash* all fruits, vegetables, and raw agricultural products (such as rice and dried beans) thoroughly before cooking.

5. *Old carpets* can hold pesticide residues. If you rent, ask your landlord about the age of the carpet and how often pesticides have been used. If you can afford it, replace your carpets, especially if they were present in the home when currently banned termiticides were applied.

6. Reduce or eliminate *nicotine* exposure during pregnancy. Women should try to quit smoking before pregnancy or as soon as possible after pregnancy is discovered. Avoid secondhand smoke if at all possible.

7. *School pesticides.* Find out whether your child's school uses pesticides on a schedule. Scheduled spraying is less effective than integrated pest management.[15]

8. *Herbicides are also pesticides.* Research shows that herbicides on lawns get tracked into the house. Dogs in families that use herbicides have a higher rate of cancer. The behavioral effects of herbicides on children have not yet been researched.

9. *Advocate and Educate.* Integrated pest management (IPM) can be used successfully to control pests while reducing pesticide use, even in inner cities. Support use of IPM and least-toxic alternatives in parks, schools, and other public facilities.[16]

Noise

If noise is bad for children's development, why do some children like noisy toys? Noise can be fun because it raises our arousal. And we all need a positive jolt for fun now and then. Get the arousing thrill without either harm to hearing or the stress of chronic noise.

1. *Promote language development and reading.* If you have your own children, grandchildren, nieces, or nephews, one of the most important things you can do to foster their language development is to spend time reading children's storybooks together.[17] Even if you live in a high-noise area, find time to sit with your child and read books together. Living in high noise might deter parents from such important quiet-time activities. But storybook reading can

serve a double purpose of building closeness and fostering intellectual development. Knowledge of this will allow you to counteract at least some of the negative effects of noise.

2. *Protect your child's hearing.* Urge the Consumer Product Safety Commission to develop noise regulations for children's toys, to label toy decibel levels, and to set decibel limits on devices that play through earphones. Avoid purchasing toys that sound loud to you.[18]

Some other sources of loud sounds for children and teens include playing with loud toys, riding snowmobiles, ATVs, or dirt bikes, mowing lawns, using leaf blowers and other power tools, practicing with firearms, attending rock concerts, and playing in music groups. Proper hearing protection should be worn during these activities. If, after exposure to a noise source, your child complains of ringing in the ears or a feeling of tightness in the ears, or says that things sound muffled, explain that those symptoms are indicators of a temporary hearing loss. Usually the symptoms will subside in about 24 hours if the ears are given rest from noise, but repeated exposures damages hearing.[19]

3. *Protect infant hearing.* Infants in critical care need to be protected from constant noise. The American Academy of Pediatrics[20] recommendations should be followed by health care personnel in neonatal intensive care units. If you have a child who was hospitalized as a newborn in a NICU, talk to your health provider about screening the child's hearing. Some states have universal screening of infant hearing. Early diagnosis of hearing problems is important for normal language development.

4. *Avoid high noise and stress during pregnancy.* Are you are a pregnant woman who works in a setting with high noise or vibrations? Not just in factories but with loud music, such as when tending bar or waiting tables in music clubs. If so, consider arranging a work assignment that will reduce your noise exposure. Of course, you must weigh economic factors and your relationship with your supervisors in making such requests. Also avoid other sources of high stress if you can.

5. *Reduce home noise.* Make noise one factor when you choose your apartment or house. Can you hear a factory, major highway, or airport when you look at the apartment or house? Also try to reduce in-home noise. Purchase quiet models of appliances such as snow blowers, lawn mowers, air conditioners, furnaces, vacuums, and dishwashers. Noise inside the home from TV and other sources is related to lower scores on standardized tests.[21] Researchers think that constant noise reduces the amount of verbal interaction with the children, as well as the children's ability to concentrate and hear what is said to them.

6. *Reduce school noise.* Children cannot learn if they cannot hear speech. Schools and day care centers should not be located within earshot of busy highways, airports, and commuter trains. Good classroom acoustics are essential

for a quality education for all children. An inadequate acoustical environment is an educational barrier to children who have a slight hearing loss (even a temporary one) or who are bilingual. The percentage of U.S. schoolchildren who are bilingual has increased dramatically in the last decade. Suggest that your school district voluntarily adopt the classroom acoustical standards for new construction and retrofit older classrooms to improve acoustics. Some retrofits can be accomplished inexpensively.[22] Construction and maintenance noise should be reduced during school hours. School remodeling projects should be scheduled when adjoining classrooms are not in use. I know a teacher who once marched his class of 7th graders outside onto the lawn to protest the noisy lawn mowing that was disrupting his class.

7. *Airport noise.* Your local airport probably has a noise policy. File a complaint when the policy is violated. The airport near where I live is supposed to direct planes to northerly landings and take-offs if weather permits. Airports have to track complaints. Complaints can influence airport noise policies.

8. *Road traffic noise* depends on the pavement, traffic speed, and the mix of trucks and cars. At 35 mph, one large truck makes the noise of 18 cars. Truck routes should avoid residential areas. The Arizona Department of Transportation has found that rubberized asphalt on highways reduces noise in nearby neighborhoods by 5–8 dBA.[23] Asphalt is quieter than cement. Ask authorities to pave roads near residential areas with asphalt rather than cement.

9. *Be a good neighbor.* Consider the time of day when you do something noisy like mowing your lawn, using a leaf blower, or turning up your stereo. Your fun might be noise to your neighbor. If you don't enjoy your neighbor's choice of music when it is turned up, the feeling is probably mutual. Can you unlock your car door remotely without beeping the horn?

10. *Advocate and educate.* Inform others about the effects of noise. Advocate for changes in laws and rules that will provide stronger protection from noise exposure for children at home and in school. Advocate for stronger rules for protection from noise and vibrations at work.[24] Promote land use planning in your community that prevents the construction of homes and schools near major highways and airports. Advocate against highway and airport expansions that will increase noise levels near residential areas. Advocate rules that require the military to comply with environmental policies such as airport noise except in true emergencies. Advocate for the establishment of quiet areas in wildlife refuges, national parks, and wilderness areas.

Community Pollution Disasters

1. *Stress and coping.* If you have the bad luck of being enmeshed in a pollution crisis, knowing that a pollution crisis is stressful can help you cope. Be deliberate

about taking care of your own psychological well-being and that of your family. When the family is stressed, very small children may regress in their developmental skills (toileting, etc.). Families and communities can become divided over pollution issues. Do your problem-oriented coping. Seek good information about the situation, and band together with your neighbors for action. But do your emotional coping too. Seek emotional support from friends and family outside the situation. Try to avoid alcohol, tobacco, and other drug use as a coping strategy. The extra stress of a pollution crisis can be very trying to anyone with a chronic mental health condition, or current or past addiction problem. People should be in close contact with their psychotherapists, physicians, and support groups. Be aware that suicide rates often increase after disasters. If someone talks about ending it all, take it seriously. Contact crisis hotlines, and convince the person to go to the nearest hospital Emergency Room.

2. *Natural disasters* often become pollution disasters. Floodwaters usually contain unprocessed sewage. Floods can mobilize wastes out of landfills, and can cause storage tanks for gasoline, pesticides, or other chemicals to leak. Similar things can happen as a result of tornados and earthquakes. Debris after natural disasters can contain toxic substances from buildings such as asbestos, lead, PCBs, mercury, and very small particle dust that can penetrate the deepest parts of the lungs.

3. *Promote and practice safe disposal of toxic wastes.* Some "Superfund" sites are former municipal dumps, as at Agriculture Street in New Orleans. I live two blocks from a former municipal dump that is now a school athletic field. This dump, like all other closed dumps, has to be monitored regularly by health officials. Municipal dumps contain an unpredictable toxic mix of household and industrial waste. We need to separate our toxics. Dispose of pesticides, old paints, mercury thermometers, and other hazardous household chemicals in your community's "clean sweep" program. Do not just put it in the trash bag at the curb. Make a couple of phone calls or check the Web to find out the appropriate disposal method. If your community does not have a "clean sweep" program to collect toxic household items, call the county or city health department and discuss how to discard your wastes. Suggest that the agency start a toxic collection program. Reduce the purchase of obviously toxic household products to the greatest extent possible.

4. *Practice energy conservation at home and at work.* All electricity generation creates pollution, whether nuclear, coal, natural gas, wind, or even solar (through the manufacture and transportation of the devices). Support energy efficiency regulations for heating, cooling, appliance, and computer manufacturers. Pay attention to the energy demands of major items you purchase, such as furnaces, air conditioners, refrigerators, dehumidifiers, computers, and other items.

5. *Avoid unnecessary exposure to radioactivity.* Whether the effects of exposure to low-level radiation cumulate over one's lifetime is highly debated in

scientific circles. Given this ongoing debate and the history of overuse of radioactive technologies, the prudent course is to be cautious. If you live "downwind" (or downstream) of a nuclear facility, try to become informed about whether the levels of radioactive isotopes present in soils, foods, and fish are elevated. *Radon in homes* and other buildings is an important source of exposure to radioactivity in the United States. Test your home for radon.[25] If your basement has high radon, do not use it as a living area, and consider radon abatement. If you are considering renting a basement-level apartment, ask the owner if it has been tested for radon. *Medical and dental uses of radioactivity* are another major source of exposure for most of us, and some of those x-rays are unnecessary. Recent research suggests this is a concern, especially with children. Make sure your dentist uses a thyroid collar to shield you or your child during dental x-rays. If you work in a lab or industry that uses radioactive isotopes, wear your "rad badge," follow safety procedures carefully to minimize your personal exposure, and discard radioactive substances in the correct manner. Consider your other sources of radiation in addition to what you receive on the job.[26]

6. *Advocate and educate.* Educate others and advocate for better control of toxic wastes and reduction at the source.

Afterword: A Few Remarks on Global Warming

Global warming poses a hazard to the entire earth. Because it affects us all, it will affect our children as well. In Chapter 8 many of the suggestions for reducing pollution will also reduce greenhouse gas emissions. Anything that is more energy efficient will help reduce global warming.

Scientific consensus is that gaseous waste products created by the human race are altering the earth's climate. Carbon dioxide (CO_2) from using fossil fuels is a strong contributor to global warming. As the climate warms, tropical diseases may move north. There will also be effects on ecosystems and agriculture that may affect food supplies. The ocean levels will rise as arctic ice is reduced.

What can we do in industrialized nations to help slow or stop global warming? One key is to cut back on our use of fossil fuels. This can be done by policy and by individual action. As I argued by using air pollution as an example in Chapter 8, individual action alone won't work. We need effective policies to reduce emissions of CO_2 and other greenhouse gases.

Policies should promote sustainable low-impact sources of energy as well as energy efficiency. By subsidizing solar electric panels on buildings, Germany has avoided constructing at least one large new power plant. Such microgeneration now exceeds global nuclear electricity generation.[1] There are technologies for solar hot water that have existed for over 30 years. Energy efficiency does not have to be uncomfortable or inconvenient, and it is cheaper than building new energy generation of any kind. Nuclear is not the solution because of cost, extremely long-lived pollutants, the complexities of handling the waste, susceptibility to terrorism, and potential for creating catastrophes.[2] The efficiency of electrical transmission needs to be improved. Approximately 7 to 15% of electrical power is lost in transmitting it from the power plant to the consumer.

Personal action can help. The largest personal contributors to global warming are transportation, home heating and appliances, and family.

Choose energy efficient transportation. A car with higher fuel efficiency is great, but how much you drive it is just as important. If you live where public transportation is available, give it a try. If you can, choose housing closer to your job. Bike or walk if you can.

Home energy efficiency can be improved by choosing a high efficiency furnace and air conditioner when you replace them, and insulating your home. Efficiency of lighting (fluorescent) helps, as does turning off lights and electric appliances such as TVs when they are not being used. But a larger reduction in your carbon footprint will come from efficient home heating and air conditioning.

Family contributes to global warming. Many biologists think the earth has more people than it can sustain, and the population is still growing. Each person born in the United States adds over 100 times the greenhouse gas emissions of a person born in most nonindustrialized parts of the world. Consider voluntarily limiting your family size. At a policy level we need to make resources for family planning and contraception available universally around the world.[3]

Appendix: Understanding Two Important Statistical Concepts

Throughout this book I have tried to describe the results in relatively nontechnical terms so that the issues would be understandable to readers who have only a passing familiarity with statistics. Some further elaboration may be helpful for some readers. The concepts here are found in most introductory statistics texts.[1] The two important concepts that I use repeatedly throughout are the proportion of variance in one variable accounted for by another, and the probabilities of Type I and Type II errors.

Type I And Type II Errors in Statistical Inference

I have used the terms "false positive error" and "false negative error" to describe what are known as the Type I and Type II errors, respectively, in statistical inference. The basic question addressed in statistical inference is how to use data to make a decision about the truth of a hypothesis. Inferential statistics dichotomizes the world into two states: a hypothesis is either true or false. For example, either there is no association between lead exposure and IQ test scores or there is. We do not ever know which is the case, but we collect data, analyze them, and make a decision about whether to believe or not believe the hypothesis that lead is associated with IQ test scores. The framework is shown in Figure A.1. There are two ways to be wrong, and two ways to be correct. We never actually know for sure whether we are correct or in error, and each decision risks one of the errors.

Current environmental policy in the United States emphasizes formal risk assessments based on research findings. One bias that is built into risk assessments is the false positive error rate that is required in order to interpret a research finding as "significant"—less than 1 chance in 20, or a

ACTUAL (but forever unknown)
State of Reality

	Lead & IQ Scores Not associated	Lead & IQ Scores ARE Associated
Lead & IQ NOT Associated	Correct Decision Probability = 1 − alpha	Type II Error Probability = beta
Lead & IQ Associated	Type I error Probability = alpha (usually alpha < .05)	Correct Decision Probability = 1 − beta (also known as 'statistical power')

Decision to believe that:

FIGURE A–1. Framework for decision making in statistical inference

probability (*p*) of .05 or less. One reason for keeping the false positive error low is to make sure that the "bricks" that we use to build the edifice of science are all solid, because scientific research builds on previous findings.[2] But most interpretations of the Precautionary Principle in environmental policy call for having a low chance of a false negative error, the error of failing to find a harmful effect when there is harm. Within any given research study, the two probabilities are linked. If $p = .05$ is changed to $p = .01$, the likelihood that the study will miss a harmful effect will increase. On the other hand, if the chance of a false positive error is set at .10 (1 chance in 10), or even higher, then the study will have a better chance of finding a harmful effect if it actually exists.

Figure A.2 shows how the probabilities of the two errors are linked.[3] Each curve in Figure A.2 shows the relationship between Type I and Type II error for a different hypothetical study. If Type I error is held constant at .05, then the probability of a Type II error for each study can be changed only by changing something about the study—collecting data on more participants, collecting data on participants who were exposed to higher levels of the pollutant, refining the laboratory procedures for measuring either exposure or the behavior that is being measured, and so on. These kinds of modifications can increase the statistical power of a study, which is just the opposite of the Type II error (statistical power = 1 − the probability of Type II error).

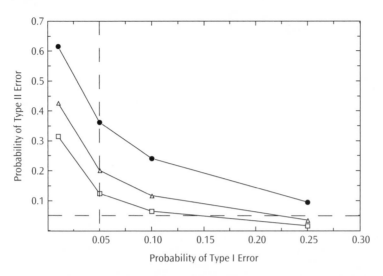

FIGURE A–2. Trade-off between Type I and Type II error. Each curve represents a hypothetical study with a differed effect size. Dashed lines show a .05 a priori chance of error. *Source:* Values calculated from methods in Murphy & Myors, 1998.

If the burden of proof of safety were shifted to the proponents of an activity, as proposed in the Precautionary Principle, one interpretation is that the probability of Type II error (beta in Figure A.1) should be set at a maximum of .05. That is, proponents of an activity would have to provide safety evidence of the same strength required for other conclusions based on scientific research. Setting Type II error at .05 would usually require allowing the Type I error to be much larger than .05, perhaps even as high as .30 or .40. If the probability of a Type I error is allowed to be that high, it would be very difficult to demonstrate safety because it would be likely that the obtained results would lead to the decision in the bottom row of Figure A.1—the decision to believe that the activity or pollutant is associated with negative effects. Setting Type II error at a level as stringent as Type I error would be a radical shift in toxicology research that is used in regulating substances and is not likely to happen in the near future.

Overlapping Variance

The majority of research that I summarize in this book relies on multiple linear regression methods to discover whether exposure to a pollutant is associated with an outcome variable. But an association between the pollutant and the outcome variable could occur either because the pollutant causes

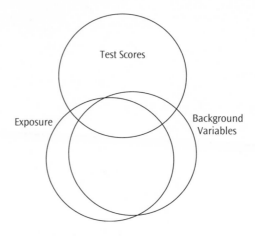

FIGURE A-3. Illustration of overlapping variance

negative effects on people or because the pollutant occurs in conjunction with other events that are associated with negative effects on people. Researchers normally deal with such "confounding" variables by including them in the statistical analyses.

Figure A.3 shows how the overlap among three variables can make interpretation of results depend on whether an important confounder is included in the analyses. I have drawn the figure so that the two circles representing Exposure and Background Variables overlap quite extensively. If the effects of Exposure on Test Scores is judged by the degree to which those two circles overlap, ignoring Background Variables, we see that Exposure accounts for a considerable proportion of test scores. But the Background Variables circle overlaps not just with Test Scores but also with Exposure. If we judge the effect of Exposure by looking at just the part of the Exposure circle that overlaps with Test Scores but not with Background Variables, then the effect of Exposure is reduced dramatically. In Chapter 1 I argued that the theories that researchers bring to their data can strongly influence the results because those theories determine when a Background Variable is considered a confounder as opposed to a cause of the exposure.

The situation in Figure A.3 is oversimplified to illustrate the idea of how overlapping variance influences the results. With more variables, the situation quickly becomes very complex.

Endnotes

Prologue

[1] Fein et al., 1984

[2] For an overview of viewpoints in philosophy of science see Suppe (1977). The effects of cultural and historical setting on scholarship have been stated eloquently by Said (1978): "No one has ever devised a method for detaching the scholar from the circumstances of life, from the fact of his involvement (conscious or unconscious) with a class, a set of beliefs, a social position, or from the mere activity of being a member of society. These continue to bear on what he does professionally, even though naturally enough his research and its fruits do attempt to reach a level of relative freedom from the inhibitions and the restrictions of brute, everyday reality. For there is such a thing as knowledge that is less, rather than more, partial than the individual (with his entangling and distracting life circumstances) who produced it. Yet this knowledge is not therefore automatically nonpolitical" (p. 10).

[3] Gergen, 1985; Howard, 1985

[4] Landrigan et al., 2002

[5] Jacobson & Jacobson, 2000

Chapter 1

[1] Lippmann, 1990

[2] Quoted from Lippmann, 1990, p. 3

[3] Rosner & Markowitz, 1985

[4] Rosner & Markowitz, 1985

[5] Rosner & Markowitz, 1985

[6] Rosner & Markowitz, 1985

[7] Rosner & Markowitz, 1985, p. 350

[8] Rosner & Markowitz, 1985

[9] Nriagu, 1998

[10] Nriagu, 1998, p. 73

[11] Quoted in Nriagu, 1998, p. 73

[12] Quoted in Nriagu, 1998

[13] Nriagu, 1998

[14] Nriagu, 1998; Patterson, 1965

[15] Nriagu, 1998

[16] Patterson, quoted in Nriagu, 1998, p. 76; see also Nriagu, 1998; Flegal, 1998; Needleman, 1998a, 2000

[17] Markowitz & Rosner, 2000

[18] Byers & Lord, 1943

[19] Micrograms per deciliter is a measure of the concentration of lead, usually measured in a blood sample. A microgram is 1/1,000,000 of a gram. A deciliter is one-tenth of a liter. The important thing is for the reader to notice how the values that are considered harmful will decrease as we go through the history of lead.

[20] See Byers & Lord, 1943, for case reports of childhood lead poisoning victims

[21] p. 484

[22] Markowitz & Rosner, 2000

[23] Mellins & Jenkins, 1955; see Berney, 1993, for a synopsis

[24] Needleman, 1998a

[25] Quoted from Markowitz & Rosner, 2000, p. 43

[26] Markowitz & Rosner, 2000

[27] Needleman et al., 1979

[28] Perino & Ernhart, 1974

[29] Ernhart et al., 1981

[30] Needleman, 1992, 2000; Scarr & Ernhart, 1996

[31] Marshall, 1983

[32] 1993a

[33] 1998b

[34] Marshall, 1983

[35] Ernhart, 1986, 1987; Needleman, 1986, 1987

[36] Ernhart et al., 1985

[37] p. 30

[38] Perino, 1973

[39] p. 478

[40] To those readers familiar with Pearsonian inferential statistics, this section describes Type I and Type II errors, statistical power, and their implications. The appendix presents a more technical version of statistical decision making.

[41] Ernhart could have concluded that her results provided weak or tentative evidence that lead harms children's intellectual development: that is, she could have explicitly acknowledged that the results were close to meeting the 5 in 100 standard. In such situations many researchers rely on a comparison to other findings in drawing a conclusion.

[42] Cairns, 1999

[43] "Superfund" is the term used for pollution cases that fall under the Comprehensive Environmental Response, Compensation, and Liability Act

(CERCLA), originally passed by the U.S. Congress in 1980 and signed into law by President Carter.

[44] Needleman, 1994

[45] Quoted in Needleman, 1992, p. 979

[46] Cited in Needleman, 1992

[47] Pediatrics, 1992

[48] Ernhart, 1993a, 1994, 1995a, 1995b, 1996, 1998, 2001

[49] 1998a, 1998b, 2000

[50] Morrow et al., 2007; Lanphear et al., 2005

[51] Needleman 1998a, 1998b, 2000, 2004, 2006

[52] See Bellinger, 2000; Berney, 1993; Schwartz, 1994; Pocock et al., 1994; Smith, 1985

[53] Canfield et al., 2003

[54] Canfield et al., 2003; Lanphear et al., 2005

[55] Fergusson et al., 1988, 1997

[56] Fergusson et al., 1997, p. 477

[57] Burns et al., 1999

[58] Achenbach, 1991

[59] Burns et al., 1999

[60] Dietrich et al., 2001

[61] Needleman et al., 1996

[62] Braun et al., 2006

[63] Dietrich et al., 2004

[64] Liu et al., 2002

[65] Pirkle et al., 1994

[66] Iqbal et al., 2008

[67] Lanphear & Roghmann, 1997

[68] Lanphear & Roghmann, 1997; Lanphear et al., 1998

[69] Bornschein, 1985

[70] Brody et al., 1994

[71] EPA, 2008

[72] Lanphear & Roghmann, 1997

[73] Hubbs-Tait et al., 2005

[74] Brown et al., 2001

[75] Brown et al., 2001, pp. 623–624

[76] U.S. Department of Housing and Urban Development, unpublished data, 2001

[77] CDC, 2001

[78] CDC, 2001

[79] Dilworth-Bart & Moore, 2006

[80] Needleman, 1998a, p. 1875

[81] Quoted in Needleman, 1998a, p. 1875

[82] Brown 2002

[83] See Brown & Fee, 1999; Jacobs, 1999; Mushak, 1999; Needleman, 1999; Piomelli & Schoenbrod, 1999; Rosen, 1999; Ryan, 1999; Vernon, 1999

[84] Jacobs, 1127

[85] Brown & Fee, 1999, p. 1131

[86] p. 323

[87] See Neisser et al., 1996, for a readable review of issues surrounding intelligence

[88] Bradley & Caldwell, 1977

[89] White, 1982

[90] McLoyd, 1998

[91] Jensen, 1985

[92] Ernhart et al., 1989

[93] Burns et al., 1999

[94] Bornschein, 1985

[95] Hubbs-Tait et al., 2005

[96] Bornschein, 1985, Ernhart et al., 1989

[97] Milar et al., 1982

[98] Ernhart et al., 1989; Yamins, 1977; Perino & Ernhart, 1974

[99] Lanphear & Roghmann, 1997; Bornschein, 1985

[100] Rice, 1993

[101] Moore et al., 2008

[102] Parush et al., 2007

[103] Dilworth-Bart & Moore, 2006

[104] See Scarr, 1985, for an excellent discussion of how scientists' beliefs affect the type of research they do

[105] CDC, 2001

[106] Minnesota Department of Health, 2006

[107] Bellinger & Mathews, 1998

[108] p. 30

[109] Brown 2002

[110] Needleman, 1998a

[111] McMichael, 1995

[112] For a review see Lippmann, 1990

[113] Stewart et al., 1999

[114] Muldoon et al., 1996; Shih et al., 2006; Stewart & Schwartz, 2007

[115] Min et al., 1996

[116] National Wildlife Magazine, 2001, p. 64

[117] Nadis, 2001

[118] University of Minnesota, 2008

[119] Kennedy, 2004

Chapter 2

[1] Zelman et al., 1991

[2] Zayas & Ozuah, 1996

[3] Halbach, 1995

[4] Bellinger et al., 2007, 2008

[5] Bender & Williams, 1999

6 Weiss, 2000
7 Bakir et al., 1973
8 Minamata City, 2001
9 Harada, 1995
10 Harada, 1995
11 Harada, 1995
12 Harada, 1995
13 Askari & Cummings, 1978
14 Bakir et al., 1973
15 Weiss, 1983
16 Marsh et al., 1987
17 NAS, 2000c
18 WDNR, 2008
19 Neergaard, 2001
20 Knobeloch, 2005
21 CIA Factbook, 2008
22 Grandjean et al., 1998, 1997b
23 Debes et al., 2006
24 Murata et al., 2004
25 Grandjean et al., 1998, p. 170
26 CIA Factbook, 2008
27 Davidson et al., 1998, p. 706
28 Davidson et al., 2000
29 Report of the Methylmercury Workshop, 1998
30 Davidson et al., 1998
31 Grandjean & White, 1999
32 Grandjean et al., 1999a
33 Grandjean & White, 1999, p. 896
34 Grandjean & White, 1999
35 Davidson et al., 1999
36 Oken et al., 2005
37 Joseph Fagan developed a way of testing infant memory in 1970. He presented two identical pictures to an infant while an observer behind the screen pushed a button to record how long the infant looked at each picture. After the infants become familiar with the two identical pictures, then one of the pictures is changed. The amount of time that the infant looks at the new picture compared to the unchanged picture is related to IQ test scores at 3 years of age (Thompson et al., 1991), and to academic achievement and verbal IQ scores at 21 years of age (Fagan et al., 2007).
38 Strain et al., 2008
39 Grandjean et al., 1999b
40 Lebel et al., 1996
41 Rice, 1996
42 Altmann et al., 1998
43 Grandjean et al., 2001
44 NAS, 2000c

[45] NAS, 2000b

[46] Burbacher et al., 2005

[47] Schecter & Grether, 2008

[48] Hviid et al., 2003; Parker et al., 2004

[49] Ip et al., 2004

[50] Parker et al., 2004

[51] Hviid et al., 2003; Institute of Medicine, 2001, 2004

[52] Geier & Geier, 2004

[53] NAS, 2000b

[54] DeSoto & Hitlan, 2007

[55] Hornig et al., 2004

[56] Newschaffer et al., 2007

[57] Risher & Amler, 2005

[58] Risher & Amler, 2005

[59] Stangle et al., 2007

[60] Stokstad, 2008

[61] Rice, 1996

[62] Yorifuji et al., 2008

[63] An excellent overview of the risk assessment process can be found in a report by another NAS panel, titled "Risk assessment in the federal government" (National Academy of Sciences [NAS], 1983). On pages 29–33 the report lists the components of the risk assessment process that require judgment calls. The panel refers to these as "inference options" and discusses some of the ways in which the inference options can be controversial. The NAS panel comments: "That a scientist makes the choices does not render the judgments devoid of policy implications. Scientists differ in their opinions of the validity of various options, even if they are not consciously choosing to be more or less conservative." (NAS, 1983, p. 36.)

[64] Murphy, 2001

[65] Crump et al., 1995

[66] Crump, 1995; Crump et al., 1998

[67] The chosen definitions of the BMR and the adverse event used in a risk analysis can be translated into what social scientists call "effect size." Effect size is a standardized index of how much a treatment (or toxic exposure) affects a measured outcome. Looking at the chosen definitions for the risk assessment in terms of effect size provides another way of thinking about what those definitions mean to us. If a risk assessor defines an adverse event as the score that the bottom 1% of unexposed children attain, and the BMR is defined as 5% more children having such low scores, this would be called a large effect size. Specifically, the calculated benchmark dose would allow an average difference of 0.77 standard deviations between those exposed and those not exposed (Crump, 1995). A large effect is usually designated as a difference of 0.80 standard deviations, a medium effect as 0.50, and a small effect as 0.20 (Cohen, 1988; Murphy & Myors, 1998). For a BMR of 10% and an adverse event defined as the lower 5% of the population, the effect size would be 0.61, between "medium" and "large." The values chosen by the NAS committee yield an effect size of 0.36, halfway between a "small" and "medium" effect size.

The main point is this: a risk assessment based on assuming a certain effect size will not protect the public from smaller effects of the toxic exposure. The NAS committee chose values for the BMR and the adverse event that will regulate mercury on development. If mercury has what social scientists call a "small" effect, then the calculated risk assessment will not protect the public. A small effect of mercury will fall in the zone between the definition of a negative event and the BMR.

What effect of exposure to methylmercury do you want children to be protected from: a small, medium, or large effect? This is a value question about which the public can be asked. But, this question is answered for you when a risk assessment team chooses a BMR and a definition of an adverse event.

[68] NAS, 2000c.

[69] NAS, 2000c, p. 299

[70] NAS, 2000c, p. 293

[71] NAS, 2000c, p. 291

[72] Neergaard, 2001, U.S. FDA, 2006

[73] Helland et al., 2003

[74] Gochfeld & Burger, 2005

[75] Pless & Risher, 2000; Stajich et al., 2000

[76] See a 1983 report by an NAS committee for another discussion of judgment calls in risk assessment.

[77] See Shrader-Frechette, 1985, for elaboration of value and ethical issues embedded in risk assessments

[78] Loew, 2001; Statz, 1991

[79] Clark, 2001

[80] Friedmanna et al., 1996

[81] Schwindt et al., 2008

[82] Wood et al., 1968

[83] Wood et al., 1968, p. 174

Chapter 3

[1] EPA, 2008

[2] Fries, 1972

[3] Broadhurst, 1972

[4] McKinney & Waller, 1994

[5] Jensen, 1966

[6] Jensen et al., 1969; Holmes et al., 1967; Risebrough et al., 1968

[7] Kalantzi et al., 2001

[8] McKinney & Waller, 1994; Tryphonas, 1995

[9] Higuchi, 1976

[10] Swanson et al., 1995

[11] Higuchi, 1976

[12] Meigs et al., 1954

[13] See Swanson et al., 1995, for a review

[14] Collier, 1943

[15] Meigs et al., 1954; Swanson et al., 1995
[16] Holmes et al., 1967; Higuchi, 1976
[17] Flick et al., 1965
[18] Higuchi, 1976
[19] Kuratsune, 1972
[20] Masuda, 1985
[21] Aoki, 2001
[22] Rogan et al., 1988
[23] Rogan, 1989
[24] Hsu et al., 1984
[25] Rogan et al., 1988
[26] Yu et al., 1991
[27] Chen et al., 1994
[28] Kashimoto et al., 1981
[29] Kashimoto et al., 1981
[30] Jacobson et al., 1984
[31] Jacobson et al., 1984
[32] Jacobson et al., 1985
[33] Jacobson et al., 1990a, 1990b; Jacobson et al., 1992
[34] Jacobson & Jacobson, 1996b
[35] Jacobson & Jacobson, 1996a
[36] Jacobson et al., 1990a, pp. 43–44
[37] Jacobson & Jacobson, 1996a, p. 788
[38] Rogan et al., 1986b
[39] See Jacobson & Jacobson, 1996b
[40] Huisman et al., 1995a, 1995b; Winneke et al., 1998
[41] Stewart et al., 2000
[42] Huisman et al., 1995a
[43] Lanting et al., 1998
[44] Patandin et al., 1999b
[45] Vreugdenhil et al., 2002, p. 51
[46] Vreugenhil et al., 2004a
[47] Vreugdenhil, 2004b
[48] Chen & Hsu, 1994
[49] Anderson et al., 1999
[50] Rogan & Gladen, 1993
[51] Anderson et al., 1999
[52] Jensen et al., 1992
[53] Lucas et al., 1992
[54] Koletzko et al., 2008
[55] Fergusson & Woodward, 1999
[56] Walkowiak et al., 2001
[57] Jacobson et al., 1989
[58] Jacobson et al., 1990a
[59] Jacobson et al., 1992
[60] Jacobson et al., 1990b

[61] See Seegal, 1996, for discussion of this point; see also Walkowiak et al., 2001; Jacobson & Jacobson, 2001

[62] Koopman-Esseboom et al., 1996

[63] Huisman et al., 1995a

[64] Patandin et al., 1999b

[65] Patandin et al., 1999b, p. 40

[66] Vreugdenhil et al., 2002

[67] Vreugdenhil, 2004a

[68] Weisglas-Kuperus et al., 2000, p. 1207

[69] Lillienthal & Winneke, 1991

[70] Rice, 1998

[71] Rice & Hayward, 1997a, 1997b, 1999; Rice, 1998

[72] Institute of Medicine, 2003

[73] Hudson Voice, 2001

[74] Hudson Voice, 2001

[75] 1996a

[76] Darvill et al., 1996

[77] Darvill et al., 1996

[78] Fein et al., 1984, p. 316

[79] Buck, 1996

[80] Jacobson & Jacobson, 1996a

[81] Jacobson & Jacobson, 1996a, p. 788

[82] Longnecker et al., 2000

[83] Jacobson & Jacobson, 1996a, p. 788

[84] Fein et al., 1984; Jacobson et al., 1984; Jacobson & Jacobson, 1996a

[85] Gladen et al., 1988

[86] Rogan et al., 1986a

[87] Huisman et al., 1995a

[88] Stewart et al., 2000

[89] Tilson et al., pp. 244–245

[90] p. 245

[91] Porterfield, 1994, p. 129

[92] McKinney & Waller, 1994

[93] for a review of environment-gene interactions, as well as the importance of retinoids, see NAS, 2000b

[94] Goodwill et al., 2007; Seegal, 1999

[95] Seegal, 1999, p. S43

[96] 1999

[97] 1995

[98] 1995

[99] p. 136

[100] Swanson et al., 1995, p. 144

[101] 1995, p. 145

[102] James et al., 1993

[103] James et al., 1993, p. 137

[104] Drinker et al., 1937

[105] See Jacobson & Jacobson, 2000, for a discussion of this point

[106] Chlorinated diphenyl is another term for chlorinated biphenyl.

[107] discussion following Drinker et al., 1937, pp. 304–305

[108] Drinker et al., p. 307

[109] Discussion following Drinker et al., 1937, p. 303, emphasis added

[110] Aoki, 2001, p. 8, emphasis added

[111] Lorber, 2008

[112] EPA, 2008

[113] Cone, 2008

[114] EPA, 2008

[115] Rogan et al., 1986b

[116] Robinson et al., 1990; Orban et al., 1994

[117] Dewailly et al., 1993

[118] Dewailly et al., 1993

[119] Dewailly et al., 1993

[120] Dewailly et al., 1999

[121] Bjerregaard et al., 2001

[122] Radio Canada International, 2001

[123] Robinson et al., 1990

[124] CDC, 2005

[125] Orban et al., 1994

[126] Robinson et al., 1990

[127] 1972

[128] Manchester-Neesvig et al., 2001

[129] Fries, 1972, p. 55

[130] Institute of Medicine, 2003

[131] Bernard, 2000

[132] Schepens et al., 2001

[133] Aoki, 2001

[134] Bent et al., 2000

[135] Wassermann et al., 1979; for a summary of research see Colborn et al., 1996

[136] Moccia et al., 1977

[137] "Tainted geese," 2001

[138] EPA, 2001a

[139] Roos et al., 2001

[140] Hickie et al., 2007

[141] General Electric, 2007

[142] Welch, 2001

[143] Johnson et al., 1999; Schantz et al., 1999

[144] Tilden et al., 1997

[145] Johnson et al., 1999

[146] Pacific salmon, steelhead, and brown trout are not native to the Great Lakes but are planted annually by various state governments to provide sport angling opportunities. Also, in the 1960s these fish were planted in the hopes of controlling the exploding population of alewife, a non-native species that hitchhiked into the Great Lakes in the bilge water of large ships that came from Europe. There are

several ironies in the current situation: non-native species were planted to control another non-native species; consumption of the salmon, steelhead, and large brown trout is hazardous to people and other wildlife; and salmon and trout show illnesses that may be related to pollutants in the Great Lakes. The yellow perch population in Green Bay has also dropped in recent years, and scientists are not sure why.

[147] Johnson et al., 1999

[148] Manchester-Neesvig et al., 2001

[149] EPA, 2001c

[150] Wisconsin Department of Natural Resources (DNR), 2008

[151] (http://www.apl.org/community/pcb.html)

[152] Kilburn, 2000

[153] Kilburn, 2000, p. 497

[154] Horn et al., 1979

[155] Horn et al., 1979

[156] EPA, 2001d

[157] New York Department of Health, 2001

Chapter 4

[1] NAS, 2000a

[2] Grossman, 1995

[3] "Rachel Carson dies," 1964, p. 25

[4] Carson, 1962

[5] "Rachel Carson dies," 1964, p. 25

[6] Udall, 1964, pp. 23, 59, emphasis in original

[7] For an excellent history of pesticide policy in the United States, see Wargo 1996.

[8] Khurana & Prabhakar, 2000

[9] Satoh & Hosokawa, 2000

[10] O'Malley, 1997

[11] Abou-Donia & Lapadula, 1990

[12] Abou-Donia & Lapadula, 1990

[13] Bidstrup et al., 1953

[14] Abou-Donia & Lapadula, 1990

[15] Abou-Donia & Lapadula, 1990

[16] Wang et al., 2000

[17] Abou-Donia & Lapadula, 1990

[18] Healy, 1959

[19] Aiuto et al., 1993

[20] Kaplan et al., 1993, p. 2196

[21] Chlorpyrifos was voluntarily withdrawn from many home and garden uses starting in 2000. In 2003 a Dow Chemical subsidiary paid $2 million to the State of New York for false advertising claims about safety of pesticides, including chlorpyrifos (New York Office of Attorney General, 2003).

[22] Lauder & Schambra, 1999

[23] Brimijoin & Koenigsberger, 1999

[24] Anonymous, "Smoking hazard," 1973; Butler & Goldstein, 1973

[25] Fried & Makin, 1987; Fried & Watkinson, 1988

[26] Butler & Goldstein, 1973

[27] Fried et al., 1998

[28] Naeye & Peters, 1984

[29] Brennan et al., 1999; Fergusson, 1999; Fergusson et al., 1998

[30] 1984

[31] Koren, 1999; Fergusson, 1999

[32] Slotkin, 1998

[33] See Slotkin, 1998, 1999, for reviews

[34] Slotkin, 1999, p. 74

[35] Slotkin, 1999

[36] The rat brain at one day after birth is much less mature than the human brain at birth, so exposure of the rat at one day of age is considered analogous to prenatal exposure in the human.

[37] This phenomenon is called a "sensitive period," a time during development when the animal or person is more sensitive to a certain kind of event, such as exposure to a toxin, than at other times.

[38] Slotkin, 1999, p. 77

[39] Barone et al., 2000, pp. 18–19

[40] Schardein & Scialli, 1999

[41] p. 10

[42] 1999

[43] 1999

[44] 1999

[45] 1998, 1999

[46] 1999

[47] See Eriksson & Talts, 2000, for a review

[48] Eriksson & Talts, 2000, p. 44

[49] Guillette et al., 1998

[50] Guillette et al., 1998, p. 348

[51] Guillette et al., 1998, p. 353

[52] CHAMACOS stands for "Center for the Health Assessment of Mothers and Children of Salinas." The word "Chamacos" means "little child" in Chicano Spanish.

[53] Young et al., 2005

[54] Rauh et al., 2006

[55] D'Amelio et al., 2005

[56] NAS, 1993, p. 7

[57] Whitmore et al., 1994

[58] An increase in cancer by 1 in 100,000 over 70 years of exposure.

[59] NAS, 1993

[60] NAS, 2000a, p. 74

[61] Whitmore et al., 1994

[62] Whitmore et al., 1994, p. 51

[63] Whitmore et al., 1994, p. 57

[64] Whitmore et al., 1994, p. 58, emphasis added

[65] EPA, 2000a

[66] O'Rourke et al., 2000

[67] O'Rourke et al., 2000

[68] Weppner et al., 2006

[69] Wargo, 1996

[70] Adgate et al., 2001

[71] Fleutsch & Sparling, 1994

[72] Lu et al., 2000

[73] Fenske et al., 2000, p. 519

[74] Curt et al., 2003

[75] Lu et al., 2006

[76] Lu et al., 2006, p. 262

[77] Charnley, 2003

[78] Curl et al., 2003

[79] Mitchell et al., 2007

[80] Sofer et al., 1989; Lifshitz et al., 1999

[81] Sofer et al., 1989

[82] O'Malley, 1997

[83] He, 2000

[84] Casey et al., 1994; Thompson et al., 1994

[85] Grossman, 1995

[86] He, 2000

[87] Lifshitz et al., 1994

[88] Gelbach & Williams, 1975

[89] Wolfe et al., 1961

[90] CDC, 1980

[91] CDC, 1984

[92] Anonymous, "Environews," 1997

[93] Ferrer & Cabral, 1995

[94] CDC, 1986

[95] Goldman et al., 1990, p. 146

[96] Goh et al., 1990

[97] Chaudhry et al., 1998, p. 269, emphasis added

[98] CDC, 1996; NAS, 2000a

[99] CDC, 1985

[100] CDC, 1999b

[101] Das et al., 2006

[102] CDC, 1996

[103] Sutton et al., 2007

[104] Calvert et al., 2007

[105] "Nonexposed" is a slightly misleading term for substances to which most Americans are exposed. A survey of a random sample of U.S. adults showed that the mean level of chlorpyrifos metabolites was 4.5 μg/dL and that the substance

was detectable in 82% of the sample. Chlorpyrifos is widely used on food crops, so most of us are exposed to it to some extent. The nonexposed sample in the North Carolina study had a mean level of 6.2 μg/dL, and the termiticide applicators who said they had not applied chlorpyrifos in the last week had a mean of 119 μg/dL. The applicators who had used the chemical within the week had a mean of 629.5 μg/dL.

106 Steenland et al., 2000

107 Steenland et al., 2000, pp. 299–300

108 EPA, 2001e

109 Lockwood et al., 1994

110 Lockwood et al., 1994

111 Landrigan et al., 1999

112 Landrigan et al., 1999

113 Rosenstreich et al., 1997

114 Bryant, 1985; Deschamps et al., 1994; O'Malley, 1997; Senthilselvan et al., 1992

115 NAS, 2000a, p. 78

116 EPA, 2001b

117 See Vyas, 1999

118 Rondeau & Desgranges, 1995

119 Vyas, 1999

120 NAS, 2000a

121 Posner et al., 2008

122 Fluetsch & Sparling, 1994

123 Ehrlich et al., 1988

Chapter 5

1 For an example, see Rabbitt, 1966

2 See Jones & Broadbent, 1979

3 Rutter, 1979; Sameroff & Chandler, 1975; Sameroff et al., 1998

4 A third characteristic of sound is its impulsive characteristic (rise time and decay time). Impulsive noises are those with sudden onset and rapid decay, such as thunderclaps, explosions, the blasts of the horn of a train or semi truck, and some noises in industry, such as drop forges, sonar "pings," or the sound of a big truck hitting a bump in the road. Impulsive noise can be more damaging to hearing in some frequencies and less damaging in other frequencies than steady-state noise (see Sulkowski & Lipowczan, 1982, for an example).

5 Kryter, 1994

6 The statement that one noise is twice as loud as another implies that it is possible to measure ratios of perceived loudness. There is evidence that people may not be able to make such comparisons. See Birnbaum (1982a, 1982b).

7 Federal Highway Administration (FHWA), 2000

8 Santa & Hoien, 1999; see Stanovich, 1986, for a review

9 1986

[10] Cohen et al., 1973

[11] Wightman & Kistler, 2005

[12] Poskiparta et al., 1999; for reviews see Wagner & Torgesen, 1987, and Adams, 1990, chap. 4

[13] Poskiparta et al., 1999

[14] In southern California, most schools are single-story, and classrooms open directly outside rather than onto interior hallways. Therefore, it is possible for some, but not all, classrooms within a school to have been sound-insulated and air-conditioned. Double-pane glass is rarely used in construction in California. Air-conditioning would generally be necessary for classroom comfort if the windows were closed.

[15] Cohen et al., 1980

[16] Cohen et al., 1981a

[17] Lane & Meecham, 1974

[18] Lane and Meecham (1974) reported that Felton, Jefferson, Whelan, Lockhaven, and Buford schools all had airport noise above 100 dBC on the playgrounds. They also said that at Felton School in a 1-year period each child would be exposed to about 3,000 jet flyovers of 115 dBC, and about 10,000 of more than 100 dBC. They concluded that noise levels outdoors at Felton School "violate the occupational health and safety codes, and constitute a high risk of permanent hearing damage" (p. 129). Cohen et al. (1981a) do not name the schools involved in their study.

[19] This is sometimes called "learned helplessness"; see Dweck, 1975

[20] Blood pressure also varies with ponderosity, an index of body mass which is usually calculated as body weight divided by height cubed. The researchers included ponderosity in their analyses of blood pressure.

[21] Department of Education, 2000

[22] Carlisle & Beeman, 2000

[23] Green et al., 1982

[24] Green et al., 1982

[25] Green et al., 1982, p. 30

[26] Another implication of the results is that it is unfair to evaluate schools or teachers by using the standardized test performances of the pupils unless important factors in the teaching environment are also taken into consideration. For example, linking teacher merit pay raises to percentage of students reading below grade level would be unfair to teachers working in high-noise locales.

[27] Bronzaft & McCarthy, 1975

[28] Bronzaft, 1981

[29] See Kryter, 1994, chap. 8, for a review

[30] Evans et al., 1995

[31] Evans et al., 1995, p. 337

[32] Bullinger et al., 1998; Evans et al., 1998

[33] Evans et al., 1998

[34] Hygge et al., 2002

[35] Stansfeld et al., 2005

[36] For an overview see Nelson & Soli, 2000; Kryter, 1994, chap. 6; Wightman & Kistler, 2005

37 Wightman & Kistler, 2005

38 Wightman & Kistler, 2005

39 Mills, 1975, p. 770

40 Niskar et al., 1998

41 Nelson, 2000

42 Niskar et al., 1998, pp. 1073–1074

43 Nelson, 2000

44 Kryter, 1994, p. 626

45 Ko, 1979

46 Crook & Langdon, 1974

47 Siebein et al., 2000

48 Lee & Khew, 1992, p. 168

49 Institute for Environment and Health, 1997; see also Passchier-Vermeer & Passchier, 2000, for a summary

50 Evans et al., 2001

51 Evans & Maxwell, 1997; Sanz et al., 1993; see Evans & Lepore, 1993, for a review that includes some unpublished studies

52 Evans et al., 2001

53 Schneider & Moore, 2000

54 Etzel et al., 1997

55 Lalande et al., 1986; Tharpe & Bess, 1991

56 Forbes & Forbes, 1927

57 Etzel et al., 1997

58 Abrams et al., 1998

59 Gerhardt & Abrams, 2000

60 Abrams et al., 1995

61 Lalande et al., 1986

62 Etzel et al., 1997, p. 725

63 Northern & Downs, 1984

64 Gottfried et al., 1981

65 Gerhardt & Abrams, 2000, p. S26; see also Pierson, 1996

66 See Smith, 1995, for a review

67 Etzel et al., 1997, p. 724

68 Kisilevsky, 1995

69 Abrams et al., 1995

70 for reviews see Lobel, 1994; Paarlberg et al., 1995; Wadhwa, 1998

71 Clarke & Schneider, 1993; Clarke et al., 1994; Schneider et al., 1999

72 See Schneider et al., 2001, for an overview

73 It is possible that the fetal monkeys were also startled and stressed by the noise blasts that startled the pregnant monkey. According to Gerhardt and Abrams's (2000) model of sound transmission, the sound received by the fetal monkey would be about 45 dB less than when measured in air, or about 70 dB. The hearing thresholds of the monkey offspring in this study have not been tested to see if this level of noise affected their hearing.

74 Sobrian et al., 1997

75 Ando, 1988; see Etzel et al., 1997, for a brief overview

[76] Knipschild et al., 1981

[77] Fisk et al., 1991

[78] Elliott, 1971

[79] See Job, 1999, for a review

[80] Weinstein, 1978

[81] Weinstein, 1982

[82] See Kryter, 1994, chap. 10, for a review

[83] Ohrstrom et al., 1990

[84] Psychologists use the term "trait anger" to refer to the characteristic of feeling and acting chronically angry. "State anger" refers to anger that is centered on a particular event or situation.

[85] Zimmer & Ellermeier, 1999

[86] People who score high on neuroticism describe themselves on questionnaires as losing sleep over their worries, being touchy about things, and not being able to handle stress well.

[87] Ohrstrom et al., 1988

[88] Stansfeld, 1992

[89] Nivison & Endresen, 1993; Ohrstrom et al., 1988

[90] Kryter, 1990; Stansfeld, 1992

[91] Tarnopolsky et al., 1978

[92] van Kempen et al., 2002

[93] American National Standards Institute, 2002

[94] ANSI, 2002

[95] Bess, 2000, p. 10

[96] ANSI, 2002; Bess, 2000, p. 10

[97] FAA, 2000; General Accounting Office (GAO), 2000

[98] FHWA, 2008

[99] GAO, 2000, p. 99

[100] GAO, 2000, p. 49

[101] FAA, 2000, p. 43819

[102] FHWA, 2008

[103] Schultz, 1978

[104] FAA, 2000, p. 43819

[105] The value of 12% for DNL 65 is given in the FAA's (2000) policy document. The FAA is using a curve from a study by Finegold et al. (1994) that differs from Schultz's original curve. Schultz (1978) used a third-order polynomial function to fit the data; Fidell et al. (1991) used a quadratic equation; Finegold et al. (1994) used a logistic function; and Miedema and Vos (1998) used a quadratic function. The predicted percentages of people who are highly annoyed at DNL 65 are 15, 18.8, 11.6, and 27.3 for the Schultz, Fidell, Finegold, and Miedema curves, respectively. Finegold (1994), who worked for the U.S. Air Force, and his colleagues excluded some data because the relationship between noise and annoyance did not meet the .05 criterion of statistical significance. Statistical significance is not normally used as a criterion for including data in a meta-analysis. The FAA has adopted Finegold et al.'s (1994) curve, which estimates lower levels of annoyance than the other curves in the range of contentious aircraft noise exposure, approximately 55 to 70 DNL.

[106] 2000
[107] 1978
[108] 1978, p. 379
[109] 1978, p. 379
[110] 1978, p. 386
[111] Kryter, 1982a
[112] 2000, 2007
[113] 1982
[114] 1982a
[115] 1982, p. 1251
[116] 1982, p. 1251
[117] 1982b, 1983
[118] 1983
[119] 1982b
[120] pp. 1255–1256
[121] 1982b
[122] 1983
[123] p. 1068
[124] von Gierke, 1982
[125] 1982a
[126] Fidell et al., 1991
[127] Fidell et al., 1991
[128] This distribution would be expected to occur by chance with a probability of less than .005 by a chi-squared test. I tabulated the numbers from the published graphs in Fidell et al. (1991), omitting data points on the curve as "too close to call." A more sophisticated analysis would examine the deviations from the curve statistically and would graph the deviations. It is routinely recommended in statistical curve-fitting that deviations be examined to see if they show any pattern (Box et al., 1978). Had Fidell et al. (1991) examined deviations from the curve as a function of noise source, they would have rediscovered the difference in annoyance between aircraft and other noise sources presented by Kryter (1982a).
[129] Finegold et al., 1994
[130] Miedema & Vos, 1998
[131] 1982a
[132] Fidell et al., 1991
[133] 1978
[134] 1978
[135] FAA, 2000
[136] Health Council of the Netherlands, 1999; Institute for Environment and Health, 1997
[137] GAO, 2000, 2001
[138] 2000
[139] p. 43
[140] FAA, 2000, p. 43819
[141] GAO, 2000
[142] 1978

[143] p. 390

[144] Schultz, 1978, p. 377

[145] FAA, 2000

[146] FICAN, 2007

[147] GAO, 2001

[148] Bradley, 1993

[149] Taylor et al., 1982

[150] As was found by Weinstein, 1982

[151] 1994, p. 620

[152] Haines et al., 2001, p. 274

[153] Green et al., 1982, p. 29

[154] Bronzaft et al., 1998

[155] Evans & Lepore, 1993

[156] Cohen et al., 1980

[157] Green et al., 2004

[158] GAO, 2007

[159] Niskar et al., 1998

[160] See Niskar et al., 1998, for an overview

[161] Lippmann, 1990

[162] de la Torre et al., 2000, p. 162

[163] Schlundt et al., 2000, p. 3496

[164] Marten et al., 2001

[165] See Richardson et al., 1995, or Tyack, 2008, for a review Frankel & Clark, 2000, Madsen & Mohl 2000

[166] Schlundt et al., 2000

[167] Erbe & Farmer, 2000

[168] Erbe & Farmer, 2000, p. 1340

[169] Maier et al., 1998, p. 753

[170] Ward et al., 1999

[171] For examples see Conomy et al., 1998a, 1998b; Krausman et al., 1998

[172] See the special section on the effects of roads on wildlife in the January 2000 issue of Conservation Biology, and also the Web site www.wildlandscpr.org

[173] Rheindt, 2003

[174] Here are a few Web sites dealing with noise in natural areas: www.hmmh.com/aviat15.html; www.fican.org/download/SCHMIDT2.PDF; www.fican.org/pages/speakers.html; www.earthisland.org/takeaction/new=action.cfm?aaID=91; www.nonoise.org/library/drowning/; www.wildmontana.org/quietsky.htm.

Chapter 6

[1] Slovic, 1987

[2] Gerusky, 1981

[3] Houts et al., 1988

[4] Trunk & Trunk, 1981

[5] Gerusky, 1981, p. 55

[6] Houts et al., 1988

[7] Houts et al., 1988

[8] Houts et al., 1988

[9] Trunk & Trunk, 1981

[10] Dohrenwend et al., 1981

[11] Flynn, 1988

[12] President Carter had been trained as a nuclear engineer during his service in the U.S. Navy.

[13] Houts et al., 1988

[14] Mynatt, 1982

[15] Ginzburg & Reis, 1991

[16] Field et al., 1981

[17] Kirk, 1983; Field et al., 1983

[18] A certain amount of radioactive gas emission is allowed from nuclear power plants. The normal procedure is to store the gas in tanks until the radioactivity of the short-lived isotopes has declined to a permissible level. In the accident, too much radioactive gas was generated and had to be vented immediately. Also, during the cleanup after the accident there were occasional ventings of radioactive gas: "Between 1979 and 1982 there were many ventings, controlled and uncontrolled, of radioactive gases from the damaged reactor. The safety of decontamination methods was hotly debated as was the restart of the undamaged reactor" (Goldsteen et al., 1989, p. 391).

[19] Mynatt, 1982

[20] Wahlen et al., 1980

[21] Gerusky, 1981

[22] Pasciak et al., 1981

[23] Fabrikant, 1979

[24] Fabrikant, 1979

[25] Radioactive exposures are often compared to estimates of exposure from background radiation due to cosmic rays and naturally radioactive elements in the earth (such as potassium-40, uranium, and radium). Measurements of natural background radiation are problematic because it is difficult to separate bomb test fallout from natural sources (Oakley, 1972). Natural background radiation near TMI in the 1960s was estimated to be 89 mrem per year in a publication cited by the President's Committee that investigated the accident (Oakley, 1972).

[26] Upton, 1981, p. 69

[27] 1981

[28] p. 156

[29] 1981

[30] p. 157

[31] Wing et al., 1997a, 1997b

[32] Bodansky, 1980

[33] Gears et al., 1980; Wasserman & Solomon, 1982

[34] Rambo, 1996

[35] Rambo, 1996; see Shrader-Frechette, 1987, for a discussion of ethical issues related to probabilistic harm

[36] Berg, 1997; Hatch et al., 1991; Hatch et al., 1997; Levin, 2008; Susser, 1997; Talbott et al., 2000a, 2000b; Wing et al., 1997b; Wing & Richardson, 2000; Wing, 2003

[37] Hatch et al., 1997, p. 12

[38] Wing et al., 1997b, pp. 56–57

[39] Susser, 1997; Wing et al., 1997a

[40] 1997b

[41] Wing et al., 1997a; Wing, 2003

[42] Wing, 2003

[43] Wing, 2003

[44] Wing et al., 1997a

[45] see Hatch et al., 1997; Wing et al., 1997a; Wing & Richardson, 2000; Talbott et al., 2000a, 2000b

[46] see Baum & Fleming, 1993, for an overview of the special challenges that face people who are coping with technological accidents

[47] Flynn, 1988

[48] Baum et al., 1983; Bromet et al., 1990; Dew & Bromet, 1993; Dohrenwend et al., 1981

[49] Bromet et al., 1990

[50] Bromet et al., 1990

[51] Dew et al., 1987

[52] Bromet et al., 1990, p. 58

[53] Dew & Bromet, 1993, p. 54

[54] Solomon et al., 1987, pp. 1107–1108

[55] Dohrenwend et al., 1981

[56] Bromet et al., 1984

[57] Bromet et al., 1984, p. 298

[58] Freeman, 1981

[59] Hohenemser et al., 1977; Ahearne, 1987

[60] Aptekar & Boore, 1990

[61] See Kleinke, 1991, for a readable overview

[62] See Lazarus & Folkman, 1984

[63] Dew et al., 1987; Goldsteen et al., 1989

[64] Goldsteen et al., 1989

[65] Bromet et al., 1990

[66] Kleinke, 1991; Lazarus & Folkman, 1984

[67] Baum et al., 1983

[68] Cavalini et al., 1991

[69] Wing et al., 1997a

[70] Baum et al., 1983

[71] Seligman, 1975, Alloy et al., 2006, Dweck, 1975, 2006

[72] Davidson et al., 1982

[73] Davidson et al., 1982

[74] Baum et al., 1983

[75] Baum et al., 1983, p. 134

[76] See Kiecolt-Glaser et al., 2002, for an overview

77 Wright, 2005
78 Cohen et al., 1993
79 Dimsdale, 2008
80 Dimsdale, 2008
81 Chida et al., 2008
82 Spelt, 1948
83 Stewart et al., 1958
84 See obituary in the Bulletin of the Atomic Scientists, 2002, or a biography of Stewart by Gayle Green, 1999
85 See Hohenemser et al., 1977, for an overview of some aspects of the controversy among scientists; Freeman, 1981, contains interviews with Gofman and Sternglass in which they relate their personal experiences
86 Maugh, 2007
87 Hohenemser et al., 1977, p. 33; see also Freudenburg, 1988, and Slovic et al., 1991, for overviews of some of the disputed aspects of nuclear risk assessments and risk perceptions
88 See Clarke, 1999; Fairlie & Sumner, 2000; Koblinger, 2000; and de Brouwer & Lagasse, 2001, for discussions of low-level radiation policies; see Birchard, 1999, for a summary of one policy argument over thresholds; see Shrader-Frechette, 2007 for gaps in radiation exposure records of workers
89 NAS, 1990, p. 7
90 Nelkin, 1977, pp. 61–65
91 Davidson & Freudenburg, 1996
92 The Ukrainian preferred spelling is "Chornobyl." Under the Soviet regime, the preferred spelling was "Chernobyl." American usage prefers the spelling "Chernobyl," so I have used that throughout rather than "Chornobyl."
93 United Nations, 2002
94 Marples, 1988
95 United Nations Scientific Committee on the Effects of Atomic Radiation (UNSCEAR), 2001b
96 Marples, 1988
97 1988
98 p. 31
99 Marples, 1997
100 Shcherbak, 1989, p. 70
101 Medvedev, 1991 pp. 187–188
102 Shcherbak, 1989, pp. 86–87
103 Many books were written between the Chernobyl accident and dissolution of the Soviet Union in 1991 that alleged Soviet minimization of the accident (Chernousenko, 1991; Marples, 1988; Medvedev, 1991 Shcherbak, 1989).
104 Marples, 1988
105 Filyushkin, 1996
106 Marples, 1997
107 Marples, 1996; see also UNSCEAR, 2001a, for maps of the polluted areas
108 Marples, 1988, p. 36

[109] UNSCEAR, 2001a

[110] 1996

[111] p. 23

[112] 1996

[113] Otake & Schull, 1998; Schull et al., 1990; Yamazaki & Schull, 1990

[114] Otake & Schull, 1998

[115] Kolominsky et al., 1999

[116] Kolominsky et al. (1999) reported r = .15, and r = .17 between IQ score and estimated iodine dose at ages 6–7 and 10–11 years. This means that radiation could account for 2%–3% of the variance in IQ scores. They report p > .10 for each. For 138 observations, the two-tailed significance test of the hypothesis that the correlation is zero meets p < .10 for r = .15 and p < .05 for r = .17.

[117] Kolominsky et al., 1999, p. 302

[118] Nyagu et al., 1998

[119] Nyagu et al., 1998

[120] Nyagu et al., 1998, p. 309

[121] Bromet et al., 2000; Litcher et al., 2000

[122] The vascular dystonia diagnosis is used in former Soviet areas as a diagnosis for "pseudoneurotic disorders" related to the autonomic nervous system (Yevelson et al., 1997). The autonomic nervous system is sensitive to psychological stress and controls blood pressure and other aspects of the so-called "flight or fight" response.

[123] Bromet et al., 2000

[124] Litcher et al., 2000

[125] Litcher et al., 2000

[126] Bromet et al., 2000

[127] Litcher et al., 2000, p. 298

[128] Bromet et al., 2000, p. 569

[129] Litcher et al. (2000) used a false positive error cutoff of .005, or 25 chances in 500, rather than 5 in 100. This would make the chance of a false negative error quite high. The three attention measures each had a false positive error rate of 2 in 100, a level that would be interpreted as significant by most researchers. See the appendix for a discussion of how the use of a more stringent false positive error increases the likelihood of a false negative error.

[130] Havenaar et al., 1997

[131] Havenaar et al., 1996

[132] Havenaar et al., 1997

[133] Havenaar et al., 1996

[134] Cwikel et al., 1997

[135] Rahu, 2006

[136] Viel et al., 1997

[137] Viel et al., 1997, p. 1543

[138] Bromet et al., 2000, p. 569

[139] Medvedev, 1991 Shcherbak, 1989

[140] Havenaar et al., 1996

[141] Yevelson et al., 1997

[142] Depleted uranium bombs are constructed with conventional, not nuclear, explosive material. Depleted uranium is used because it is cheap and dense enough to pierce tanks and other objects that are difficult for munitions to penetrate. Depleted uranium has approximately 60% of the radioactivity of "natural" uranium. Like plutonium, it emits alpha particles. Alpha particles are minimally harmful unless they are inhaled or ingested or come into direct contact with tissues (McDiarmid et al., 2000). However, depleted uranium bombs used in the 1999 Kosovo conflict contained at least trace amounts of plutonium and two other nuclear reactor by-products ("Pentagon admits," 2001). Plutonium is more highly radioactive than depleted uranium and is a more serious health hazard if inhaled. Bomb explosions create large quantities of dust, raising the potential for inhalation.

[143] McDiarmid et al., 2000

[144] Bard et al., 1997; Holm, 2000; Lomat et al., 1997; WHO, 2006

[145] WHO, 2006

[146] Holm, 2000

[147] Balter, 1996

[148] Wing et al., 1991

[149] See Preston, 1998, or Schull, 1995, for overviews

[150] The bombs dropped on Japan created very little fallout because they were exploded in the air. Whether the results from bomb survivors can be extrapolated to situations involving radioactivity in soils and the food chain is a subject of ongoing debate.

[151] Goldsmith et al., 1999; Pacini et al., 1999

[152] Bard et al., 1997; Bromet & Havenaar, 2007

[153] Doyle et al., 2000, 2001; Parker et al., 1999; Parker, 2001

[154] Otake et al., 1990, p. 10

[155] Dubrova et al., 1996

[156] UNSCEAR, 2001b, p. 2

[157] UNSCEAR, 2001b

[158] Zimbrick, 1998

[159] UNSCEAR, 2001b; WHO, 2006

[160] Williams, 1995

[161] Ellegren et al., 1997

[162] Kovalchuk et al., 2000

[163] Voitsekhovitch et al., 1996

[164] Burger et al., 2001; Nezu et al., 1962; Revelle et al., 1956

[165] Bugai et al., 1996

[166] Micronesia Support Committee, 1981, p. 5

[167] Kiste, 1974

[168] Republic of Marshall Islands, 2002

[169] Republic of Marshall Islands, 2002

[170] International Atomic Energy Agency, 1998

[171] Cronkite et al., 1997; Micronesia Support Committee, 1981

[172] Cronkite et al., 1997

[173] Republic of the Marshall Islands, 2002

[174] Sydney Morning Herald, 2008

[175] NAS, 1982; Noshkin & Robison, 1997

[176] Republic of Marshall Islands, 2002

[177] Republic of Marshall Islands, 2002

[178] The tribunal Web site also gives some history of U.S. nuclear testing in the Marshall Islands and has a link where you can purchase a video: www.nuclear-claimstribunal.com.

[179] 1974

[180] p. 75

[181] Andrews, 1955

[182] Clarke, 1954; Meinke, 1951; McCarthy, 1997

[183] Oakley, 1972, p. 21

[184] Martell, 1964

[185] Miller, 1986

[186] www.JSTOR.org

[187] 1958

[188] 1958

[189] Archer, 1987, p. 269

[190] Archer, 1999; Dworkin, 1994; Gilbert et al., 1998; Kerber et al., 1993; Lapp, 1994

[191] Contact the Department of Justice at www.usdoj.gov:80/civil/torts/const/reca/index.htm.

[192] Gallagher, 1993

[193] Hardison, 1990

[194] Gallagher, 1993

[195] Garcia, 1998; Parrish, 1998

[196] Johnson et al., 1997

[197] Miller, 1986; Parrish, 1998

[198] Garcia, 1994

[199] Stender & Walker, 1974

[200] Murphy et al., 1990

[201] Gensler, 1998

[202] Patel, 1987

[203] Parrish, 1998

[204] The term "dirty" is applied colloquially to atomic or nuclear explosions that produce a large amount of radioactive debris or fallout.

[205] Caldwell et al., 1980

[206] Hansen & Schreiner, 2005; Johnson et al., 1997

[207] Brady, 1998

[208] Gensler, 1998; Hansen & Schreiner, 2005; Volante, 1998

[209] Unidentified U.S. atomic veteran, quoted in Murphy et al., 1990

[210] EPA, 2000a

[211] Marx, 1979

[212] Burger et al., 2001; Olsen et al., 1981

[213] Mangano et al., 2000

214 Mangano et al., 2000
215 Goldsmith et al., 1999
216 EPA, 2000a, p. 23
217 Freeman, 1981; Owens, 1988
218 Brugge et al., 2007
219 Brugge et al., 2007
220 The Web address is www.usdoj.gov:80/civil/torts/const/reca/index.htm.
221 McKay, 2002
222 2002, 2007
223 Goldman, 1997
224 Warner & Kirchmann, 2000
225 Goldman, 1997
226 Arnold, 1992
227 Copplestone et al., 2000
228 Walker, 2007
229 Warner & Kirchmann, 2000
230 Warner & Kirchmann, 2000
231 Copplestone et al., 2000
232 EPA, 2000a
233 Copplestone et al., 2000; Donaldson et al., 1997; Walker et al., 1997
234 Warner & Kirchmann, 2000, p. 95
235 See Brown, 1979; Gibbs, 1982
236 Office of Technology Assessment, 1983
237 Paigen et al., 1987
238 EPA, 2002
239 Levine, 1982; Stone & Levine, 1985
240 Blum 2008; Paigen et al., 1985
241 Levine, 1982
242 Interested readers will find resources at the University of Buffalo's Web site on Love Canal, at ublib.buffalo.edu/libraries/units/sel/exhibits/lovecanal.html. The site includes a link to a retrospective health study by the New York Department of Health. That study was funded by a legal settlement with Occidental Chemical Company.
243 EPA, 2002
244 Levine, 1982
245 New York Department of Health (NYDOH), 1978, pp. 13–15
246 For example, see Office of Technology Assessment, 1983
247 Vianna & Polan, 1984
248 Paigen et al., 1987
249 Goldman et al., 1985; Paigen et al., 1985
250 Heath et al., 1984
251 Janerich et al., 1981
252 Levine, 1982
253 Dolk et al., 1998
254 Levine, 1982; Mazur, 1998
255 NYDOH, 1978

[256] Roswell Park Memorial Institute is part of the New York Department of Health.

[257] Paigen, 1982

[258] Levine, 1982, p. 133

[259] Paigen, 1982

[260] Levine, 1982, p. 128

[261] Scientific data collected from individuals are required to be held in confidence. However, this does not imply that data cannot be shared with other scientists. Names, addresses, and other personal identification information can be deleted from the data. It is best not to enter personal identification information in data files but to identify participants by a code number that is recorded only in a key kept in a separate location. This could easily have been done with the Love Canal data and should be done with any data set.

[262] Levine, 1982, pp. 110–111; Paigen, 1982

[263] Levine, 1982, p. 167

[264] NYDOH, 1981, p. 52

[265] Levine, 1982, pp. 159–160

[266] Bross, 1980

[267] A hypospadia is an abnormal position of the opening of the urethra. In males with hypospadia the urethra opens on the underside of the penis (Dox et al., 1993).

[268] Dolk et al., 1998

[269] See Hammond, 1996, for an overview

[270] Levine, 1982

[271] Blum, 2008

[272] 1982

[273] p. 34

[274] pp. 83–84

[275] p. 101

[276] p. 109

[277] p. 29

[278] p. 37

[279] p. 37

[280] p. 32

[281] p. 103

[282] p. 102

[283] p. 108

[284] p. 105

[285] p. 45

[286] See Rawls, 1971, for a philosophical treatise that includes procedural justice

[287] Levine, 1982, pp. 108–109, emphasis added

[288] Mazur, 1998, p. 100

[289] Mazur, 1998

[290] Levine, 1982, p. 102

[291] Levine, 1982, p. 81

[292] Levine, 1982, p. 98

[293] Levine, 1982, p. 21; Mazur, 1998, p. 91

[294] See Rawls, 1971, for discussion of the failure of procedural justice to invariably create justice

[295] Levine, 1982

[296] EPA, 2002; see table 6.3

[297] Gupta, 1991; Shrivastava, 1987

[298] Landi et al., 1997; Needham et al., 1999

[299] Bertazzi et al., 1998

[300] Mocarelli et al., 2000

[301] Baccarelli et al., 2008

[302] EPA, 1999c

[303] Carter et al., 1975

[304] Missouri Department of Natural Resources (MoDNR), 2002b

[305] Carter et al., 1975

[306] MoDNR, 2002a

[307] MoDNR, 2002a

[308] EPA, 2007a

[309] Missouri DNR, 2008

[310] Current information on Superfund (CERCLA) waste sites can be found in the EPA Web site (www.epa.gov).

[311] EPA, 1999a

[312] EPA, 1999a

[313] EPA, 2007b

[314] EPA, 2007b

[315] Fitzgerald et al., 2000; Hwang et al., 2001

[316] Hwang et al., 2001

[317] Fitzgerald et al., 2007

[318] Denham et al., 2005

[319] Adeola, 2000

[320] EPA, 2008b

[321] NRDC, 2007

[322] EPA, 2008b

[323] EPA, 2000b

[324] EPA, 2000b

Chapter 7

[1] Shrader-Frechette, 1985

[2] Broome, 1994

[3] See Rayner & Cantor, 1987, for discussion of the role of perceived fairness in perceived risk

[4] See Landrigan et al., 2002, for monetary estimates of some environmentally induced conditions in children

[5] The term "compensating wage differential" is used to describe the idea that dangerous or unhealthy jobs should pay more to compensate for the danger. One

interpretation is that taking such a job amounts to selling one's health. The compensating wage differential assumes that people are fully informed about the risks and are free to choose other occupations in a free market. These conditions are rarely met. See Shrader-Frechette (2002) for an explication of these and other ethical issues with respect to compensating wage differentials.

6 Slovic, 1987

7 Sjoberg, 2000

8 Slovic, 1987

9 1987

10 See Freuedenberg, 1996; see also Shrader-Frechette, 1983, for a critique of risk assessments of nuclear power

11 Zeckhauser & Viscusi, 1990

12 Bazerman et al., 2001

13 See Shrader-Frechette, 1991, for an in-depth analysis of the ethical and philosophical issues embedded in criteria for rational decision making in situations involving risk and uncertainty

14 National Research Council (NRC), 1996, chap. 3; Shrader-Frechette, 1985, 1991

15 Shrader-Frechette, 1994, p. 112

16 United Nations, 1992

17 Kriebel & Tickner, 2001

18 Kriebel & Tickner, 2001

19 Cranor, 2001; O'Riordan & Jordan, 1995; Soule, 2000

20 Robert Wardrop, personal communication, 1997

21 The last two sources of scientific uncertainty are linked to the roles of theory in science and how scientists evaluate their research findings and theories. The viewpoint I have presented is derived from a mix of classic sources in philosophy of science, such as Kuhn (1970), Quine and Ullian (1970), Suppe (1977), and Kaplan (1964). My own ideas were shaped in part by Professor Don E. Dulany during my graduate studies at the University of Illinois.

22 See Shrader-Frechette, 1991, chap. 9

23 Meta-analysis is another way to resolve differences in research outcomes. In meta-analysis the results of many different studies are examined and statistically combined. Instead of looking at whether each study met the p = .05 cutoff, in meta-analysis the effect sizes are combined. Effect size is an index of how much of a difference a pollutant makes in the outcome variable, compared to the standard deviation of the outcome variable. One issue with meta-analysis is that the standard deviation is likely to be larger for studies that are poorly conducted. This implies that the results of good research studies can be watered down by poorly conducted studies. Nevertheless, where there are many studies available on a topic, meta-analysis is useful for summarizing the strength of the findings.

24 For further discussions of the burden of proof see Cranor, 1990, 2001, and Hull, 1999; see the appendix of this volume for further explanation of the linkage between the probabilities of false positive and false negative errors in research.

25 Kaplan, 1964; Shrader-Frechette, 1991

26 Buhl-Mortensen & Welin, 1998

27 An important influence on the likelihood of a false negative error is how much the pollutant actually affects the outcome, called effect size (see chapter 2;

see also Cohen, 1988). The probability of a false negative error (or the statistical power) can be calculated for the observed effect size in a particular study. A problematic aspect of effect size is that it is related to the variability of the outcome. It can be argued that a simple way to lower the observed effect size is to conduct a sloppy piece of research in which the variability of the outcome measures is high. Because of this, interpretations of effect sizes are not always straightforward.

[28] Goldstein, 2001

[29] 2001

[30] Gerstein, 2002

[31] Buhl-Mortensen & Welin, 1998

[32] NRC, 1996

[33] See NRC, 1996; Raynor & Cantor, 1987; Shrader-Frechette, 1991, for discussions of citizen negotiating panels

[34] A coalition of fishing organizations and Native Americans opposed the mine. This coalition is an unlikely combination because there were many protests, some of them violent, when Ojibwe people began exercising their treaty rights to spearfish in the late 1980s (Loew, 2001).

[35] Wisconsin Stewardship Network, 2002; Wood, 2001

[36] Other sources of information on the Crandon mine are the Wisconsin Department of Natural Resources, www.dnr.state.wi.us/org/es/science/mining/crandon/crandon.htm, or any newspaper in Wisconsin.

Chapter 8

[1] Hardin, 1968; Kortenkamp & Moore, 2001, 2006

[2] California Air Resources Board, 2008

[3] Bellinger & Matthews, 1998

[4] CDC, 1998, 1999a

[5] CDC, 1998

[6] Many of them are listed at http://www.epa.gov/ost/fish.

[7] http://www.noharm.org/library/docs/Mercury=Thermometers=and=Your=Familys=Health.htm.

[8] Saper et al., 2004

[9] Tsuruta et al., 2004

[10] Boyd et al., 2000

[11] Patandin et al., 1999a

[12] Koppe, 1995

[13] 2003

[14] Williams et al., 2006

[15] See the National Institute for Occupational Safety and Health publication on reducing pesticide exposure in schools: http://www.cdc.gov/niosh/docs/2007-150/

[16] Some environmental organizations, such as the National Audubon Society, have links to least-toxic alternatives: www.audubon.org/bird/at=home/alternatives.html.

[17] Senechal & LeFevre, 2001

[18] For further information see the Web site of the League for the Hard of Hearing at www.lhh.org/noise/children/toys.htm. The League for the Hard of Hearing lists a number of toys that exceed 100 dB. Other information can be found at the Clearinghouse for Noise Pollution, www.nonoise.org.

[19] Mills, 1975

[20] Etzel et al., 1997

[21] See Wachs & Gruen, 1982, chap. 3

[22] See the following Web sites for further information: www.access-board.gov/publications/acoustic-factsheet.htm; www.fican.org/pages/sympos02.html; asa.aip.org/classroom.html, www.cefpi.org:80/pdf/issue9.pdf.

[23] Dennis, 2006

[24] It is illegal in the United States for employers to discriminate against women by assigning them different jobs from men against their wishes even when they are pregnant (Bertin, 1994). There is not enough research on the potential negative reproductive effects of pollutants on men, and it should not be assumed that men are somehow "immune" from those effects. An ethical argument can be made that work safety rules for exposure should be no different from the exposure standards for the public and certainly should be stringent enough to protect the most vulnerable worker; sometimes the "most vulnerable" is the category of women who are pregnant. See Shrader-Frechette (1985, chap. 4) for elaboration of arguments regarding the ethics of compensating wage differentials, i.e., paying more for dangerous work.

[25] Test kits are available from some state laboratories, such as the Wisconsin State Laboratory of Hygiene, www.sph.wisc.edu/index.shtml

[26] Shrader-Frechette, 2007

Afterword

[1] Lovins & Sheikh, 2008

[2] For one view of the costs, see Lovins and Sheikh 2008 in the biology journal Ambio.

[3] Guillebaud, 2008

Appendix

[1] for example, Glenberg, 1996

[2] Cranor, 1990

[3] See Murphy & Myors, 1998

References

Abou-Donia, M. B., & Lapadula, D. M. (1990). Mechanisms of organophosphorus ester-induced delayed neurotoxicity: Type I and type II. *Annual Review of Pharmacology and Toxicology, 30,* 405–440.

Abrams, R. M., Gerhardt, K. J., & Peters, A. J. M. (1995). Transmission of sound and vibration to the fetus. In J-P. Lecanuet, W. P. Fifer, N. A. Krasnegor, & W. P. Smotherman (Eds.), *Fetal development: A psychobiological perspective.* Hillsdale, NJ: Erlbaum.

Abrams, R. M., Griffiths, S. K., Huang, X., Sain, J., et al. (1998). Fetal music perception: The role of sound transmission. *Music Perception, 15*(3), 307–317.

Abramson, J. S., Baker, C. J., Fisher, M. C., & Gerber, M. A. (1999). Thimerosal in vaccines: An interim report to clinicians. *Pediatrics, 104*(3), 570–574.

Achenbach, T. M. (1991). *Child behavior checklist.* Burlington: University of Vermont.

Adams, M. J. (1990). *Beginning to read: Thinking and learning about print.* Cambridge, MA: MIT.

Adeola, F. O. (2000). Endangered community, enduring people: Toxic contamination, health, and adaptive responses in a local context. *Environment and Behavior, 32*(2), 209–249.

Adgate, J. L., Barr, D. B., Clayton, C. A., Eberly, L. E., et al. (2001). Measurement of children's exposure to pesticides: Analysis of urinary metabolite levels in a probability-based sample. *Environmental Health Perspectives, 109,* 583–590.

Agency for Toxic Substances and Disease Registry, and U.S. Environmental Protection Agency. (1996, December). *Public Health Implications of PCB Exposures.* Accessed online at www.epa.gov/region5/foxriver/lower=fox=river=PCB=Exposures.htm on May 1, 2001.

Ahearne, J. F. (1987). Nuclear power after Chernobyl. *Science, 236,* 673–679.

Aiuto, L. A., Pavlakis, S. G., & Boxer, R. A. (1993). Life-threatening organophosphate-induced delayed polyneuropathy in a chld after accidental chlorpyrifos ingestion. *Journal of Pediatrics, 122,* 658–660.

"Alice Stewart" (2002, September/October). *Bulletin of the Atomic Scientists,* 11.

Alloy, L. B., Abramson, L. Y., Whitehouse, W. G., Hogan, M. E., Panzarella, C., & Rose, D. T. (2006). Prospective incidence of first onsets and recurrences of depression in individuals at high and low cognitive risk for depression. *Journal of Abnormal Psychology, 115*, 145–156.

Altmann, L., Sveinsson, K., Kraemer, U., Weishoff-Houben, M., et al. (1998). Visual functions in 6-year-old children in relation to lead and mercury levels. *Neurotoxicology and Teratology, 20*, 9–17.

Amdur, M. O., Doull, J., & Klaassen, C. D. (1991). *Casarett and Doull's toxicology: The basic science of poisons* (4th ed.). New York: Pergamon.

American National Standards Institute (2002). Acoustical performance criteria, design requirements, and guidelines for schools. ANSI S12.60–2002. http://asastore.aip.org/ accessed on August 23, 2008.

Anderson, J. W., Johnstone, B. M., & Remley, D. T. (1999). Breast-feeding and cognitive development: A meta-analysis. *American Journal of Clinical Nutrition, 70*, 525–535.

Ando, Y. (1988). Effects of daily noise on fetuses and cerebral hemisphere specialization in children. *Journal of Sound and Vibration, 127*(3), 411–417.

Andrews, H. L. (1955). Radioactive fallout from bomb clouds. *Science, 122*, 453–456.

Anonymous. (1997). Environews forum: Methyl parathion comes inside. *Environmental Health Perspectives, 105*, 690–691.

Anonymous. (1973). Smoking hazard to the fetus [Editorial]. *British Medical Journal*, no. 5830, 369–370.

Aoki, Y. (2001). Polychlorinated biphenyls, polychlorinated dibenso-p-dioxins, and polychlorinated dibenzofurans as endocrine disrupters—what we have learned from yusho disease. *Environmental Research, Section A, 86*, 2–11.

Appleton Public Library (2008). PCBs and the Fox River. http://www.apl.org/community/pcb.html Accessed on November 19, 2008.

Aptekar, L., & Boore, J. A. (1990). The emotional effects of disaster on children: A review of the literature. *International Journal of Mental Health, 19*(2), 77–90.

Archer, V. E. (1987). Association of nuclear fallout with leukemia in the United States. *Archives of Environmental Health, 42*(5), 263–271.

Archer, V. E. (1999). Re: Thyroid cancer rates and 131-I doses from Nevada atmospheric nuclear bomb tests. *Journal of the National Cancer Institute, 91*(16), 1423–1424.

Arnold, L. (1992). *Windscale 1957: Anatomy of a nuclear accident.* Dublin, Ireland: Gill and Macmillan.

Askari, H. & Cummings, J. (1978). Food shortages in the Middle East. *Middle Eastern Studies, 14*, 326–351.

Avery, A. (2006). Organic diets and children's health. *Environmental Health Perspectives, 114*, A210.

Baccarelli, A., Giacomini, S. M., Corbetta, C., Landi, M. T., et al. (2008). Neonatal thyroid function in Seveso 25 years after maternal exposure to dioxin. *PloS Medicine, 5*, e161.

Bakir, F., Damluji, L., Amin-Zaki, M. M., et al. (1973). Methylmercury poisoning in Iraq: An interuniversity report. *Science, 181,* 230–241.

Balter, M. (1996). Poor dose data hamper study of cleanup workers [News]. *Science, 272,* 360.

Bard, D., Verger, P., & Hubert, P. (1997). Chernobyl, 10 years after: Health consequences. *Epidemiologic Reviews, 19*(2), 187–204.

Barone, S., Jr., Das, K. P., Lassiter, T. L., & White, L. D. (2000). Vulnerable processes of nervous system development: A review of markers and methods. *NeuroToxicology, 21*(1–2), 15–36.

Bauer, N. (1958). Fallout near Nevada test site. *Science, 128,* 40.

Baum, A., & Fleming, R. (1993). Implications of psychological research on stress and technological accidents. *American Psychologist, 48*(6), 665–672.

Baum, A., Fleming, R., & Singer, J. E. (1983). Coping with victimization by technological disaster. *Journal of Social Issues, 39*(2), 117–138.

Bazerman, M. H., Baron, J., & Shonk, K. (2001). *You can't enlarge the pie: Six barriers to effective government.* New York: Basic Books.

Bellinger, D. C. (2000). Effect modification in epidemiologic studies of low-level neurotoxicant exposures and health outcomes. *Neurotoxicology and Teratology, 22,* 133–140.

Bellinger, D. C., & Matthews, J. A. (1998). Social and economic dimensions of environmental policy: Lead poisoning as a case study. *Perspectives in Biology and Medicine, 41,* 307–326.

Bellinger, D. C., Trachtenberg, F., Zhang, A., Tavares, M., Daniel, D., & McKinlay, S. (2008). Dental amalgam and psychosocial status: The New England children's amalgam trial. *Journal of Dental Research, 87*(5), 470–474.

Bellinger, D. C., Trachtenberg, F., Daniel, D., Zhang, A., Tavares, M., & McKinlay, S. (2007). A dose-effect analysis of children's exposure to dental amalgam and neuropsychological function: The New England children's amalgam trial. *Journal of the American Dental Association, 138*(9), 1210–1216.

Bender, M. T., & Williams, J. M. (1999). A real plan of action on mercury. *Public Health Reports, 114,* 416–420.

Bent, S., Rachor-Ebbinghaus, R., & Schmidt, C. (2000). Decontamination of highly polychlorinated biphenyl contaminated indoor areas by complete removal of primary and secondary sources. [German; abstract in English]. *Gesundheitswesen, 62*(2), 86–92.

Berg, G. G. (1997). Radiation exposure and cancer: A simpler view of Three Mile Island. *Environmental Health Perspectives, 105*(6), 566.

Bernard, A. (2000). Food contamination by PCBs/dioxins in Belgium: Analysis of an accident with improbable health consequences. [French; abstract in English]. *Bulletin et Memoires de l Academie Royale de Medecine de Belgique, 155,* 195–201.

Berney, B. (1993). Round and round it goes: The epidemiology of childhood lead poisoning, 1950–1990. *Milbank Quarterly, 71,* 3–38.

Bertazzi, P. A., Bernucci, I., Brambilla, G., Consonni, D., & Pesatori, A. (1998). The Seveso studies on early and long-term effects of dioxin exposure: A review. *Environmental Health Perspectives, 106*(Supplement 2), 625–633.

Bertin, J. E. (1994). Reproductive hazards in the workplace: Lessons from UAW v. Johnson Controls. In H. L. Needleman & D. Bellinger (Eds.), *Prenatal exposure to toxicants: Developmental consequences.* Johns Hopkins series in environmental toxicology. Baltimore, MD: Johns Hopkins University Press

Bess, F. H. (2000). Classroom acoustics: An overview. *Volta Review, 101*(5), 1–14.

Bidstrup, P. L., Bonnell, J. A., & Beckett, A. G. (1953). Paralysis following poisoning by a new organic phosphorus insecticide ("Mipafox"). *British Medical Journal, 1,* 1068.

Birchard, K. (1999). Experts still arguing over radiation doses. *Lancet, 354,* 400.

Birnbaum, M. H. (1982a). Controversies in psychological measurement. In B. Wegener (Ed.), *Social attitudes and psychophysical measurement* (pp. 401–485). Hillsdale, NJ: Erlbaum.

Birnbaum, M. H. (1982b). Problems with so-called "direct" scaling. In J. T. Kuznicki, R. A. Johnson, & A. F. Rutkiewic (Eds.), *Selected sensory methods: Problems and approaches to hedonics, ASTM STP 773* (pp. 34–48). New York: American Society for Testing and Materials.

Bjerregaard, P., Dewailly, E., Ayotte, P., Pars, T., Ferron, L., & Mulvad, G. (2001). Exposure of Inuit in Greenland to organochlorines through the marine diet. *Journal of Toxicology & Environmental Health, Part A, 62*(2), 69–81.

Bjorklund, I., Borg, H., & Johansson, K. (1984). Mercury in Swedish lakes—Its regional distribution and causes. *Ambio, 13*(2), 118–121.

Bleier, R. (1984). *Science and gender: A critique of biology and its theories on women.* New York: Penguin.

Blum, E. D. (2008). *Love canal revisited: Race, class and gender in environmental activism.* Lawrence, KS: University Press of Kansas.

Bodansky, D. (1980). Electricity generation choices for the near term. *Science, 207,* 721–728.

Bornschein, R. L. (1985). Influence of social factors on lead exposure and child development. *Environmental Health Perspectives, 62,* 343–351.

Box, G. E. P., Hunter, W. G., & Hunter, J. S. (1978). *Statistics for experimenters: An introduction to design, data analysis, and model building.* New York: Wiley.

Boyd, A. S., Seger, D., Vannucci, S., Langley, M., Abraham, J. L., & King, L. E. (2000). Mercury exposure and cutaneous disease. *Journal of the American Academy of Dermatology, 43,* 81–90.

Bradley, J. S. (1986). Speech intelligibility in classrooms. *Journal of the Acoustical Society of America, 80*(3), 846–854.

Bradley, J. S. (1993). Disturbance caused by residential air conditioner noise. *Journal of the Acoustical Society of America, 93*(4), 1978–1986.

Bradley, R. H., & Calwell, B. M. (1977). Home observation for measurement of the environment: A validation study of screening efficiency. *American Journal of Mental Deficiency, 81,* 417–420.

Brady, W. J. (1998). Statement of William J. Brady. *Ionizing radiation, veterans health care, and related issues: Hearing before the Committee on Veterans' Affairs,* 105th Congress, 2nd sess. (April 21, 1998). Washington, D.C.: U.S. Government Printing Office.

Braun, J. M., Kahn, R. S., Froehlich, T., Auinger, P., & Lanphear, B. P. (2006). Exposures to environmental toxicants and attention deficit hyperactivity disorder in U.S. children. *Environmental Health Perspectives, 114,* 1904–1909.

Brennan, P. A., Grekin, E. R., & Mednick, S. A. (1999). Maternal smoking during pregnancy and adult male criminal outcomes. *Archives of General Psychiatry, 56,* 215–219.

Brimijoin, S., & Koenigsberger, C. (1999). Cholinesterases in neural development: New findings and toxicologic implications. *Environmental Health Perspectives, 107*(Supplement 1), 59–64.

Broadhurst, M. G. (1972). Use and replaceability of polychlorinated biphenyls. *Environmental Health Perspectives, 1*(October), 81–102.

Brody, D. J., Pirkle, J. L., Kramer, R. A., Flegal, K. M., et al. (1994). Blood lead levels in the U.S. population: Phase 1 of the third national health and nutrition examination survey (NHANES III, 1988–1991). *Journal of the American Medical Association, 272,* 277–283.

Bromet, E. J., Goldgaber, D., Carlson, G., Panina, N., Golovakha, E., Gluzman, S., et al. (2000). Children's well-being 11 years after the Chornobyl catastrophe. *Archives of General Psychiatry, 57,* 563–571.

Bromet, E. J., & Havenaar, J. M. (2007). Psychological and perceived health effects of the Chernobyl Disaster: A 20-year review. *Health Physics, 93,* 516–521.

Bromet, E. J., Hough, L., & Connell, M. (1984). Mental health of children near the Three Mile Island reactor. *Journal of Preventive Psychiatry, 2*(3/4), 275–301.

Bromet, E. J., Parkinson, D. K., & Dunn, L. O. (1990). Long-term mental health consequences of the accident at Three Mile Island. *International Journal of Mental Health, 19*(2), 48–60.

Bronzaft, A. L. (1981). The effect of a noise abatement program on reading ability. *Journal of Environmental Psychology, 1,* 215–222.

Bronzaft, A. L., Ahern, K. D., McGinn, R., O'Connor, J., & Savino, B. (1998). Aircraft noise: A potential health hazard. *Environment and Behavior, 30*(1), 101–113.

Bronzaft, A. L., & McCarthy, D. P. (1975). The effect of elevated train noise on reading ability. *Environment and Behavior, 7*(4), 517–527.

Broome, J. (1994). Discounting the future. *Philosophy and Public Affairs, 23,* 128–156.

Bross, I. D. J. (1980). Muddying the water at Niagara. *New Scientist, 88,* 728–729.

Brown, M. (1979). *Laying waste: The poisoning of America by toxic chemicals.* New York: Pantheon.

Brown, M. (2002). Costs and benefits of enforcing housing policies to prevent children's lead poisoning. *Medical Decision Making, 22,* 482–492.

Brown, M. J., Gardner, J., Sargent, J. D., Swartz, K., et al. (2001). The effectiveness of housing policies in reducing children's lead exposure. *American Journal of Public Health, 91,* 621–624.

Brown, T. M., & Fee, E. (1999). The editors comment. *American Journal of Public Health, 89,* 1131.

Brugge, D., deLemos, J. L., & Bui, C. (2007). The Sequoyal Corporation fuels release and the Churck Rock spill: Unpublicized nuclear releases in American Indian communities. *American Journal of Public Health, 97,* 1595–1600.

Bryant, D. H. (1985). Asthma due to insecticide sensitivity. *Australian & New Zealand Journal of Medicine, 5*(1), 66–68.

Buck, G. M. (1996). Epidemiologic perspective of the developmental neurotoxicity of PCBs in humans. *Neurotoxicology and Teratology, 18*, 239–241.

Bugai, D. A., Waters, R. D., Dzhepo, S. P., & Skal'skij, A. S. (1996). Risks from radionuclide migration to groundwater in the Chernobyl 30-km zone. *Health Physics, 71*(1), 9–18.

Buhl-Mortensen, L., & Welin, S. (1998). The ethics of doing policy relevant science: The Precautionary Principle and the significance of non-significant results. *Science and Engineering Ethics, 4*(4), 401–412.

Bullinger, M., Hygge, S., Evans, G. W., Meis, M., & Mackensen, S. V. (1998). The psychological cost of aircraft noise for children. *Zentralblatt fur Hygiene und Umweltmedizin, 202*, 127–138.

Burbacher, T., Shen, D., Liberato, N. Grant, K., Cernichiari, E., & Clarkson, T. (2005). Comparison of blood and brain mercury levels in infant monkeys exposed to methylmercury or vaccines containing thimerosal. *Environmental Health Perspectives, 113*, 1015–1021.

Burger, J., Gaines, K. F., Peles, J. D., Stephens, Jr., W. L., et al. (2001). Radiocesium in fish from the Savannah River and Steel Creek: Potential food chain exposure to the public. *Risk Analysis, 21*(3), 545–559.

Burns, J. M., Baghurst, P. A., Sawyer, M. G., McMichael, A. J., & Tong, S. (1999). Lifetime low-level exposure to environmental lead and children's emotional and behavioral development at ages 11–13 years. *American Journal of Epidemiology, 149*, 740–749.

Butler, N. R., & Goldstein, H. (1973). Smoking in pregnancy and subsequent child development. *British Medical Journal, 4*, 573–575.

Byers, R. K., & Lord, E. E. (1943). Late effects of lead poisoning on mental development. *American Journal of Diseases of Children, 66*, 471–494.

Cabana, G., Tremblay, A., Kalff, J., & Rasmussen, J. P. (1994). Pelagic food chain structure in Ontario lakes: A determinant of mercury levels in lake trout (*Salvelinus namaycush*). *Canadian Journal of Fisheries and Aquatic Sciences, 51*, 381–389.

Cairns, J., Jr. (1999). Absence of certainty is not synonymous with absence of risk [Editorial]. *Environmental Health Perspectives, 107*(2), A56–A57.

Caldwell, G. G., Kelley, D. B., & Heath, C. W. (1980). Leukemia among participants in military maneuvers at a nuclear bomb test: A preliminary report. *Journal of the American Medical Association, 244*, 1575–1578.

California Air Resources Board (2008). Key events in the history of air quality in California. http://www.arb.ca.gov/html/brochure/history.htm Accessed on November 19, 2008.

Calvert, J. M., Petersen, A. M., Sievert, J., Mehler, L. N., et al. (2007). Acute pesticide poisoning in the U.S. retail industry, 1998–2004. *Public Health Reports, 122*, 232–244.

Canfield, R.L., Henderson, C.R. Jr., Cory-Slechta, D.A., Cox, C., Jusko, T.A. & Lanphear, B. P. (2003). Intellectual impairment in children with blood lead concentrations below 19 microg per decliliter. *New England Journal of Medicine, 348*(16), 1517–1526.

Carlisle, J. F., & Beeman, M. M. (2000). The effects of language of instruction on the reading and writing achievement of first-grade Hispanic children. *Scientific Studies of Reading, 4*(4), 331–353.

Carson, R. (1962). *Silent Spring.* New York: Houghton-Mifflin.

Carter, C. D., Kimbrough, R. D., Liddle, J. A., Cline, R. E., et al. (1975). Tetrachlorodibenzodioxin: An accidental poisoning in horse arenas. *Science, 188,* 738–740.

Casey, P. B., Thompson, J. P., & Vale, J. A. (1994). Suspected paediatric pesticide poisoning in the UK. I—Home accident surveillance system 1982–1988. *Human & Experimental Toxicology, 13,* 529–533.

Caution required with the precautionary principle [Editorial]. (2000). *Lancet, 356,* 265.

Cavalini, P. M., Koeter-Kemmerling, L. G., & Pules, M. P. J. (1991). Coping with odour annoyance and odour concentrations: Three field studies. *Journal of Environmental Psychology, 11,* 123–142.

Center for Online Ethics. (2002). Current events at Love Canal. Accessed online at onlineethics.org/cases/l.canal/recent.html on January 25, 2002.

Centers for Disease Control. (1980). Pesticide poisoning in an infant—California. *Morbidity & Mortality Weekly Report,* 254–255.

Centers for Disease Control. (1984). Epidemiologic notes and reports: Organophosphate insecticide poisoning among siblings—Mississippi. *Morbidity & Mortality Weekly Report, 33*(42), 592–594.

Centers for Disease Control (1985). Acute poisoning following exposure to an agricultural insecticide—California. *Morbidity and Mortality Weekly Report, 34*(30), 464–466, 471.

Centers for Disease Control. (1986). Epidemiologic notes and reports: Aldicarb food poisoning from contaminated melons—California. *Morbidity & Mortality Weekly Report, 35*(16), 254–258.

Centers for Disease Control. (1996). Epidemiological notes and reports: Pentachlorophenol poisoning in newborn infants—St. Louis, Missouri, April–August 1967. *Morbidity & Mortality Weekly Report, 45*(25), 545–549.

Centers for Disease Control. (1997). Poisonings associated with illegal use of aldicarb as a rodenticide—New York City, 1994–1997. *Morbidity & Mortality Weekly Report, 46*(41), 961–963.

Centers for Disease Control. (1998). Lead poisoning associated with imported candy and powdered food coloring—California and Michigan. *Morbidity & Mortality Weekly Report, 47*(48), 1041–1043.

Centers for Disease Control. (1999a). Adult lead poisoning from an Asian remedy for menstrual cramps—Connecticut, 1997. *Morbidity & Mortality Weekly Report, 48*(02), 27–29.

Centers for Disease Control. (1999b). Farm worker illness following exposure to carbofuran and other pesticides—Fresno County, California, 1998. *Morbidity & Mortality Weekly Report, 48*(06), 113–116.

Centers for Disease Control (2001). Fatal pediatric lead poisoning—New Hampshire, 2000. *Morbidity & Mortality Weekly Report, 50*(22), 457–459.

Centers for Disease Control (2005). Polychlorinated dibenzo-p-dioxins, poly-chlorinated dibenzofurans, and coplanar and mono-ortho-substituted polychlorinated biphenyls. Third National Report on Human Exposure to Environmental Chemicals. http://www.cdc.gov/exposurereport/report.htm accessed on August 17, 2008.

Charnley, G. (2003). Pesticide exposures and children's risk tradeoffs. *Environmental Health Perspectives*, 111, A689.

Chaudhry, R., Mishra, B., & Dhawan, B. (1998). A foodborne outbreak of organo-phosphate poisoning. *British Medical Journal*, *317*, 268–269.

Chen, A. C., & Charuk, K. (2001). Speech interference levels at airport noise impacted schools. *Sound and Vibration*, July, 26–31.

Chen, Y. J., & Hsu, C. C. (1994). Effects of prenatal exposure to PCBs on the neu-rological function of children: A neuropsychological and neurophysiological study. *Developmental Medicine and Child Neurology*, *36*, 312–320.

Chen, Y-C. J., Yu, M-L. M., Rogan, W. J., Gladen, B. C., & Hsu, C-C. (1994). A 6-year follow-up of behavior and activity disorders in the Taiwan Yu-cheng chil-dren. *American Journal of Public Health*, *84*, 415–421.

Chernousenko, V. M. (1991). *Chernobyl: Insight from the inside*. (J. G. Hine, Trans.). Heidelberg, Germany: Springer-Verlag.

Chia, L-G., & Chu, F-L. (1984). Neurological studies on polychlorinated biphy-enyl (PCB)-poisoned patients. *American Journal of Industrial Medicine*, *5*, 117–126.

Chida, Y., Hamer, M., Wardle, J., & Steptoe, A. (2008). Do stress-related psycho-social factors contribute to cancer incidence and survival? *Nature Clinical Practice: Oncology*, *5*, 466–475.

CIA Factbook (2008). Available online at www.cia.gov/cia/publications/factbook/. Accessed November 2008.

Clark, M. J. (2001). EPA, tribes to study local lakes. *Lander Journal*, *117*(65), August 22, pp. A1, A8.

Clarke, A. S., & Schneider, M. L. (1993). Prenatal stress has long-term effects on behavioral responses to stress in juvenile rhesus monkeys. *Developmental Psychobiology*, *26*, 293–304.

Clarke, A. S., Wittwer, D. J., Abbott, D. H., & Schneider, M. L. (1994). Long-term effects of prenatal stress on HPA axis activity in juvenile rhesus monkeys. *Developmental Psychobiology*, *27*, 257–269.

Clarke, H. M. (1954). The occurrence of an unusually high-level radioactive rain-out in the area of Troy, NY. *Science*, *119*, 619–622.

Clarke, R. (1999). Control of low-level radiation exposure: Time for a change? *Journal of Radiological Protection*, *19*, 107–115.

Cohen, J. (1988). *Statistical power analysis for the behavioral sciences* (2nd ed.). Hillsdale, NJ: Erlbaum.

Cohen, S., Evans, G. W., Krantz, D. S., & Stokols, D. (1980). Physiological, moti-vational, and cognitive effects of aircraft noise on children: Moving from the laboratory to the field. *American Psychologist*, *35*, 231–243.

Cohen, S., Evans, G. W., Krantz, D. S., Stokols, D., & Kelly, S. (1981a). Aircraft noise and children: Longitudinal and cross-sectional evidence on adaptation to noise and the effectiveness of noise abatement. *Journal of Personality and Social Psychology, 40*, 331–345.

Cohen, S., Glass, D. C., & Singer, J. E. (1973). Apartment noise, auditory discrimination, and reading ability in children. *Journal of Experimental Social Psychology, 9*, 407–422.

Cohen, S., Krantz, D. S., Evans, G. W., & Stokols, D. (1981b). Cardiovascular and behavioral effects of community noise. *American Scientist, 69*, 528–535.

Cohen, S., Tyrrell, D. A., & Smith, A. P. (1993). Negative life events, perceived stress, negative affect, and susceptibility to the common cold. *Journal of Personality and Social Psychology, 64*, 131–140.

Colborn, T., Dumanoski, D., & Myers, J. P. (1996). *Our stolen future.* New York: Dutton.

Collier, E. (1943). Poisoning by chlorinated naphthalene. *Lancet*, January 16, 72–74.

Cone, M. (2008). Outspoken scientist dismissed from panel on chemical safety. Los Angeles Times, February 29, 2008. http://www.latimes.com/news/science/environment/la-me-epa29feb29,1,2980474.story?page=1 accessed on August 18, 2008.

Conomy, J. T., Collazo, J. A., Dubovsky, J. A., & Fleming, W. J. (1998a). Dabbling duck behavior and aircraft activity in coastal North Carolina. *Journal of Wildlife Management, 62*(3), 1127–1134.

Conomy, J. T., Dubovsky, J. A., Collazo, J. A., & Fleming, W. J. (1998b). Do black ducks and wood ducks habituate to aircraft disturbance? *Journal of Wildlife Management, 62*(3), 1135–1142.

Copplestone, D., Toal, M. E., Johnson, M. S., Jackson, D., & Jones, S. R. (2000). Environmental effects of radionuclides—observations on natural ecosystems. *Journal of Radiological Protection, 20*, 29–40.

Cranor, C. F. (1990). Some moral issues in risk assessment. *Ethics, 101*, 123–143.

Cranor, C. F. (2001). Learning from the law to address uncertainty in the Precautionary Principle. *Science and Engineering Ethics, 7*(3), 313–326.

Cronkite, E. P., Conard, R. A., & Bond, V. P. (1997). Historical events associated with fallout from Bravo shot—Operation Castle and 25 Y of medical findings. *Health Physics, 73*(1), 176–186.

Crook, M. A., & Langdon, F. J. (1974). The effects of aircraft noise in schools around London airport. *Journal of Sound and Vibration, 34*(2), 221–232.

Crump, K. (1995). Calculation of benchmark doses from continuous data. *Risk Analysis, 15*(1), 79–89.

Crump, K. S., Kjellstrom, T., Shipp, A. M., Silvers, A., & Stewart, A. (1998). Influence of prenatal mercury exposure upon scholastic and psychological test performance: Benchmark analysis of a New Zealand cohort. *Risk Analysis, 18*(6), 701–713.

Crump, K. S., Van Landingham, C., Shamlaye, C., Cox, C., Davidson, P. W., et al. (2000). Benchmark concentrations for methylmercury obtained from the Seychelles Child Development Study. *Environmental Health Perspectives, 108*(3), 257–263.

Crump, K., Viren, J., Silvers, A., Clewell, H., Gearhart, J., & Shipp, A. (1995). Reanalysis of dose-response data from the Iraqi methylmercury poisoning episode. *Risk Analysis, 15*(4), 523–532.

Curl, C. L., Fenske, R. A., & Elgethun, K. (2003). Organophosphorus pesticide exposure of urban and suburban preschool children with organic and conventional diets. *Environmental Health Perspectives, 111*, 377–382.

Curl, C. L., Fenske, R. A., & Elgethun, K. (2003). Organophosphate exposure: Response to Krieger et al. and Charnley [Letter]. *Environmental Health Perspectives, 111*, A689–691.

Cutler, A. R., Wilkerson, A. E., Gingras, J. L., & Levin, E. D. (1996). Prenatal cocaine and/or nicotine exposure in rats: Preliminary findings on long-term cognitive outcome and genital development at birth. *Neurotoxicology and Teratology, 18*(6), 635–643.

Cwikel, J., Abdelgani, A., Goldsmith, J. R., Quastel, M., & Yevelson, I. I. (1997). Two-year follow-up study of stress-related disorders among immigrants to Israel from the Chernobyl area. *Environmental Health Perspectives, 105*(Supplement 6), 1545–1550.

D'Amelio, M., Ricci, I., Sacco, R., Liu, X., et al. (2005). Paraoxonase gene variants are associated with autism in North America, but not in Italy: possible regional specificity in gene-environment interactions. *Molecular Psychiatry, 10*, 1006–1016.

Darvill, T., Lonky, E., Reihman, J., & Daly, H. (1996). Critical issue for research on the neurobehavioral effects of PCBs in humans. *Neurotoxicology and Teratology, 18*, 265–270.

Das, R., Materna, B., Windham, G., Beckman, J., et al. (2006). Worker illness related to ground application of pesticide – Kern County, California, 2005. *Morbidity and Mortality Weekly Review, 55*(17), 486–488.

David, O. J., Hoffman, S., Clark, J., Grad, G., & Sverd, J. (1983). Penicillamine in the treatment of hyperactive children with moderately elevated lead levels. In M. Rutter & R. Russell-Jones (Eds.), *Lead versus health*. Chichester: John Wiley.

Davidson, D. J., & Freudenburg, W. R. (1996). Gender and environmental risk concerns: A review and analysis of available research. *Environment and Behavior, 28*, 302–339.

Davidson, L. M., Baum, A., & Collins, D. L. (1982). Stress and control-related problems at Three Mile Island. *Journal of Applied Social Psychology, 12*(5), 349–359.

Davidson, P. W., Myers, G. J., Cox, C., Axtell, C., et al. (1998). Effects of prenatal and postnatal methylmercury exposure from fish consumption on neurodevelopment: Outcomes at 66 months of age in the Seychelles Child Development Study. *Journal of the American Medical Association, 280*(8), 701–707.

Davidson, P. W., Myers, G. J., Cox, C., Cernichiari, E., Clarkson, T. W., & Shamlaye, C. (1999). In reply [Letter]. *Journal of the American Medical Association, 281*(10), 897.

Davidson, P., Palumbo, D., Myers, G. J., Cox, C., Shamlaye, C. F., et al. (2000). Neurodevelopmental outcomes of Seychellois children from the pilot cohort at 108 months following prenatal exposure to methylmercury from a maternal fish diet. *Environmental Research, Section A, 84*(1), 1–11.

Debes, F., Budtz-Jorgensen, E., Weihe, P., White, R. F., & Grandjean, P. (2006). Impact of prenatal methylmercury exposure on neurobehavioral function at age 14 years. *Neurotoxicology and Teratology, 28*, 536–547.

deBrouwer, C., & Lagasse, R. (2001). The risk linked to ionizing radiation: An alternative epidemiologic approach. *Environmental Health Perspectives, 109*(9), 877–880.

de la Torre, S., Snowdon, C. T., & Bejarano, M. (2000). Effects of human activities on wild pygmy marmosets in Ecuadorian Amazonia. *Biological Conservation, 94*, 153–163.

Denham, M., Schell, L. M., Deane, G., Gallo, M. V., et al. (2005). Relationship of lead, mercury, mirex, dichlorodiphenldichloroethylene, hexachorobenzene, and polychlorinated biphenyls to timing of menarche among Akwesasne Mohawk girls. *Pediatrics, 115*, 127–134.

Dennis, M. (2006). Overview of ADOT's quiet pavement pilot program. http://www.azdot.gov/highways/EEG/EEG_common/documents/files/workshops/brown_bag/quiet_roads_presentation.pdf accessed on August 23, 2008.

Department of Consumer and Employment Protection, Government of Western Australia. (2002). Noise management data sheets: Vacuum cleaner. Available online at www.safetyline.wa.gov.au/pagebin/noisegen10044.htm.

Department of Education. (2000). *Trends in academic progress: Three decades of student performance.* U.S. Department of Education, Office of Educational Research and Improvement, Publication Number 2000–469. Available online at www.ed.gov. Accessed February 2002.

Deschamps, D., Questel, F., Baud, F. J., Gervais, P., & Dally, S. (1994). Persistent asthma after acute inhalation of organophosphate insecticide [Letter]. *Lancet, 344*(8938), 1712.

DeSoto, M. C., & Hitlan, R. T. (2007). Blood levels of mercury are related to diagnosis of autism: A reanalysis of an important data set. *Journal of Child Neurology, 22*(11), 1308–1311.

DeSoto, M. C., & Hitlan, R. T. (2008). Concerning blood mercury levels and autism: A need to clarify. *Journal of Child Neurology, 23*, 463–465.

Dew, M. A., Bromet, E. J., & Schulberg, H. C. (1987). Application of a temporal persistence model to community residents' long-term beliefs about the Three Mile Island nuclear accident. *Journal of Applied Social Psychology, 17*, 1071–1091.

Dew, M. A., & Bromet, E. J. (1993). Predictors of temporal patterns of psychiatric distress during 10 years following the nuclear accident at Three Mile Island. *Social Psychiatry and Psychiatric Epidemiology, 28*, 49–55.

Dewailly, E., Ayotte, P., Bruneau, S., Laliberte, C., et al. (1993). Inuit exposure to organochlorines through the aquatic food chain in arctic Quebec. *Environmental Health Perspectives, 101*, 618–620.

Dewailly, E., Mulvad, G., Pedersen, H. S., Ayotte, P., et al. (1999). Concentration of organochlorines in human brain, liver, and adipose tissue autopsy samples from Greenland. *Environmental Health Perspectives, 107*, 823–828.

Dietrich, K. N., Ris, M. D., Succop, P. A., Berger, O. G., & Bornschein, R. L. (2001). Early exposure to lead and juvenile delinquency. *Neurotoxicology & Teratology, 23*, 511–518.

Dietrich, K. N., Ware, J. H., Salganik, M., Radcliffe, J., et al. (2004). Effect of chelation therapy on the neuropsychological and behavioral development of lead-exposed children after school entry. *Pediatrics, 114*, 19–26.

Dilworth-Bart, J. E., & Moore, C. F. (2006). Mercy mercy me: Social injustice and the prevention of environmental pollutant exposures among ethnic minority and poor children. *Child Development, 77,* 247–265.

Dimsdale, J. E. (2008). Psychological stress and cardiovascular disease. *Journal of the American College of Cardiology, 51,* 1237–1246.

Dohrenwend, B. P., Dohrenwend, B. S., Warheit, G. J., Bartlett, G. S., et al. (1981). Stress in the community: A report to the president's commission on the accident at Three Mile Island. *Annals of the New York Academy of Sciences, 365,* 159–174.

Dolk, H., Vrijheid, M., Armstrong, B., Abramsky, L., et al. (1998). Risk of congenital anomalies near hazardous-waste landfill sites in Europe: The EUROHAZCON study. *Lancet, 352,* 423–427.

Donaldson, L. R., Seymour, A. H., & Nevissi, A. E. (1997). University of Washington's radioecological studies in the Marshall Islands, 1946–1977. *Health Physics, 73*(1), 214–222.

Dox, I. G., Melloni, B. J., & Eisner, G. M. (1993). *The HarperCollins illustrated medical dictionary.* New York: HarperCollins.

Doyle, P., Maconochie, N., & Roman, E. (2001). Author's reply [Letter]. *Lancet, 357,* 556–557.

Doyle, P., Maconochie, N., Roman, E., Davies, G., et al. (2000). Fetal death and congenital malformation in babies born to nuclear industry employees: Report from the nuclear industry family study. *Lancet, 356,* 293–299.

Drinker, C. K., Warren, M. F., & Bennett, G. A. (1937). The problem of possible systemic effects from certain chlorinated hydrocarbons. *Journal of Industrial Hygiene and Toxicology, 19,* 283–311.

Dubrova, Y. E., Nesterev, V. N., Krouchinsky, N. G., Ostapenko, V. A., Neumann, R., et al. (1996). Human minisatellite mutation rate after the Chernobyl accident. *Nature, 380,* 683–686.

Dweck, C. S. (1975). The role of expectations and attributions in the alleviation of learned helplessness. *Journal of Personality and Social Psychology, 31*(4), 674–685.

Dweck, C. S. (2006). *Mindset: The new psychology of success.* New York: Random House.

Dworkin, H. J. (1994). Nuclear weapons testing fallout and thyroid disease [Letter]. *Journal of the American Medical Association, 271*(11), 825–826.

Ehrlich, P. R., Dobkin, D. S., & Wheye, D. (1988). *The birder's handbook: A field guide to the natural history of North American birds.* New York: Simon & Schuster.

Ellegren, H., Lindgren, G., Primmer, C. R., & Meller, A. P. (1997). Fitness loss and germline mutations in barn swallows breeding in Chernobyl [Letter]. *Nature, 389,* 593–596.

Elliott, C. D. (1971). Noise tolerance and extraversion in children. *British Journal of Psychology, 62,* 375–380.

Environmental Protection Agency. (1974). *Information on levels of environmental noise requisite to protect public health and welfare with an adequate margin of safety.* Washington, D.C.: U.S. Government Printing Office. Document 550/9-74-004. Available online at www.nonoise.org/library/levels74/levels74.htm. Accessed January 2002.

Environmental Protection Agency. (1999a). EPA selects cleanup action for contaminated soils and sediments at the General Motors Superfund Site in Massena, New York [Press release]. Accessed online at www.epa.gov/region2/epd/99042.htm on February 13, 2002.

Environmental Protection Agency. (1999b). GM to start removal of contaminated sediments and soils next week at federal Superfund site in Massena, New York [Press release]. Accessed online at www.epa.gov/region2/epd/99091.htm on February 13, 2002.

Environmental Protection Agency. (1999c). *Superfund permanent relocations, October 1999.* Accessed online at www.epa.gov on January 27, 2002.

Environmental Protection Agency. (2000a). *Radiation protection at the EPA: The first 30 years.* EPA 402-B-00–001, August 2000. Accessed online at www.epa.gov in July 2002.

Environmental Protection Agency. (2000b). *Superfund national relocation policy dialogue: International City/County Management Association.* Washington, D.C., March 2–3, 2000. Accessed online at www.epa.gov on January 27, 2002.

Environmental Protection Agency. (2001a). Areas of concern home page. Accessed online at www.epa.gov/grtlakes/aoc/greenbay.html in March 2001.

Environmental Protection Agency. (2001b). *Diazinon revised risk assessment and agreement with registrants.* (7506C). Accessed online at www.epa.gov/pesticides/op/diazinon.htm in August 2001.

Environmental Protection Agency. (2001c). Fox River, Wisconsin, home page (www.epa.gov/region5/foxriver).

Environmental Protection Agency. (2001d). Hudson River home page. Accessed online at www.epa.gov/hudson in March 2001.

Environmental Protection Agency (2001e). Interim reregistration eligibility decision for chlorpyrifos. Accessed online at http://www.epa.gov/oppsrrd1/REDs/chlorpyrifos=ired.pdf in December 2001.

Environmental Protection Agency. (2002). *Love Canal national priority list fact sheet.* (EPA ID# NYD000606947). Accessed online at www.epa.gov on January 25, 2002.

Environmental Protection Agency (2007a). Progress report on site activities, Herculaneum lead smelter site, Herculaneum, Missouri. July 2007 Fact Sheet. http://www.epa.gov/region07/factsheets/2007/fs_comm_advsry_mtg_lead_smelter_herculaneium_mo0707.htm accessed on August 25, 2008.

Environmental Protection Agency (2007b). General Motors (Central Foundry Division) New York, EPA ID#: NYD091972554. 7/20/2007. http://www.epa.gov/region02/superfund/npl/0201644c.pdf accessed on August 25, 2008.

Environmental Protection Agency (2008). America's Children and the Environment, Body burdens. http://www.epa.gov/economics/children/body_burdens/index.htm accessed on August 30, 2008.

Environmental Protection Agency (2008). PCBs in fluorescent light fixtures. http://yosemite.epa.gov/R10/OWCM.NSF/88fa11a23f885ef3882565000062d635/d053fb2a8fcba715882569ed00782e8a!OpenDocument accessed on August 30, 2008.

Environmental Protection Agency (2008). Toxicological review of decabromodiphenyl ether (BDE-209). June 2008. http://nlquery.epa.gov/epasearch/

epasearch?fld=nceawww1%2Cnceaprtp&areaname=National+Center+ for+Environmental+Assessment&areacontacts=http%3A%2F%2Fcfpub. epa.gov%2Fncea%2Fcfm%2Fnceacontact.cfm&areasearchurl=&result_ template=epafiles_default.xsl&filter=samplefilt.hts&typeofsearch=epa&q uerytext=toxicological+review+bde-209&submit=Go accessed on August 18, 2008.

Environmental Protection Agency (2008b). Second five-year review for the Agriculture Street Landfill superfund site. April, 2008. Region 6, EPA. http:// www.epa.gov/region6/6sf/pdffiles/asl_5yr_final_apr08.pdf accessed on August 25, 2008.

Erbe, C., & Farmer, D. M. (2000). Zone of impact around icebreakers affecting beluga whales in the Beaufort Sea. *Journal of the Acoustical Society of America, 108*(3), 1332–1340.

Eriksson, P., & Talts, U. (2000). Neonatal exposure to neurotoxic pesticides increases adult susceptibility: A review of current findings. *NeuroToxicology, 21*(1–2), 37–48.

Ernhart, C. B. (1986). Reply to Needleman's comment. *Journal of Learning Disabilities, 19,* 322–323.

Ernhart, C. B. (1987). Lead levels and child development: Statements by Claire B. Ernhart and Herbert L. Needleman. *Journal of Learning Disabilities, 20,* 262–265.

Ernhart, C. B. (1993a). Declining blood lead levels and cognitive change in children [Letter]. *Journal of the American Medical Association, 270,* 827–828.

Ernhart, C. B. (1993b). Deliberate misrepresentations. *Pediatrics, 91,* 171–172.

Ernhart, C. B. (1994). Letter to the editor. *Archives of Environmental Health, 49,* 77–78.

Ernhart, C. B. (1995a). Cleveland study hypothesis was not confirmed. *British Medical Journal, 310,* 397.

Ernhart, C. B. (1995b). Inconsistencies in the lead-effects literature exist and cannot be explained by "effect modification." *Neurotoxicology and Teratology, 17,* 227–234.

Ernhart, C. B. (1996). Bone lead levels and delinquent behavior [Letter]. *Journal of the American Medical Association, 275,* 1726.

Ernhart, C. B. (1998). Lead effects research [Letter]. *American Journal of Public Health, 88,* 1879.

Ernhart, C. B. (2001). Effects of lead exposure (letter). *Science, 293* (5529), 426–427.

Ernhart, C. B., Landa, B., & Schell, N. B. (1981). Subclinical levels of lead and developmental deficit: A multivariate follow-up reassessment. *Pediatrics, 67,* 911–919.

Ernhart, C. B., Landa, B., & Wolf, A. W. (1985). Subclinical lead level and developmental deficit: Re-analyses of data. *Journal of Learning Disabilities, 18,* 475–479.

Ernhart, C. B., Morrow-Tlucak, M., Wolf, A. W., Super, D., & Drotar, D. (1989). Low level lead exposure in the prenatal and early preschool periods: Intelligence prior to school entry. *Neurotoxicology and Teratology, 11,* 161–170.

Etzel, R. A., et al. (1997). Noise: A hazard for the fetus and newborn: Committee on Environmental Health. *Pediatrics, 100*(4), 724–727.

Evans, G. W., Bullinger, M., & Hygge, S. (1998). Chronic noise exposure and physiological response: A prospective study of children living under environmental stress. *Psychological Science, 9*, 75–77.

Evans, G. W., Hygge, S., & Bullinger, M. (1995). Chronic noise and psychological stress. *Psychological Science, 6*, 333–338.

Evans, G. W., & Lepore, S. J. (1993). Nonauditory effects of noise on children: A critical review. *Children's Environments, 10*(1), 31–51.

Evans, G. W., Lercher, P., Meis, M., Ising, H., & Kofler, W. W. (2001). Community noise exposure and stress in children. *Journal of the Acoustical Society of America, 109*(3), 1023–1027.

Evans, G. W., & Maxwell, L. (1997). Chronic noise exposure and reading deficits: The mediating effects of language acquisition. *Environment and Behavior, 29*(5), 638–656.

Fabrikant, J. I. (1979). *Reports of the public health and safety task force to the President's Commission on the accident at Three Mile Island.* Washington, D.C.: U.S. Government Printing Office.

Fabrikant, J. I. (1981). Health effects of the nuclear accident at Three Mile Island [Guest editorial]. *Health Physics, 40*, 151–161.

Fagan, J. F. (1970). Memory in the infant. *Journal of Experimental Child Psychology, 9*(2), 217–226.

Fagan, J. F., Holland, C. R., & Wheeler, K. (2007). The prediction, from infancy, of adult IQ and achievement. *Intelligence, 35*, 225–231.

Fairlie, I., & Sumner, D. (2000). In defence of collective dose. *Journal of Radiological Protection, 20*, 9–19.

Federal Aviation Administration. (2000). Aviation noise abatement policy 2000. *Federal Register, 65*(136), 43802–43824.

Federal Highway Administration. (2000). *Highway traffic noise in the United States: Problem and response.* U.S. Department of Transportation, April 2000. Accessed online at www.fhwa.dot.gov in December 2001.

Federal Highway Administration (2006). Highway traffic noise in the United States: Problem and response. FHWA-HEP-06–020. http://www.fhwa.dot.gov/environMent/probresp.htm#t9 accessed on August 23, 2008.

Federal Highway Administration (2008). What are the FHWA noise abatement criteria (NAC)? http://www.fhwa.dot.gov/environment/noise/faq_nois.htm#note17 accessed on November 19, 2008.

Federal Interagency Committee on Aviation Noise (2007). Findings of the FICAN pilot study on the relationship between aircraft noise reduction and changes in standardized test scores. http://www.fican.org/pages/findings.html accessed on August 23, 2008.

Fein, G. G., Jacobson, J. L., Jacobson, S. W., Schwartz, P. M., & Dowler, J. K. (1984). Prenatal exposure to polychlorinated biphenyls: Effects on birth size and gestational age. *Journal of Pediatrics, 105*, 315–320.

Fenske, R. A., Kissel, J. C., Lu, C., Kalman, D. A., et al. (2000). Biologically based pesticide dose estimates for children in an agricultural community. *Environmental Health Perspectives, 108*, 515–520.

Fergusson, D. M. (1999). Prenatal smoking and antisocial behavior. *Archives of General Psychiatry, 56*, 223–224.

Fergusson, D. M., Fergusson, J. E., Horwood, L. J., & Kinzett, N. G. (1988). A longitudinal study of dentine lead levels, intelligence, school performance and behaviour: Part II: Dentine lead levels and cognitive ability. *Journal of Child Psychology and Psychiatry, 29,* 793–809.

Fergusson, D. M., Horwood, L. J., & Lynskey, M. T. (1997). Early dentine lead levels and educational outcomes at 18 years. *Journal of Child Psychology and Psychiatry, 38,* 471–478.

Fergusson, D. M., & Woodward, L. J. (1999). Breast feeding and later psychosocial adjustment. *Paediatric & Perinatal Epidemiology, 13,* 144–157.

Fergusson, D. M., Woodward, L. J., & Horwood, J. (1998). Maternal smoking during pregnancy and psychiatric adjustment in late adolescence. *Archives of General Psychiatry, 55,* 721–727.

Ferrer, A., & Cabral, R. (1995). Recent epidemics of poisoning by pesticides. *Toxicology Letters, 82/83,* 55–63.

Fidell, S., Barber, D. S., & Schultz, T. J. (1991). Updating a dosage-effect relationship for the prevalence of annoyance due to general transportation noise. *Journal of the Acoustical Society of America, 89*(1), 221–233.

Field, R. W., Field, E. H., Zegers, D. A., & Steucek, G. L. (1981). Iodine-131 in the thyroids of the meadow vole (*Microtus pennsylvanicus*) in the vicinity of the Three Mile Island nuclear generating plant. *Health Physics, 41,* 297–301.

Field, R. W., Zegers, D. A., Steucek, G. L., & Field, E. A. (1983). Regarding I-131 in meadow vole thyroids. *Health Physics, 44*(2), 177–180.

Filyushkin, I. V. (1996). The Chernobyl accident and the resultant long-term relocation of people. *Health Physics, 71*(1), 4–8.

Finegold, L. S., Harris, C. S., & von Gierke, H. E. (1994). Community annoyance and sleep disturbance: Updated criteria for assessing the impacts of general transportation noise on people. *Noise Control Engineering Journal, 42*(1), 25–30.

Fisk, N. M., Nicolaidis, P. K., Arulkumaran, S., Weg. M. W., Tannirandorn, Y., et al. (1991). Vibroacoustic stimulation is not associated with sudden fetal catecholamine release. *Early Human Development, 25,* 11–17.

Fitzgerald, E. F., Brix, K. A., Deres, D. A., Hwang, S., et al. (2000). Polychlorinated biphenyl (PCB) and dichlorodiphenyl dichloroehtylese (DDE) exposure among Native American men from contaminated Great Lakes fish and wildlife. *Toxicology and Industrial Health, 12,* 361–368.

Fitzgerald, E. F., Hwang, S. A., Gomez, M., Bush, B., Yang, B. Z., & Tarbell, A. (2007). Environmental and occupational exposures and serum PCB concentrations and patterns among Mohawk men at Akwesasne. *Journal of Exposure Science and Environmental Epidemiology, 17,* 269–278.

Flegal, A. R. (1998). Clair Patterson's influence on environmental research. *Environmental Research, Section A, 78,* 65–70.

Flick D. F., O'Dell, R. G., & Childs, V. A. (1965). *Poultry Science, 44,* 1460.

Fluetsch, K. M., & Sparling, D. W. (1994). Avian nesting success and diversity in conventionally and organically managed apple orchards. *Environmental Toxicology and Chemistry, 13,* 1651–1659.

Flynn, C. B. (1988). Reactions of local residents to the accident at Three Mile Island. In D. L. Sills, et al. (Eds.), *Accident at Three Mile Island: The human dimension.* Boulder, CO: Westview Press.

Food and Drug Administration (2004). What you need to know about mercury in fish and shellfish. http://www.cfsan.fda.gov/~dms/admehg3.html accessed on November 19, 2008.

Forbes, H. S., & Forbes, H. B. (1927). Fetal sense reaction: Hearing. *Journal of Comparative Psychology, 7,* 353–355.

Forman, R. T. T., & Deblinger, R. D. (2000). The ecological road-effect zone of a Massachusetts (U.S.A.) suburban highway. *Conservation Biology, 14,* 36–46.

Foulke, J. E. (1995, May). Mercury in fish: Cause for concern? *FDA Consumer.*

Frankel, A. S., & Clark, C. W. (2000). Behavioral responses of humpback whales (*Megaptera novaeanglieae*) to full-scale ATOC signals. *Journal of the Acoustical Society of America, 108*(4), 1930–1937.

Freeman, L. A. (1981). *Nuclear witnesses: Insiders speak out.* New York: Norton.

Freiman, A., Al-Layali, A., & Sasseville, D. (2003). Patch testing with thimerosal in a Canadian center: An 11-year experience. *American Journal of Contact Dermatitis, 14*(3), 138–143.

Freudenburg, W. R. (1988). Perceived risk, real risk: Social science and the art of probabilistic risk assessment. *Science, 242,* 44–49.

Freudenberg, W. R. (1996). Risky thinking: Irrational fears about risk and society. *Annals of the American Academy of Political and Social Sciences, 545,* 44–53.

Fried, P. A., & Makin, J. E. (1987). Neonatal behavioural correlates of prenatal exposure to marijuana, cigarettes, and alcohol in a low risk population. *Neurotoxicology and Teratology, 9,* 1–7.

Fried, P. A., & Watkinson, B. (1988). 12- and 24-month neurobehavioral follow-up of children prenatally exposed to marijuana, cigarettes, and alcohol. *Neurotoxicology and Teratology, 10,* 305–313.

Fried, P. A., Watkinson, B., & Gray, R. (1998). Differential effects on cognitive functioning in 9- to 12-year-olds prenatally exposed to cigarettes and marijuana. *Neurotoxicology and Teratology, 20,* 293–306.

Friedmanna, A. S., Watzinb, M. C., Brinck-Johnsenc, T., & Leitera, J. C. (1996). Low levels of dietary methylmercury inhibit growth and gonadal development in juvenile walleye (*Stizostedion vitreum*). *Aquatic Toxicology, 35*(3–4), 265–278.

Fries, G. F. (1972). Polychlorinated biphenyl residues in milk of environmentally and experimentally contaminated cows. *Environmental Health Perspectives, 1* (Experimental Issue No. 1), 55–59.

Gallagher, C. (1993). *American ground zero: The secret nuclear war.* New York: Random House.

Garcia, B. (1994). Social-psychological dilemmas and coping of atomic veterans. *American Journal of Orthopsychiatry, 64*(4), 651–655.

Garcia, T. (1998). Statement of Tidoro Garcia, New Mexico atomic veteran. *Ionizing radiation, veterans health care, and related issues: Hearing before the Committee on Veterans' Affairs,* 105th Congress, 2nd sess. (April 21, 1998). Washington, D.C.: U.S. Government Printing Office.

GE ad trumpets company's government-ordered environmental cleanup. (2002). *The Onion, 38*(5), 1. Available online at www.theonion.com.

Gears, G. E., LaRoche, G., Cable, J., Jaroslow, B., & Smith, D. (1980). *Investigations of reported plant and animal health effects in the Three Mile Island area.* Document

NUREG-0738, EPA 600/4–80–049. Washington: U.S. Nuclear Regulatory Commission.

Geier, D., & Geier, M. R. (2004). Neurodevelopmental disorders following thimerosal-containing immunizations: A follow-up analysis. *International Journal of Toxicology, 23*(6), 369–376.

Gelbach, S. H., & Williams, W. A. (1975). Pesticide containers: Their contribution to poisoning. *Archives of Environmental Health, 30,* 49–50.

General Accounting Office. (2000, April). *Aviation and the environment: FAA's role in major airport noise programs.* GAO/RCED 00–98. Available online at www.access.gpo.gov/su=docs/aces/aces160.shtml.

General Accounting Office. (2001, September). *Aviation and the environment: Transition to quieter aircraft occurred as planned, but concerns about noise persist.* GAO-01–1053. Available online at www.access.gpo.gov/su=docs/aces/aces160.shtml.

General Electric. (1999). *Annual Report 1999.* Available online at www.ge.com.

General Electric (2007). GE Annual Report 2007. http://www.ge.com/ar2007/pdf/ge_ar2007_full_book.pdf accessed on November 19, 2008.

Gensler, M. D. (1998). Staff study on atomic veterans issues for Senator Paul D. Wellstone by Dr. Martin D. Gensler, legislative fellow. *Ionizing radiation, veterans health care, and related issues: Hearing before the Committee on Veterans' Affairs,* 105th Congress, 2nd sess. (April 21, 1998). Washington, D.C.: U.S. Government Printing Office.

Gergen, K. J. (1985). The social constructivist movement in psychology. *American Psychologist, 40*(3), 266–275.

Gerhardt, K. J., & Abrams, R. M. (2000). Fetal exposures to sound and vibroacoustic stimulation. *Journal of Perinatology, 20,* S20–S29.

Gerstein, E. R. (2002). Manatees, bioacoustics, and boats. *American Scientist, 90*(2), 154–163.

Gerusky, T. M. (1981). Three Mile Island: Assessment of radiation exposures and environmental contamination. *Annals of the New York Academy of Sciences, 365,* 54–62.

Gibbs, L. (1982). *The Love Canal: My story.* Albany, NY: SUNY Press.

Gilbert, E. S., Tarone, R., Bouville, A., & Ron, E. (1998). Thyroid cancer rates and 131-I doses from Nevada atmospheric nuclear bomb tests. *Journal of the National Cancer Institute, 90*(21), 1654–1660.

Ginzburg, H. M., & Reis, E. (1991). Consequences of the nuclear power plant accident at Chernobyl. *Public Health Reports, 106*(1), 32–40.

Gladen, B. C., Rogan, W.J., Hardy, P., Thullen, J., Tingelstad, J. & Tully, M. (1988). Development after exposure to polychlorinated biphenyls and dichlorodiphenyl dichloroethene transplacentally and through human milk. *Journal of Pediatrics, 113,* 991–995.

Glenberg, A. M. (1996). *Learning from data: An introduction to statistical reasoning* (2nd ed.). Mahwah, NJ: Erlbaum.

Glynn, A. W., Wernroth, L., Atuma, S., Linder, C. E., et al. (2000). PCB and chlorinated pesticide concentrations in swine and bovine adipose tissue in Sweden 1991–1997: Spatial and temporal trends. *Science of the Total Environment, 246,* 195–206.

Gochfeld, M., & Burger, J. (2005). Good fish/bad fish: A composite benefit-risk by dose curve. *NeuroToxicology, 26,* 511–520.

Goh, K. T., Yew, F. S., Ong, K. H., & Tan, I. K. (1990). Acute organophosphorus food poisoning caused by contaminated green leafy vegetables. *Archives of Environmental Health, 45,* 180–184.

Goldman, I. R., Paigen, B., Magnant, M. M., & Highland, J. (1985). Low birth weight, prematurity and birth defects in children living near the hazardous waste site, Love Canal. *Hazardous Waste and Hazardous Materials, 2*(2), 209–223.

Goldman, L. R., Beller, M., & Jackson, R. J. (1990). Aldicarb food poisonings in California, 1985–1988: Toxicity estimates for humans. *Archives of Environmental Health, 45,* 141–147.

Goldman, M. (1997). The Russian radiation legacy: Its integrated impact and lessons. *Environmental Health Perspectives, 105*(Supplement 6), 1385–1391.

Goldsmith, J. R., Grossman, C. M., Morton, W. E., Nussbaum, R. H., Kordysh, E. A., Quastel, M. R., Sobel, R. B., & Nussbaum, F. D. (1999). Juvenile hypothyroidism among two populations exposed to radioiodine. *Environmental Health Perspectives, 107*(4), 303–308.

Goldsteen, R., Schorr, J. K., & Goldsteen, K. S. (1989). Longitudinal study of appraisal at Three Mile Island: Implications for life event research. *Social Science and Medicine, 28*(4), 389–398.

Goldstein, B. D. (2001). The precautionary principle also applies to public health actions. *American Journal of Public Health, 91*(9), 1358–1361.

Goodwill, M. H., Lawrence, D. A., & Seegal, R. F. (2007). Polychlorinated biphenyls induce proinflammatory cytokine release and dopaminergic dysfunction: protection in interleukin-6 knockout mice. *Journal of Neuroimmunology, 183*(1–2), 125–132.

Gottfried, A. W., Wallace, P., Sherman-Brown, S., King, J., & Coen, C. (1981). Physical and social environment of newborn infants in special care. *Science, 214*(4521), 673–675.

Gould, S. J. (1981). *The mismeasure of man.* New York: Norton.

Government Accountability Office (2007). Aviation and the environment: Impact of aviation noise on communities presents challenges for airport operations and future growth of the national airspace system. GAO-08–216T. http://www.gao.gov/new.items/d08216t.pdf accessed on August 23, 2008.

Grandjean, P., Budtz-Jorgensen, E., White, R. F., Jorgensen, P. J., et al. (1999a). Methylmercury exposure biomarkers as indicators of neurotoxicity in children aged 7 years. *American Journal of Epidemiology, 150*(3), 301–305.

Grandjean, P., Guldager, B., Larsen, I. B., Jorgensen, P. J., & Holmstrup, P. (1997a). Placebo response in environmental disease: Chelation therapy of patients with symptoms attributed to amalgam fillings. *Journal of Occupational and Environmental Medicine, 39,* 707–714.

Grandjean, P., Weihe, P., White, R. F., et al. (1997b). Cognitive deficit in 7-year-old children with prenatal exposure to methylmercury. *Neurotoxicology and Teratology, 20,* 417–428.

Grandjean, P., Weihe, P., White, R. F., & Debes, F. (1998). Cognitive performance of children prenatally exposed to "safe" levels of methylmercury. *Environmental Research, Section A, 77,* 165–172.

Grandjean, P., & White, R. F. (1999). Effects of methylmercury exposure on neurode-velopment [Letter]. *Journal of the American Medical Association, 281*(10), 896.

Grandjean, P., White, R. F., Nielsen, A., Cleary, D., & de Oliveira Santos, E. C. (1999b). Methylmercury neurotoxicity in Amazonian children downstream from gold mining. *Environmental Health Perspectives, 107,* 587–591.

Grandjean, P., White, R. F., Sullivan, K., Debes, F., Murata, K., et al. (2001). Impact of contrast sensitivity performance on visually presented neurobe-havioral tests in mercury-exposed children. *Neurotoxicology and Teratology, 23,* 141–146.

Green, G. (1999). *The woman who knew too much: Alice Stewart and the secrets of radi-ation.* Ann Arbor: University of Michigan Press.

Green, K. B., Pasternack, B. S., & Shore, R. E. (1982). Effects of aircraft noise on reading ability of school-age children. *Archives of Environmental Health, 37,* 24–31.

Green, R. S., Smorodinsky, S., Kim, J. J., McLaughlin, R., & Ostro, B. (2004). Proximity of California public schools to busy roads. *Environmental Health Perspectives, 112,* 61–66.

Grossman, J. (1995). Focus. What's hiding under the sink: Dangers of household pesticides. *Environmental Health Perspectives, 103,* 550–554.

Guillebaud, J. (2008). Editorials: Population growth and climate change. *BMJ, 337,* a576.

Guillette, E. A., Meza, M. M., Aquilar, M. G., Soto, A. D., & Garcia, I. E. (1998). An anthropological approach to the evaluation of preschool children exposed to pesticides in Mexico. *Environmental Health Perspectives, 106*(6), 347–353.

Gupta, A. (1991). *Ecological nightmares and the management dilemma: The case of Bhopal.* Delhi, India: Ajanta Publications.

Haines, M. M., Stansfeld, S. A., Job, R. F. S., Berglund, B., & Head, J. (2001). Chronic aircraft noise exposure, stress responses, mental health and cognitive perfor-mance in school children. *Psychological Medicine, 31,* 265–277.

Halbach, S. (1995). Estimation of mercury dose by a novel quantitation of elemen-tal and inorganic species released from amalgam. *International Archives of Occupational and Environmental Health, 67,* 295–300.

Hall, F. L., Birnie, S., Tayler, S. M., & Palmer, J. (1981). Direct comparison of com-munity response to road traffic noise and to aircraft noise. *Journal of the Acoustical Society of America, 70,* 1690–1698.

Hammond, K. R. (1996). *Human judgment and social policy: Irreducible uncertainty, inevitable error, unavoidable injustice.* New York: Oxford University Press.

Hansen, D., & Schreiner, C. (2005). Unanswered questions: The legacy of atomic veterans. *Health Physics, 89*(2), 155–163.

Harada, M. (1995). Minamata disease: Methylmercury poisoning in Japan caused by environmental pollution. *Critical Reviews in Toxicology, 25,* 1–24.

Hardin, G. (1968). The tragedy of the commons. *Science, 162,* 1243–1248.

Hardison, J. D. (1990). *The megaton blasters: Story of the 4925th test group (atomic).* Arvada, CO: Boomerang Publishers.

Hatch, M., Susser, M., & Beyea, J. (1997). Comments on "A reevaluation of can-cer incidence near the Three Mile Island nuclear plant." *Environmental Health Perspectives, 105*(1), 12.

Hatch, M. C., Wallenstein, S., Beyea, J., Nieves, J. W., & Susser, M. (1991). Cancer rates after the Three Mile Island nuclear accident and proximity of residence to the plant. *American Journal of Public Health, 81*(6), 719–724.

Havenaar, J. M., Rumyantzeva, G., Kasyanenko, A. P., Kaasjager, K., et al. (1997). Health effects of the Chernobyl disaster: Illness or illness behavior? A comparative general health survey in two former Soviet regions. *Environmental Health Perspectives, 105*(Supplement 6), 1533–1537.

Havenaar, J. M., Van den Brink, W., Van den Bout, J., Kasyanenko, A. P., et al. (1996). Mental health problems in the Gomel region (Belarus): An analysis of risk factors in an area affected by the Chernobyl disaster. *Psychological Medicine, 26,* 845–855.

He, F. (2000). Neurotoxic effects of insecticides—Current and future research: A review. *NeuroToxicology, 21,* 829–836.

Health Council of the Netherlands: Committee on the Health Impact of Large Airports. (1999). *Public health impact of large airports.* The Hague: Health Council of the Netherlands.

Healy, J. K. (1959). Ascending paralysis following malathion intoxication: A case report. *Medical Journal of Australia, 1,* 765–767.

Heath, C. W., Nadel, M. R., Zack, M. M., Chen, A. T. L., et al. (1984). Cytogenetic findings in persons living near the Love Canal. *Journal of the American Medical Association, 251*(11), 1437–1440.

Helland, I. B., Smith, L., Saarem, K., Saugstad, O. D., & Drevon, C. A. (2003). Maternal supplementation with very-long-chain n-3 fatty acids during pregnancy and lactation augments children's IQ at 4 years of age. *Pediatrics,* 2003, *111,* e39–e44.

Hickie, B. E., Ross, P. S., MacDonald, R. W., & Ford, J. K. B. (2007). Killer whales (Orcinus orca) face protracted health risks associated with lifetime exposure to PCBs. *Environmental Science and Technology, 41,* 6613–6619.

Higuchi, K. (Ed.). (1976). *PCB poisoning and pollution.* Tokyo: Kodansha Ltd.

Hohenemser, C., Kasperson, R., & Kates, R. (1977). The distrust of nuclear power. *Science, 196,* 25–34.

Holm, L-E. (2000). Chernobyl effects [Letter]. *Lancet, 356,* 344.

Holmes, D. C., Simmons, J. H., & Tatton, J. O. G. (1967). Chlorinated hydrocarbons in British wildlife. *Nature, 216,* 227–229.

Horn, E. G., Hetling, L. J., & Tofflemire, J. (1979). The problem of PCBs in the Hudson River system. *Annals of the New York Academy of Sciences, 320,* 591–609.

Hornig, M., Chian, D., & Lipkim, W. I. (2004). Neurotoxic effects of postnatal thimerosal are mouse strain dependent. *Molecular Psychiatry, 9,* 833–845.

Houts, P. S., Cleary, P. D., & Hu, T-W. (1988). The Three Mile Island crisis: Psychological, social, and economic impacts on the surrounding population. *Pennsylvania State University Studies,* no. 49.

Howard, G. S. (1985). The role of values in the science of psychology. *American Psychologist, 40*(3), 255–265.

Hsu, S-T., Ma, C-I., Hsu, S. K-H., Wu, S-S., et al. (1984). Discovery and epidemiology of PCB poisoning in Taiwan. *American Journal of Industrial Medicine, 5,* 71–79.

Hubbs-Tait, L., Nation, J. R., Krebs, N. F., & Bellinger, D. C. (2005). Neurotoxicants, micronutrients and social environments: Individual and combined effects on children's development. *Psychological Science in the Public Interestt*, 6(3), 57–121.

HudsonVoice. (2001). *GE's Analysis on Health*. Accessed online at www.hudson-voice.com/auxiliary/about=pcbs/health1.html on May 2, 2001.

Huisman, M., Esseboom-Koopman, C., Lanting, C. I., van der Paauw, C. G., et al. (1995a). Neurological condition in 18-month-old children perinatally exposed to polychlorinated biphenyls and dioxins. *Early Human Development*, 43, 165–176.

Huisman, M., Koopman-Esseboom, K., Fidler, V., Hadders-Algra, M., et al. (1995b). Perinatal exposure to polychlorinated biphenyls and dioxins and its effect on neonatal neurological development. *Early Human Development*, 41, 111–127.

Hull, C. L. (1999). When something is to be done: Proof of environmental harm and the philosophical tradition. *Environmental Values*, 8, 3–25.

Hviid, A., Stellfeld, M. Wohlfahrt, J., & Melbye, M. (2003). Association between Thimerosal–containing vaccine and autism. *Journal of the American Medical Association*, 290(13), 1763–1766.

Hwang, S. A., Yang, B. Z., Fitzgerald, E. F., Bush, B., & Cook, K. (2001). Fingerprinting PCB patterns among Mohawk women. *Journal of Exposure Analysis and Environmental Epidemiology*, 11, 184–192.

Hygge, S., Evans, G. W., & Bullinger, M. (2002). A prospective study of some effects of aircraft noise on cognitive performance in schoolchildren. *Psychological Science*, 13(5), 469–474.

Institute for Environment and Health. (1997). *The non-auditory effects of noise*. Leicester, UK: Institute for Environment and Health.

Institute of Medicine (2001). *Immunization safety review: Thimerosal-containing vaccines and neurodevelopmental disorders*. Washington, D.C.: National Academies Press.

Institute of Medicine (2003). *Dioxins and dioxin-like compounds in the food supply: Strategies to decrease exposure*. Washington D.C.: National Academies Press.

Institute of Medicine (2004). *Immunization safety review: vaccines and autism*. Washington, D.C.: National Academies Press.

International Atomic Energy Agency. (1998, March). *Radiological conditions at Bikini Atoll: Prospects for resettlement*. Available online at sales.publications@ iaea.org.

Ip, P., Wong, V., Ho, M., Lee, J., & Wong, W. (2004). Mercury exposure in children with autistic spectrum disorder: Case-control study. *Journal of Child Neurology*, 19, 431–434.

Iqbal, S., Muntner, P., Batuman, V., & Rabito, F. A. (2008). Estimated burden of blood lead levels > 5 μg/dl in 1999–2002 and declines from 1988 to 1994. *Environmental Research*, 107, 305–311.

Jacobs, D. E. (1999). Jacobs re Needleman. *American Journal of Public Health*, 89, 1127–1128.

Jacobson, J. L., Humphrey, H. E. B., Jacobson, S. W., Schantz, S. L., et al. (1989). Determinants of polychlorinated biphenyls (PCBs), polybrominated biphenyls (PBBs), and dichlorodiphenyl trichloroethane (DDT) levels in the sera of young children. *American Journal of Public Health*, 79, 1401–1404.

Jacobson, J. L., & Jacobson, S. W. (1996a). Intellectual impairment in children exposed to polychlorinated biphenyls in utero. *New England Journal of Medicine, 335*, 783–789.

Jacobson, J. L., & Jacobson, S. W. (1996b). Sources and implications of interstudy and interindividual variability in the developmental neurotoxicity of PCBs. *Neurotoxicology and Teratology, 18*(3), 257–264.

Jacobson, J. L., & Jacobson, S. W. (2001). Commentary: Postnatal exposure to PCBs and child development. *Lancet, 358*(9293), 1568–1569.

Jacobson, J. L., Jacobson, S. W., Fein, G. G., Schwarts, P. M., & Dowler, J. K. (1984). Prenatal exposure to an environmental toxin: A test of the multiple effects model. *Developmental Psychology, 20*, 523–532.

Jacobson, J. L., Jacobson, S. W., & Humphrey, H. E. B. (1990a). Effects of exposure to PCBs and related compounds on growth and activity in children. *Neurotoxicology and Teratology, 12*, 319–326.

Jacobson, J. L., Jacobson, S. W., & Humphrey, H. E. B. (1990b). Effects of in utero exposure to polychlorinated biphenyls and related contaminants on cognitive functioning in young children. *Journal of Pediatrics, 116*, 38–45.

Jacobson, J. L., Jacobson, S. W., Padgett, R. J., Brumitt, G. A., & Billings, R. L. (1992). Effects of prenatal PCB exposure on cognitive processing efficiency and sustained attention. *Developmental Psychology, 28*, 297–306.

Jacobson, S. W., Fein, G. G., Jacobson, J. L., Schwartz, P. M., & Dowler, J. K. (1985). The effect of intrauterine PCB exposure on visual recognition memory. *Child Development, 56*, 853–860.

Jacobson, S. W., & Jacobson, J. L. (2000). Teratogenic insult and neurobehavioral function in infancy and childhood. In C. A. Nelson (Ed.), *The effects of early adversity on neurobehavioral development: The Minnesota symposium on child psychology*, Vol. 31. Mahwah, NJ: Erlbaum.

James, R. C., Busch, H., Tamburro, C. H., Roberts, S. M., et al. (1993). Polychlorinated biphenyl exposure and human disease. *Journal of Occupational Medicine, 35*, 136–148.

Janerich, D. T., Burnett, W. S., Feck, G., Hoff, M., et al. (1981). Cancer incidence in the Love Canal area. *Science, 212*, 1404–1407.

Jank, B., & Rath, J. (2000). The precautionary principle [Letter]. *Nature Biotechnology, 18*, 697.

Jensen, A. (1985). The nature of the black-white difference on various psychometric tests: Spearman's hypothesis. *Behavioral and Brain Sciences, 8*, 193–263.

Jensen, R. G., Ferris, A. M., & Lammi-eefe, C. J. (1992). Lipids in human milk and infant formulas. *Annual Review of Nutrition, 12*, 417–441.

Jensen, S. (1966). Report of a new chemical hazard. *New Scientist, 32*, 612.

Jensen, S., Johnels, A. G., Olsson, M., & Otterlind, G. (1969). DDT and PCB in marine animals from Swedish waters. *Nature, 224*, 247–250.

Job, R. F. S. (1999). Noise sensitivity as a factor influencing human reaction to noise. *Noise & Health, 3*, 57–68.

Johnson, B. L., Hicks, H. E., & DeRosa, C. T. (1999). Introduction: Key environmental human health issues in the Great Lakes and St. Lawrence River basins. *Environmental Research, Section A, 80*, S2–S12.

Johnson, J. C., Thaul, S., Page, W. F., & Crawford, H. (1997). Mortality of veteran participants in the Crossroads nuclear test. *Health Physics, 73*(1), 187–189.

Jones, D., & Broadbent, D. (1979). Side-effects of interference with speech by noise. *Ergonomics, 22*, 1073–1081.

Kalantzi, O. I., Alcock, R. E., Johnston, P. A., Snatillo, D., et al. (2001). The global distribution of PCBs and organochlorine pesticides in butter. *Environmental Science & Technology, 35*, 1013–1018.

Kaplan, A. (1964). *The conduct of inquiry.* San Francisco: Chandler.

Kaplan, J. G., Kessler, J., Rosenberg, N., Pack, D., & Schaumburg, H. H. (1993). Sensory neuropathy associated with Dursban (chlorpyrifos) exposure. *Neurology, 43*, 2193–2196.

Kashimoto, T., Miyata, H., Kunita, S., Tung, T-C., et al. (1981). The role of poly-chlorinated dibenzofuran in Yusho (PCB poisoning). *Archives of Environmental Health, 36*, 321–326.

Kennedy, R. F.Jr. (2004). *Crimes against nature.* New York: Harper Collins.

Kerber, R. A., Till, J. E., Simon, S. L., Lyon, J. L., et al. (1993). A cohort study of thy-roid disease in relation to fallout from nuclear weapons testing. *Journal of the American Medical Association, 270*(17), 2076–2082.

Khurana, D., & Prabhakar, S. (2000). Organophosphorus intoxication. *Archives of Neurology, 57*, 600–602.

Kiecolt-Glaser, J. K., McGuire, L., Robles, T. F., & Glaser, R. (2002). Emotions, morbidity, and mortality: New perspectives from psychoneuroimmunology. *Annual Review of Psychology, 53*, 83–107.

Kilburn, K. H. (2000). Visual and neurobehavioral impairment associated with polychlorinated biphenyls. *NeuroToxicology, 21*, 489–500.

Kirk, W. P. (1983). I-131 in meadow voles near the Three Mile Island nuclear gen-erating plant [Letter]. *Health Physics, 44*(2), 175–177.

Kisilevsky, B. S. (1995). The influence of stimulus and subject variables on human fetal responses to sound and vibration. In J-P. Lecanuet, W. P. Fifer, N. A. Krasnegor, & W. P. Smotherman (Eds.), *Fetal development: A psychobiological perspective.* Hillsdale, NJ: Erlbaum.

Kiste, R. C. (1974). *The Bikinians: A study in forced migration.* Menlo Park, CA: Cummings Publishing Company.

Kleinke, C. L. (1991). *Coping with life challenges.* Belmont, CA: Brooks/Cole.

Knipschild, P., Meijer, H., & Salle, H. (1981). Aircraft noise and birth weight. *International Archives of Occupational and Environmental Health, 48*, 131–136.

Knobeloch, L. (2005). Assessing MeHg exposure in Wisconsin. Lecture given at the University of Wisconsin-Madison.

Ko, N. W. M. (1979). Responses of teachers to aircraft noise. *Journal of Sound and Vibration, 62*(2), 277–292.

Koblinger, L. (2000). Can we put aside the LNT dilemma by the introduction of the controllable dose? *Journal of Radiological Protection, 20*, 5–8.

Kolbert, E. (2000). The river: Will the EPA finally make GE clean up its PCBs? *New Yorker, 76*(37), 56–62.

Koletzko, B., Lien, E., Agostoni, C., Bohles, H., et al. (2008). The roles of long-chain polyunsaturated fatty acids in pregnancy, lactation and infancy: Review of current knowledge and consensus recommendations. *Journal of Perinatal Medicine, 36*(1), 5–14.

Kolominsky, Y., Igumnov, S., & Drozdovitch, V. (1999). The psychological development of children from Belarus exposed in the prenatal period to radiation from the Chernobyl atomic power plant. *Journal of Child Psychology and Psychiatry, 40*(2), 299–305.

Koopman-Esseboom, C., Weisglas-Kuperus, N., DeRidder, M. A., Van der Paauw, C. G., et al. (1996). Effects of polychlorinated biphenyl/dioxin exposure and feeding type on infants' mental and psychomotor development. *Pediatrics, 97,* 700–706.

Koppe, J. G. (1995). Nutrition and breast feeding. *European Journal of Obstetrics & Gynecology, 61,* 73–78.

Koren, G. (1999). The association between maternal cigarette smoking and psychiatric diseases or criminal outcome in the offspring: A precautionary note about the assumption of causation. *Reproductive Toxicology, 13,* 345–346.

Kortenkamp, K. V. & Moore, C. F. (2001). Ecocentrism and anthropocentrism: Moral reasoning about ecological commons dilemmas. *Journal of Environmental Psychology, 21,* 262–272.

Kortenkamp, K. V. & Moore, C. F. (2006). Time, uncertainty, and individual differences in decisions to cooperate in resource dilemmas. *Personality and Social Psychology Bulletin, 32,* 603–615.

Kovalcuk, O., Dubrova, Y. E., Arkhipov, A., Hohn, B., & Kovalchuk, I. (2000). Wheat mutation rate after Chernobyl. *Nature, 407,* 583–584.

Krausman, P. R., Wallace, M. C., Hayes, C. L., & DeYoung, D. W. (1998). Effects of jet aircraft on mountain sheep. *Journal of Wildlife Management, 62*(4), 1246–1254.

Kriebel, D., & Tickner, J. (2001). Reenergizing public health through precaution. *American Journal of Public Health, 91*(9), 1351–1355.

Krieger, R. I., Dinoff, T. M., Williams, R. L., Zhang, X., et al. (2003). Preformed biomarkers in produce inflate human organophosphate exposure assessments. *Environmental Health Perspectives, 111,* A688.

Kryter, K. D. (1982a). Community annoyance from aircraft and ground vehicle noise. *Journal of the Acoustical Society of America, 72*(4), 1222–1242.

Kryter, K. D. (1982b). Rebuttal by Karl D. Kryter to comments by T. J. Schultz. *Journal of the Acoustical Society of America, 72*(4), 1253–1257.

Kryter, K. D. (1983). Response of K. D. Kryter to modified comments by T. J. Schultz on K. D. Kryter's paper, "Community annoyance from aircraft and ground vehicle noise" [J. Acoust. Soc. Am. *72,* 1243–1252 (1982)]. *Journal of the Acoustical Society of America, 73*(3), 1066–1068.

Kryter, K. D. (1990). Aircraft noise and social factors in psychiatric hospital admission rates: A re-examination of some data. *Psychological Medicine, 20,* 395–411.

Kryter, K. D. (1994). *The handbook of hearing and the effects of noise.* San Diego: Academic Press.

Kuhn, T. S. (1970). *The structure of scientific revolutions* (2nd ed., enl.). Chicago: University of Chicago.

Kuratsune, M. (1972). An abstract of results of laboratory examinations of patients with Yusho and of animal experiments. *Environmental Health Perspectives, 1* (Experimental Issue No. 1), 129–136.

Kuratsune, M., & Masuda, Y. (1972). Polychlorinated biphenyls in non-carbon copy paper. *Environmental Health Perspectives, 1* (Experimental Issue No. 1), 61–62.

Lalande, N. M., Hetu, R., & Lambert, J. (1986). Is occupational noise exposure during pregnancy a risk factor of damage to the auditory system of the fetus? *American Journal of Industrial Medicine, 10,* 427–435.

Landi, M. T., Needham, L. L., Lucier, G., Mocarelli, P., et al. (1997). Concentration of dioxin 20 years after Seveso [Research letter]. *Lancet, 349,* 1811.

Landrigan, P. J., Claudio, L., Markowitz, S. B., Berkowitz, G. S., et al. (1999). Pesticides and inner-city children: Exposures, risks, and prevention. *Environmental Health Perspectives, 107*(Supplement 3), 431–437.

Landrigan, P. J., Schechter, C. B., Lipton, J. M., Fahs, M. C., & Schwartz, J. (2002). Environmental pollutants and disease in American children: Estimates of morbidity, mortality, and costs for lead poisoning, asthma, cancer, and developmental disabilities. *Environmental Health Perspectives, 110*(7), 721–728.

Lane, S. R., & Meecham, W. C. (1974). Jet noise at schools near Los Angeles International Airport. *Journal of the Acoustical Society of America, 56*(1), 127–131.

Lanphear, B. P., Burgoon, D. A., Rust, S. W., Eberly, S., & Galke, W. (1998). Environmental exposures to lead and urban children's blood lead levels. *Environmental Research, Section A, 76,* 120–130.

Lanphear, B. P., Hornung, R., Khoury, J., Yolton, K., Baghurst, P. Bellinger, D. C., Canfield, R. L., Dietrich, K. N., Bornschein, R. Greene, T., Rothernberg, S. J., Needleman, H. L., Schnaas, L., Wasserman, G., Graziano, J., & Roberts, R. (2005). Low-level environmental lead exposure and children's intellectual function: An international pooled analysis. *Environmental Health Perspectives, 113*(7), 894–899.

Lanphear, B. P., & Roghmann, K. J. (1997). Pathways of lead exposure in urban children. *Environmental Research, 74,* 67–73.

Lanting, C. I., Patandin, S., Fidler, V., Weisglas-Kuperus, N., et al. (1998). Neurological condition in 42-month-old children in relation to pre- and postnatal exposure to polychlorinated biphenyls and dioxins. *Early Human Development, 50,* 283–292.

Lapp, R. E. (1994). Nuclear weapons testing fallout and thyroid disease [Letter]. *Journal of the American Medical Association, 271*(11), 826.

Lauder, J. M., & Schambra, U. B. (1999). Morphogenetic roles of acetylcholine. *Environmental Health Perspectives, 107*(Supplement 1), 65–69.

Lazarus, R. S., & Folkman, S. (1984). *Stress, appraisal, and coping.* New York: Springer.

Lead shot ban saves waterfowl. (2001). *National Wildlife Magazine, 39*(2), 64.

Lebel, J., Mergler, D., Lucotte, M., Amorim, M., et al. (1996). Evidence of early nervous system dysfunction in Amazonian populations exposed to low levels of methylmercury. *NeuroToxicology, 17,* 157–168.

Lee, S. E., & Khew, S. K. (1992). Impact of road traffic and other sources of noise on the school environment. *Indoor Environ, 1*, 162–169.

Levin, E. D., Briggs, S. J., Christopher, N. C., & Rose, J. E. (1993). Prenatal nicotine exposure and cognitive performance in rats. *Neurotoxicology and Teratology, 15*(4), 251–260.

Levin, R. J. (2008). Incidence of thyroid cancer in residents surrounding the Three Mile Island nuclear facility. *The Laryngoscope, 118*, 618–628.

Levine, A. G. (1982). *Love Canal: Science, politics, and people*. Lexington, MA: Lexington Books.

Liem, A. K., Furst, P., & Rappe, C. (2000). Exposure of populations to dioxins and related compounds. *Food Additives & Contaminants, 17*, 241–59.

Lifshitz, M., Rotenberg, M., Sofer, S., Tamiri, T., et al. (1994). Carbamate poisoning and oxime treatment in children: A clinical and laboratory study. *Pediatrics, 93*, 652–655.

Lifshitz, M., Shahak, E., & Sofer, S. (1999). Carbamate and organophosphate poisoning in young children. *Pediatric Emergency Care, 15*(2), 102–103.

Lillienthal, H., & Winneke, G. (1991). Sensitive periods for behavioral toxicity of polychlorinated biphenyls: Determination by cross-fostering in rats. *Fundamental and Applied Toxicology, 17*, 368–375.

Lippmann, M. (1990). Lead and human health: Background and recent findings [The 1989 Alice Hamilton Lecture given to the National Institute for Occupational Safety and Health]. *Environmental Research, 51*, 1–24.

Litcher, L., Bromet, E. J., Carlson, G., Squires, N., Goldgaber, D., Panina, N., Golovakha, E., & Gluzman, S. (2000). School and neuropsychological performance of evacuated children in Kyiv 11 years after the Chornobyl disaster. *Journal of Child Psychology and Psychiatry, 41*(3), 291–299.

Liu, X., Dietrich, K. M. Radcliffe, J., Ragan, N. B., Rhoads, G. G., & Rogan, W. J. (2002). Do children with falling blood lead levels have improved cognition? *Pediatrics, 110*, 787–791.

Lobel, M. (1994). Conceptualizations, measurement, and effects of prenatal maternal stress on birth outcomes. *Journal of Behavioral Medicine, 17*, 225–272.

Lockwood, J. A., Wangberg, J. K., Ferrell, M. A., & Hollon, J. D. (1994). Pesticide labels: Proven protection or superficial safety? *Journal of the American Optometric Association, 65*(1), 18–26.

Loew, P. (2001). *Indian nations of Wisconsin: Histories of endurance and renewal*. Madison, WI: Wisconsin Historical Society Press.

Lomat, L., Galburt, G., Quastel, M. R., Polyakov, S., et al. (1997). Incidence of childhood disease in Belarus associated with the Chernobyl accident. *Environmental Health Perspectives, 105*(Supplement 6), 1529–1532.

Longnecker, M. P., Ryan, J. J., Gladen, B. C., & Schecter, A. J. (2000). Correlations among human plasma levels of dioxin-like compounds and polychlorinated biphenyls (PCBs) and implications for epidemiologic studies. *Archives of Environmental Health, 55*(3), 195–200.

Lorber, M. (2008). Exposure of Americans to polybrominated diphenyl ethers. *Journal of Exposure Science and Environmental Epidemiology, 18*, 2–19.

Lovins, A. B., & Sheikh, I. (2008). The nuclear illusion. *Ambio* in press.

Lu, C., Fenske, R. A., Simcox, N. J., & Kalman, D. (2000). Pesticide exposure of children in an agricultural community: Evidence of household proximity to farmland and take home exposure pathways. *Environmental Research, Section A, 84,* 290–302.

Lu, C., Toepel, K., Irish, R., Fenske, R. A., Barr, D. B., & Bravo, R. (2006). Organic diets significantly lower children's dietary exposure to organophosphorus pesticides. *Environmental Health Perspectives, 114,* 260–263.

Lu, C., Toepel, K., Irish, R., Fenske, R. A., Barr, D. B., & Bravo, R. (2006). Organic diets: Lu et al. Respond. *Environmental Health Perspectives, 114,* A211.

Lucas, A., Morley, R., Cole, T. J., Lister, G., & Leeson-Payne, C. (1992). Breast milk and subsequent intelligence quotient in children born preterm. *Lancet, 339,* 261–264.

Madsen, P. T., & Mohl, B. (2000). Sperm whales (*Physeter catodon* L. 1758) do not react to sounds from detonators. *Journal of the Acoustical Society of America, 107*(1), 668–671.

Maier, J. A. K., Murphy, S. M., White, R. G., & Smith, M. D. (1998). Responses of caribou to overflights by low-altitude jet aircraft. *Journal of Wildlife Management, 62*(2), 752–766.

Manchester-Neesvig, J. B., Valters, K., & Sonzogni, W. C. (2001). Comparison of polybrominated diphenyl ethers (PBDEs) and polychlorinated biphenyls (PCBs) in Lake Michigan salmonids. *Environmental Science & Technology, 35,* 1072–1077.

Mangano, J. J., Sternglass, E. J., Gould, J. M., Sherman, J. D., et al. (2000). Strontium-90 in newborns and childhood disease. *Archives of Environmental Health, 55*(4), 240–244.

Markowitz, G., & Rosner, D. (2000). "Cater to the children": The role of the lead industry in a public health tragedy, 1900–1955. *American Journal of Public Health, 90,* 36–46.

Marples, D. R. (1988). *The social impact of the Chernobyl disaster.* New York: St. Martin's Press.

Marples, D. R. (1996). The decade of despair. *Bulletin of the Atomic Scientists, 52*(3), 22[CF3]–[CF1]31.

Marples, D. R. (1997). The legacy of Chernobyl in 1997: Impact on Ukraine and Belarus. *Post-Soviet Geography and Economics, 38*(3), 163–170.

Marsh, D. O., Clarkson, T. W., Cox, C., Myers, G. J., et al. (1987). Fetal methylmercury poisoning: Relationship between concentration in single strands of maternal hair and child effects. *Archives of Neurology, 44,* 1017–1022.

Marshall, E. (1983). EPA faults classic lead poisoning study. *Science, 222,* 906–907.

Martell, E. A. (1964). Iodine-131 fallout from underground tests. *Science, 143,* 126–129.

Marten K., Herzing D., Poole, M., & Newman, A. K. (2001). The acoustic predation hypothesis: Linking underwater observations and recordings during odontocete predation and observing the effects of loud impulsive sounds on fish. *Aquatic Mammals, 27*(1), 56–66.

Marx, J. L. (1979). Low-level radiation: Just how bad is it? [News]. *Science, 204,* 160–164.

Masuda, Y. (1985). Health status of Japanese and Taiwanese after exposure to contaminated rice oil. *Environmental Health Perspectives, 60,* 321–325.

Maugh, T.H. (2007) John Gofman, 88; Physicist warmed about radiation risks. *Los Angeles Times,* August 28, 2007 Obituaries. http://articles.latimes.com/2007/aug/28/local/me-gofman28

Mazur, A. (1998). *A hazardous inquiry: The Rashomon effect at Love Canal.* Cambridge, MA: Harvard University Press.

McCarthy, M. (1997). Nuclear bomb test fallout may cause many U.S. cancers [News]. *Lancet, 350,* 415.

McConnell, R., & Hruska, A. J. (1993). An epidemic of pesticide poisoning in Nicaragua: Implications for prevention in developing countries. *American Journal of Public Health, 83*(11), 1559–1562.

McCombie, B. (2001, February 2). Poison in our lakes. *Isthmus,* 8–9.

McDiarmid, M. A., Keogh, J. P., Hooper, F. J., McPhaul, K., et al. (2000). Health effects of depleted uranium on exposed Gulf War veterans. *Environmental Research, Section A, 82,* 168–180.

McKay, J. (2002, January 8). [No title]. *Pittsburgh Post-Gazette.* Available online at www.ibew29.org/shippingport.htm).

McKinney, J. D., & Waller, C. L. (1994). Polychlorinated biphenyls as hormonally active structural analogues. *Environmental Health Perspectives, 102,* 290–297.

McLoyd, V. C. (1998). Socioeconomic disadvantage and child development. *American Psychologist, 53*(2), 185–204.

McMichael, A. J. (1995). Environmental lead and intellectual development: Strengths and limitations of epidemiological research. *Neurotoxicology and Teratology, 17,* 237–240.

Medvedev, G. (1991). *The truth about Chernobyl.* (Evelyn Rossiter, Trans.). New York: Basic Books. Original work published 1989.

Meigs, J. W., Albom, J. J., & Kartin, B. L. (1954). Chloracne from an unusual exposure to Arochlor. *Journal of the American Medical Association, 154,* 1417–1418.

Meinke, W. W. (1951). Observations on radioactive snows at Ann Arbor, Michigan. *Science, 113,* 545–546.

Mellins, R. B., & Jenkins, C. D. (1955). Epidemiological and psychological study of lead poisoning in children. *Journal of the American Medical Association, 158*(1), 15–20.

Micronesia Support Committee. (1981). *Marshall Islands. A chronology: 1944–1981.* Honolulu, HI: Micronesia Support Committee.

Miedema, H. M. E., & Vos, H. (1998). Exposure-response relationships for transportation noise. *Journal of the Acoustical Society of America, 104*(6), 3432–3445.

Milar, C. R., Schroeder, S. R., Mushak, P., Dolcourt, J. L., & Grant, L. D. (1982). Contributions of the caregiving environment to increased lead burden of children. *American Journal of Mental Deficiency, 84,* 339–344.

Miller, H., & Conko, G. (2000). Reply [Letter]. *Nature Biotechnology, 18,* 697.

Miller, R. L. (1986). *Under the cloud: The decades of nuclear testing.* New York: Free Press.

Mills, J. H. (1975). Noise and children: A review of literature. *Journal of the Acoustical Society of America, 58*(4), 767–779.

Min, Y-I., Correa-Villasenor, A., & Stewart, P. A. (1996). Paternal occupational lead exposure and low birth weight. *American Journal of Industrial Medicine, 30,* 569–578.

Minamata City. (2001). Accessed online at island.qqq.or.jp/hp/minamata.city/english/me2e.htm on August 27, 2001.

Minnesota Department of Health (2006). Child's death from lead poisoning prompts recall and warning about children's jewelry. http://www.health.state.mn.us/news/pressrel/lead032306.html accessed on August 30, 2008.

Missouri Department of Natural Resources. (2002a). *Herculaneum Lead Contamination.* Accessed online at www.dnr.state.mo.us/deq/herc.htm on February 12, 2002.

Missouri Department of Natural Resources. (2002b). *Times Beach, other sites dropped from list.* Accessed online at www.dnr.state.mo.us/magazine/2001–02=winter/News=Briefs.htm on February 1, 2002.

Missouri Department of Natural Resources (2008). Herculaneum Broad Street Site: Lead (Pb). http://www.dnr.mo.gov/env/esp/aqm/herculaneum-broad-st.htm accessed on August 25, 2008.

Mitchell, A. E., Hong, Y-J., Koh, E., Barrett, D. M., Bryant, D. E., Denison, R. F., & Kaffka, S. (2007). Ten-year comparison of the influence of organic and conventional crop management practices on the content of flavonoids in tomatoes. *Journal of Agricultural and Food Chemistry, 55,* 6154–6159.

Mocarelli, P., Gerthoux, P. M., Ferrari, E., Petterson, D. G., et al. (2000). Paternal concentrations of dioxin and sex ratio of offspring. *Lancet, 355,* 1858–1863.

Moccia, R. D., Leatherland, J. F., & Sonstegard, R. A. (1977). Increasing frequency of thyroid goiters in coho salmon (*Oncorhynchus kisutch*) in the Great Lakes. *Science, 198,* 425–426.

Moore, C.F., Gajewski, L.L., Laughlin, N. K., Luck, M. L., Larson, J. A. & Schneider, M. L. (2008). Developmental lead exposure induces tactile defensiveness in rhesus monkeys (Macaca mulatta). *Environmental Health Perspectives, 116*(10), 1322–1326.

Morrow, L., Needleman, H. L., McFarland, C., Metheny, K., & Tobin, M. (2007). Past occupational exposure to lead: Association between current blood lead and bone lead. *Archives of Environmental and Occupational Health, 62*(4), 183–186.

Muldoon, S. B., Cauley, J. A., Kuller, L. H., Morrow, L., Needleman, H. L., Scott, J., & Hooper, F. J. (1996). Effects of blood lead levels on cognitive function of older women. *Neuroepidemiology, 15,* 62–72.

Murata, K., Weihe, P., Budtz-Jorgensen, E., Jorgensen, P., & Grandjean, P. (2004). Delayed brainstem auditory evoked potential latencies in 14-year-old children exposed to methylmercury. *Journal of Pediatrics, 144,* 177–183.

Murphy, B. C., Ellis, P., & Greenberg, S. (1990). Atomic veterans and their families: Responses to radiation exposure. *American Journal of Orthopsychiatry, 60*(3), 418–427.

Murphy, C. (2001, March 3–4). 1/3 of black students here in special ed: School officials express concern, are taking steps. *Capital Times,* pp. 2A, 7A.

Murphy, K. R., & Myors, B. (1998). *Statistical power analysis*. Mahwah, NJ: Erlbaum.

Mushak, P. (1999). Mushak re Needleman. *American Journal of Public Health, 89,* 1128.

Mynatt, F. R. (1982). Nuclear reactor safety research since Three Mile Island. *Science, 216,* 131–135.

Nabalek, A. K., & Robinson, P. K. (1982). Monaural and binaural speech perception in reverberation for listeners of various ages. *Journal of the Acoustical Society of America, 71*(5), 1242–1248.

Nadis, S. (2001). Getting the lead out. *National Wildlife Magazine, 39*(5), 46–51.

Naeye, R. L., & Peters, E. C. (1984). Mental development of children whose mothers smoked during pregnancy. *Obstetrics and Gynecology, 64,* 601–607.

National Academy of Sciences. (1982). *Evaluation of Enewetak radioactivity containment*. Washington, D.C.: National Academy Press.

National Academy of Sciences. (1983). *Risk assessment in the federal government: Managing the process*. Washington, D.C.: National Academy Press.

National Academy of Sciences. (1990). *Health effects of exposure to low levels of ionizing radiation: BEIR V*. Washington, D.C.: National Academy Press.

National Academy of Sciences. (1993). *Pesticides in the diets of infants and children*. Washington, D.C.: National Academy Press.

National Academy of Sciences. (2000a). *The future role of pesticides in agriculture*. Washington, D.C.: National Academy Press.

National Academy of Sciences. (2000b). *Scientific frontiers in developmental toxicology and risk assessment*. Washington, D.C.: National Academy Press.

National Academy of Sciences. (2000c). *Toxicological effects of methylmercury*. Washington, D.C.: National Academy Press.

National Association for the Advancement of Colored People. (2001, July 10). Mfume calls lead paint poisoning "the silent epidemic," will ask for a meeting with President Bush to discuss the issue [Press release]. Accessed online at www.naacp.org/news/pressreleases.shtml on August 25, 2001.

National Research Council. (1996). *Understanding risk: Informing decisions in a democratic society*. Washington, D.C.: National Academy Press.

Natural Resources Defense Council (2007). Katrina's wake: Arsenic-laced schools and playgrounds put New Orleans at risk. August 2007. http://www.nrdc.org/health/effects/wake/wake.pdf accessed on August 25, 2008.

Needham, L. L., Gerthouz, P. M., Patterson, D. G., Jr., Brambilla, P., et al. (1999). Exposure assessment: Serum levels of TCDD in Seveso, Italy. *Environmental Research, Section A, 80,* S200–S206.

Needleman, H. L. (1986). Lead levels—Comment. *Journal of Learning Disabilities, 19,* 322.

Needleman, H. L. (1987). Lead levels and child development: Statements by Claire B. Ernhart and Herbert L. Needleman. *Journal of Learning Disabilities, 20,* 264–265.

Needleman, H. L. (1992). Salem comes to the National Institutes of Health: Notes from inside the crucible of scientific integrity. *Pediatrics, 90,* 977–981.

Needleman, H. L. (1994). Correction: Lead and cognitive performance in children. *New England Journal of Medicine, 331,* 616–617.

Needleman, H. L. (1998a). Childhood lead poisoning: The promise and abandonment of primary prevention. *American Journal of Public Health, 88,* 1871–1877.

Needleman, H. L. (1998b). Clair Patterson and Robert Kehoe: Two views of lead toxicity. *Environmental Research, Section A, 78,* 79–85.

Needleman, H. L. (1999). Needleman responds. *American Journal of Public Health, 89,* 1130–1131.

Needleman, H. L. (2000). The removal of lead from gasoline: Historical and personal reflections. *Environmental Research, Section A, 84,* 20–35.

Needleman, H. L. (2004). Values, errors, and precautions. *International Journal of Occupational Medicine and Environmental Health, 17*(1), 111–114.

Needleman, H. L. (2006). Mercury in dental amalgam: A neurotoxic risk? (Editorial). *Journal of the American Medical Association, 295*(15), 1935–1836.

Needleman, H. L., Gunnoe, C., Leviton, A., Reed, R., et al. (1979). Deficits in psychologic and classroom performance of children with elevated dentine lead levels. *New England Journal of Medicine, 300,* 689–695.

Needleman, H. L., Riess, J. A., Tobin, M. J., Biesecker, G. E., & Greenhouse, J. B. (1996). Bone lead levels and delinquent behavior. *Journal of the American Medical Association, 275*(5), 363–369.

Neergaard, L. (2001, January 13–14). FDA: Fish can be bad for unborn babies. *Capital Times,* 6A.

Neisser, U., Boodoo, G., Bouchard, T. J., Boykin, A. W., et al. (1996). Intelligence: Knowns and unknowns. *American Psychologist, 51*(2), 77–101.

Nelkin, D. (1977). *Technological decisions and democracy: European experiments in public participation.* Beverly Hills, CA: Sage Publications.

Nelson, P. (2000). The changing demand for improved acoustics in our schools. *Volta Review, 101*(5), 23–31.

Nelson, P., & Soli, S. (2000). Acoustical barriers to learning: Children at risk in every classroom. *Language, Speech, and Hearing Services in Schools, 31*(4), 356–361.

Newschaffer, C. J., Croen, L. A., Daniels, J., Giarelli, E., Grether, J. K., et al. (2007). The epidemiology of autism spectrum disorder. *Annual Review of Public Health, 28,* 235–258.

New York Department of Health. (1978). *Love Canal: Public health time bomb.* Albany, NY: New York State Department of Health.

New York Department of Health. (1981). *Love Canal: A special report to the Governor and Legislature.* Albany, NY: New York Department of Health.

New York Department of Health. (2001). Fish consumption advisory posted at www.health.state.ny.us/nysdoh/environ/fish.htm.

New York Office of the Attorney General (2003). Dow subsidiary to pay $2 million for making false safety claims in pesticide ads: Largest pesticide enforcement penalty in U.S. history. http://www.oag.state.ny.us/press/2003/dec/dec15a_03.html accessed on August 19, 2008.

Nezu, N., Asano, M., & Ouchi, S. (1962). Cerium-144 in food. *Science, 135,* 102–103.

Niskar, A. S., Kieszak, S. M., Holmes, A., Esteban, E., et al. (1998). Prevalence of hearing loss among children 6 to 19 years of age: The Third National Health and Nutrition Examination Survey. *Journal of the American Medical Association, 279*, 1071–1075.

Nivison, M. E., & Endresen, I. M. (1993). An analysis of the relationships among environmental noise, annoyance and sensitivity to noise, and the consequences for health and sleep. *Journal of Behavioral Medicine, 16*, 257–276.

Northern, J. L., & Downs, M. P. (1984). *Hearing in children* (3rd ed.). Baltimore, MD: Williams & Wilkins.

Noshkin, V. E., & Robison, W. L. (1997). Assessment of a radioactive waste disposal site at Enewetak Atoll. *Health Physics, 73*(1), 234–247.

Nriagu, J. O. (1998). Clair Patterson and Robert Kehoe's paradigm of "Show Me the Data" on environmental lead poisoning. *Environmental Research, Section A, 78*, 71–78.

Nyagu, A. I., Loganovsky, K. N., & Loanovskaja, T. K. (1998). Psychophysiologic aftereffects of prenatal irradiation. *International Journal of Psychophysiology, 30*, 303–311.

Oakley, D. T. (1972). *Natural radiation exposure in the United States*. Washington, D.C.: U.S. EPA.

Office of Technology Assessment. (1983). *Habitability of the Love Canal area*. Washington, D.C.: U.S. Congress, Office of Technology Assessment. (Document OTA-TM-M-13, June 1983).

Ohrstrom, E., Bjorkman, M., & Rylander, R. (1988). Noise annoyance with regard to neurophysiological sensitivity, subjective noise sensitivity and personality variables. *Psychological Medicine, 18*, 605–618.

Ohrstrom, E., Bjorkman, M., & Rylander, R. (1990). Effects of noise during sleep with reference to noise sensitivity and habituation. *Environment International, 16*, 477–482.

Oken, E., Radesky, J. S., Wright, R. O., Bellinger, D. C., et al. (2008). Maternal fish intake during pregnancy, blood mercury levels and child cognition at age 3 years in a U.S. cohort. *American Journal of Epidemiology, 167*(10), 1171–1181.

Oken, E., Wright, R. O ., Kleinman, K. P.,, Bellinger, D. C., et al. (2005). Maternal fish consumption, hair mercury and infant cognition in a U.S. cohort. *Environmental Health Perspectives, 113*, 1376–1380.

Olsen, C. R., Larsen, I. L., Cutshall, N. H., Donoghue, J. F., et al. (1981). Reactor-released radionuclides in Susquehanna River sediments. *Nature, 294*, 242–245.

O'Malley, M. (1997). Clinical evaluation of pesticide exposure and poisonings. *Lancet, 349*(9059), 1161–1166.

Orban, J. E., Stanley, J. S., Schwemberger, J. G., & Remmers, J. C. (1994). Dioxins and dibenzofurans in adipose tissue of the general U.S. population and selected subpopulations. *American Journal of Public Health, 84*, 439–445.

O'Riordan, T., & Jordan, A. (1995). The precautionary principle in contemporary environmental politics. *Environmental Values, 14*, 191–212.

O'Rourke, M. K., Lizardi, P. S., Rogan, S. P., Freeman, N. C., Aguirre, A., & Saint, C. G. (2000). Pesticide exposure and creatinine variation among young children. *Journal of Exposure Analysis and Environmental Epidemiology, 10*, 672–681.

Otake, M., & Schull, W. J. (1998). Review: Radiation-related brain damage and growth retardation among the prenatally exposed atomic bomb survivors. *International Journal of Radiation Biology, 74*(2), 159–171.

Otake, M., Schull, W. J., & Neel, J. V. (1990). Congenital malformations, stillbirths, and early mortality among the children of atomic bomb survivors: A reanalysis. *Radiation Research, 122,* 1–11.

Ouis, D. (2001). Annoyance from road traffic noise: A review. *Journal of Environmental Psychology, 21,* 101–120.

Owens, W. (1988). Testimony by Rep. Wayne Owens. *Hearing before the Subcommittee on Administrative Law and Governmental Relations of the Committee on the Judiciary,* 100th Cong., on H.R. 5022, Radiation Exposure Compensation Act. Washington, D.C.: U.S. Government Printing Office.

Paarlberg, K. M., Vingerhoets, J. P., Dekker, G. A. & van Geijn, H. P. (1995). Psychosocial factors and pregnancy outcome: A review with emphasis on methodological issues. *Journal of Psychosomatic Research, 39,* 563–595.

Pacini, F., Vorontsova, T., Molinaro, E., Shavrova, E., et al. (1999). Thyroid consequences of the Chernobyl nuclear accident. *Acta Paediatrica Supplement, 433,* 23–27.

Paigen, B. (1982). Controversy at Love Canal. *Hastings Center Report, 12*(3), 29–37.

Paigen, B , Goldman, I. R.., Magnant, M. M., & Highland, J. (1985). Prevalence of health problems in children living near Love Canal. *Hazardous Waste and Hazardous Materials, 2*(1), 23–43.

Paigen, B., Goldman, L. R., Magnant, M. M., Highland, J. H., & Steegmann, A. T. (1987). Growth of children living near the hazardous waste site, Love Canal. *Human Biology, 59*(3), 489–508.

Parker, L. (2001). Fetal death and radiation exposure [Letter]. *Lancet, 357,* 556.

Parker, L., Pearce, M. S., Dickinson, H. O., Aitkin, M., & Craft, A. W. (1999). Stillbirths in offspring of male radiation workers from Sellafield nuclear reprocessing plant. *Lancet, 354,* 1407–1414.

Parker, S. K., Schwartz, B., Todd, J., & Pickering, L. K. (2004). Thimerosal-containing vaccines and autistic spectrum disorder: A critical review of published original data. *Pediatrics, 114,* 793–804.

Parrish, A. G. (1998). Statement of Albert G. "Smoky" Parrish, founder of "The Forgotten 216th." *Ionizing radiation, veterans health care, and related issues: Hearing before the Committee on Veterans' Affairs,* 105th Congress, 2nd sess. (April 21, 1998). Washington, D.C.: U.S. Government Printing Office.

Parush, S., Sohmer, H., Steinberg, A., & Kaitz, M. (2007). Somatosensory function in boys with ADHD and tactile defensiveness. *Physiology & Behavior, 90,* 553–558.

Pasciak, W., Branagan, E. F., Jr., Congel, F. J., & Fairobent, J. E. (1981). A method for calculating doses to the population from Xe-133 releases during the Three Mile Island accident. *Health Physics, 40,* 457–465.

Passchier-Vermeer, W., & Passchier, W. F. (2000). Noise exposure and public health. *Environmental Health Perspectives, 108*(Supplement 1), 123–131.

Patandin, S., Dagnelie, P. C., Mulder, P. G., Op de Coul, E., et al. (1999a). Dietary exposure to polychlorinated biphenyls and dioxins from infancy until adulthood: A comparison between breast-feeding, toddler, and long-term exposure. *Environmental Health Perspectives, 107,* 45–51.

Patandin, S., Lanting, C. I., Mulder, P. G. H., Boersma, R., et al. (1999b). Effects of environmental exposure to polychlorinated biphenyls and dioxins on cognitive abilities in Dutch children at 42 months of age. *Journal of Pediatrics, 134*, 33–41.

Patel, M. H. (1987). Decision in National Association of Radiation Survivors v. Thomas K. Turnage, et al. No. C-83–1861 MHP. 115 FRD543; 1887 U.S. Dist. LEXIS 3468. Accessed online at web.lexis-nexis.com/universe/ on July 4, 2002.

Patterson, C. C. (1965). Contaminated and natural lead environments of man. *Archives of Environmental Health, 11*, 344–360.

Pauling, L. (1958). Genetic and somatic effects of Carbon-14. *Science, 128*, 1183–1186.

Pediatrics. (1992). Editor's note. *Pediatrics, 90*(6), 981.

Pentagon admits plutonium exposure: NATO shells used radioactive metals. (2001, February 3–4). *Capital Times*, 6A.

Perino, J. (1973). The relation of subclinical lead levels to cognitive and sensorimotor ability in black preschool children. *Dissertation Abstracts International, 34*(5-B), 2315–2316.

Perino, J., & Ernhart, C. B. (1974). The relation of subclinical lead level to cognitive and sensorimotor impairment in black preschoolers. *Journal of Learning Disabilities, 7*, 26–30.

Pierson, L. L. (1996). Hazards of noise exposure on fetal hearing. *Seminars in Perinatology, 20*(1), 21–29.

Piomelli, S., & Schoenbrod, D. (1999). Piomelli and Schoenbrod re Needleman. *American Journal of Public Health, 89*, 1129–1130.

Pirkle, J. L., Brody, D. J., Gunter, E. W., Kramer, R. A., et al. (1994). The decline in blood lead levels in the United States: The national health and nutrition examination surveys (NHANES). *Journal of the American Medical Association, 272*, 284–291.

Pless, R., & Risher, J. F. (2000). Mercury, infant neurodevelopment, and vaccination. *Journal of Pediatrics, 136*, 571–573.

Pocock, S., Smith, M., & Baghurst, P. (1994). Environmental lead and children's intelligence: Review of the epidemiological evidence. *British Medical Journal, 309*, 1189–1197.

Porterfield, S. P. (1994). Vulnerability of the developing brain to thyroid abnormalities: Environmental insults to the thyroid systems. *Environmental Health Perspectives, 102*(Supplement 2), 125–130.

Poskiparta, E., Niemi, P., & Vauras, M. (1999). Who benefits from training in linguistic awareness in the first grade, and what components show training effects? *Journal of Learning Disabilities, 32*, 437–446, 456.

Posner, J. L., Baldock, J. O., & Hedtcke, J. L. (2008). Organic and conventional production systems in the Wisconsin integrated cropping systems trials: I. Productivity 1990–2002. *Agronomy Journal, 100*(2), 253–260.

Potomac Electric Power Company. (2000). *Annual Report 2000*. Washington, D.C. Available online at www.pepco.com.

Preston, D. L. (1998). A historical review of leukemia risks in atomic bomb survivors. In L. E. Peterson & S. Abrahamson (Eds.), *Effects of ionizing radiation: Atomic bomb survivors and their children (1945–1995)*. Washington, D.C.: Joseph Henry Press.

Quine, W. V., & Ullian, J. S. (1970). *The web of belief.* New York: Random House.

Rabbitt, P. M. A. (1966). Channel-capacity, intelligibility and immediate memory. *Quarterly Journal of Experimental Psychology, 20*, 241–248.

Rachel Carson dies of cancer; "Silent Spring" author was 56. (1964). *New York Times, 1*, 25.

Radio Canada International. (2001). News broadcast, May 23, 2001.

Rahu, K., Rahu, M., Tekkel, M., & Bromet, E. (2006). Suicide risk among Chernobyl cleanup workers in Estonia is still increased: An updated cohort study. *Annals of Epidemiology, 16*, 917–919.

Rambo, S. H. (1996). Summary court judgment. Excerpts available online at www.pbs.org/wgbh/pages/frontline/shows/reaction/readings/tmi.html.

Rauh, V. A., Garfinkel, R., Perera, F. P., Andrews, H. F., et al. (2006). Impact of prenatal chlorpyrifos exposure on neurodevelopment in the first 3 years of life among inner-city children. *Pediatrics, 118*, 1845–1859.

Rawls, J. (1971). *A theory of justice.* Cambridge, MA: Harvard University Press.

Rayner, S., & Cantor, R. (1987). How fair is safe enough? The cultural approach to societal technology choice. *Risk Analysis, 7*(1), 3–9.

Report of the Methylmercury Workshop (1998). National Toxicology Program Press Center. http://ntp.niehs.nih.gov/index.cfm?objectid=03614B65-BC68-D231–4E915F93AF9A6872#execsumm accessed on August 14, 2008.

Republic of Marshall Islands. (2002). Accessed online at www.rmiembassyus.org/nuclear/exhibit.html on October 20, 2002.

Revelle, R., Buroughs, H., Carritt, D. E., Chipman, W. A., et al. (1956). Oceanography, fisheries, and atomic radiation. *Science, 124*, 13–16.

Rheindt, F. E. (2003). The impact of roads on birds: Does song frequency play a role in determining susceptibility to noise pollution? *Journal of Ornithology, 144*, 295–306.

Rice, D. C. (1993). Lead-induced changes in learning: Evidence for behavioral mechanisms from experimental animal studies. *NeuroToxicology, 14*, 167–178.

Rice, D. C. (1996). Sensory and cognitive effects of developmental methylmercury exposure in monkeys, and a comparison to effects in rodents. *NeuroToxicology, 17*, 139–154.

Rice, D. C. (1998). Effects of postnatal exposure of monkeys to a PCB mixture on spatial discrimination reversal and DRL performance. *Neurotoxicology and Teratology, 20*, 391–400.

Rice, D. C. (2007). Polybrominated flame retardants (PBDEs) with an emphasis on deca BDE. Maine Center for Disease Control and Prevention, August, ME. http://www.maine.gov/dep/oc/safechem/present.htm accessed on August 18, 2008.

Rice, D. C., & Hayward, S. (1997a). Effect of postnatal exposure to a PCB mixture in monkeys on multiple fixed interval–fixed ratio performance. *Neurotoxicology and Teratology, 19*, 429–434.

Rice, D. C., & Hayward, S. (1997b). Effects of postnatal exposure to a PCB mixture in monkeys on nonspatial discrimination reversal and delayed alternation performance. *NeuroToxicology, 18,* 479–494.

Rice, D. C., & Hayward, S. (1999). Effect of postnatal exposure of monkeys to a PCB mixture on concurrent random interval–random interval and progressive ratio performance. *Neurotoxicology and Teratology, 21,* 47–58.

Richardson, W. J., Greene, C. R., Malme, C. I., & Thomson, D. H. (1995). *Marine mammals and noise.* San Diego: Academic Press.

Risebrough, R. W., Rieche, P., Herman, S. G., Peakall, D. B., & Kirven, M. N. (1968). Polychlorinated biphenyls in the global ecosystem. *Nature, 220,* 1098–1102.

Risher, J. F., & Amler, S. N. (2005). Mercury exposure: Evaluation and intervention. The inappropriate use of chelating agents in the diagnosis and treatment of putative mercury poisoning. NeuroToxicology, *26,* 691–699.

Robertson, G. L., Lebowitz, M. D., O'Rourke, M. K., Gordon, S., & Moschandreas, D. (1999). The National Human Exposure Assessment Survey (NHEXAS) study in Arizona—an introduction and preliminary results. *Journal of Exposure Analysis and Environmental Epidemiology, 9,* 427–434.

Robinson, P. E., Mack, G. A., Remmers, J., Levy, R., & Mohadjer, L. (1990). Trends of PCB, hexachlorobenzene, and beta-benzene hexachloride levels in adipose tissue of the U.S. population. *Environmental Research, 53,* 175–192.

Rogan, W. J. (1989). Yu-Cheng. In R. D. Kimbrough & A. Jensen (Eds.), *Halogenated biphenyls, terphenyls, naphthalenes, dibenzodioxins and related products.* Amsterdam: Elsevier Science.

Rogan, W. J., & Gladen, B. C. (1993). Breast-feeding and cognitive development. *Early Human Development, 31,* 181–193.

Rogan, W. J., Gladen, B. C., Hung, K-L., Koong, S-L., et al. (1988). Congenital poisoning by polychlorinated biphenyls and their contaminants in Taiwan. *Science, 241,* 334–336.

Rogan, W. J., Gladen, B. C., McKinney, J. D., Carreras, N., et al. (1986a). Neonatal effects of transplacental exposure to PCBs and DDE. *Journal of Pediatrics, 109,* 335–341.

Rogan, W. J., Gladen, B. C., McKinney, J. D., Carreras, N., et al. (1986b). Polychlorinated biphenyls (PCBs) and dichlorodiphenyl dichloroethene (DDE) in human milk: Effects of maternal factors and previous lactation. *American Journal of Public Health, 76,* 172–177.

Rohlman, D. S., Bailey, S. R., Anger, W. K., & McCauley, L. (2001). Assessment of neurobehavioral function with computerized tests in a population of Hispanic adolescents working in agriculture. *Environmental Research, Section A, 85,* 14–21.

Rolvaag, O. E. (1927). *Giants in the earth: A saga of the prairie.* (L. Colcord & O. E. Rolvaag, Trans.). New York: Harper.

Rondeau, G., & Desgranges, J-L. (1995). Effects of insecticide use on breeding birds in Christmas tree plantations in Quebec. *Ecotoxicology, 4,* 281–298.

Roos, A., Greyerz, E., Olsson, M., & Sandegren, F. (2001). The otter (*Lutra lutra*) in Sweden—population trends in relation to sigma DDT and total PCB concentrations during 1968–99. *Environmental Pollution, 111,* 457–469.

Rosen, J. F. (1999). Rosen re Needleman. *American Journal of Public Health, 89,* 1129.

Rosenstreich, D. L., Eggelston, P., Kattan, M., Baker, D., et al. (1997). The role of cockroach allergy and exposure to cockroach allergen in causing morbidity among inner-city children with asthma. *New England Journal of Medicine, 336,* 1356–1363.

Rosner, D., & Markowitz, G. (1985). A "Gift of God"?: The public health controversy over leaded gasoline during the 1920s. *American Journal of Public Health, 75,* 344–352.

Ruff, H. A., Bijur, P. E., Markowitz, M., Ma, Y., & Rosen, J. F. (1993). Declining blood lead levels and cognitive changes in moderately lead-poisoned children. *Journal of the American Medical Association, 269,* 1641–1646.

Rutter, M. (1979). Maternal deprivation, 1972–1978: New findings, new concepts, new approaches. *Child Development, 50,* 283–305.

Ryan, D. (1999). Ryan re Needleman. *American Journal of Public Health, 89,* 1126–1127.

Said, E. W. (1978). *Orientalism.* New York: Vintage Books.

Sameroff, A. J., Bartko, W. T., Baldwin, A., Baldwin, C., & Seifer, R. (1998). Family and social influences on the development of child competence. In M. Lewis & C. Feiring (Eds.), *Families, risk, and competence* (pp. 161–185). Mahwah, NJ: Erlbaum.

Sameroff, A. J., & Chandler, M. J. (1975). Reproductive risk and the continuum of caretaking casualty. In F. D. Horowitz (Ed.), *Review of child development research, Vol. 4.* Chicago: University of Chicago Press.

Santa, C. M., & Hoien, T. (1999). An assessment of Early Steps: A program for early intervention of reading problems. *Reading Research Quarterly, 34,* 54–73.

Sanz, S. A., Garcia, A. M., & Garcia, A. (1993). Road traffic noise around schools: A risk for pupil's performance? *International Archives of Occupational and Environmental Health, 65,* 205–207.

Saper, R. B., Kales, S. N., Paquin, J., Burns, M. J., et al. (2004). Heavy metal content of Ayuredic herbal medicine products. *Journal of the American Medical Association, 292*(23), 2868–2873.

Satoh, T., & Hosokawa, M. (2000). Organophosphates and their impact on the global environment. *NeuroToxicology, 21,* 223–228.

Scarr, S. (1985). Constructing psychology: Making facts and fables for our times. *American Psychologist, 40,* 499–512.

Scarr, S., & Ernhart, C. B. (1996). A reply from Scarr and Ernhart. *American Journal of Public Health, 86,* 113–114.

Schantz, S. L. (1996a). Developmental neurotoxicity of PCBs in humans: What do we know and where do we go from here? *Neurotoxicology and Teratology, 18,* 217–227.

Schantz, S. L. (1996b). Response to commentaries. *Neurotoxicology and Teratology, 18,* 271–276.

Schantz, S. L., Gardiner, J. C., Gasior, D. M., Sweeney, A. M., et al. (1999). Motor function in aging Great Lakes fisheaters. *Environmental Research, Section A, 80,* S46–S56.

Schardein, J. L., & Scialli, A. R. (1999). The legislation of toxicologic safety factors: The Food Quality Protection Act with Chlorpyrifos as a test case. *Reproductive Toxicology, 13*(1), 1–14.

Schecter, R., & Grether, J. K. (2008). Continuing increases in autism reported to California's developmental services system: Mercury in retrograde. *Archives of General Psychiatry, 65*(1), 19–24.

Schepens, P. J. C., Covaci, A., Jorens, P. G., Hens, L., et al. (2001). Surprising findings following a Belgian food contamination with polychlorobiphenyls and dioxins. *Environmental Health Perspectives, 109,* 101–103.

Schlundt, C. E., Finneran, J. J., Carder, D. A., & Ridgway, S. H. (2000). Temporary shift in masked hearing thresholds of bottlenose dolphins, *Tursiops truncatus,* and white whales, *Delphinapterus leucas,* after exposure to intense noises. *Journal of the Acoustical Society of America, 197*(6), 3496–3508.

Schneider, M.L. & Moore, C. F. (2000). Effect of prenatal stress on development: A nonhuman primate model. In C.A. Nelson (Ed.), The Minnesota symposia on child psychology, Vol 31, pp. 201–244: The effects of early adversity on neurobehavioral development. Mahwah, NJ: Lawrence Erlbaum.

Schneider, M. L., Moore, C. F., Roberts, A. D., & DeJesus, O. (2001). Prenatal stress alters early neurobehavior, stress reactivity and learning in non-human primates: A brief review. *Stress, 4,* 183–193.

Schneider, M. L., Roughton, E. C., Koehler, A. J., & Lubach, G. R. (1999). Growth and development following prenatal stress exposure in primates: An examination of ontogenetic vulnerability. *Child Development, 70,* 263–274.

Schull, W. J. (1995). *Effects of atomic radiation: A half century of studies from Hiroshima and Nagasaki.* New York: Wiley.

Schull, W. J., Norton, S., & Jensh, R. P. (1990). Ionizing radiation and the developing brain. *Neurotoxicology and Teratology, 12,* 249–260.

Schultz, T. J. (1978). Synthesis of social surveys on noise annoyance. *Journal of the Acoustical Society of America, 64*(2), 377–405.

Schultz, T. J. (1982). Comments on K. D. Kryter's paper, "Community annoyance from aircraft and ground vehicle noise." *Journal of the Acoustical Society of America, 72*(4), 1243–1252.

Schwartz, J. (1994). Low-level lead exposure and children's IQ: A meta-analysis and search for a threshold. *Environmental Research, 65,* 42–55.

Schwindt, A. R., Fournie, J. W., Landers, D. H., Schreck, C. B., & Kent, M. L. (2008). Mercury concentrations in salmonids from western U.S. national parks and relationships with age and macrophage aggregates. *Environmental Science and Technology, 42,* 1365–1370.

Seegal, R. F. (1996). Epidemiological and laboratory evidence of PCB-induced neurotoxicity. *Critical Reviews in Toxicology, 26,* 709–737.

Seegal, R. F. (1999). Are PCBs the major neurotoxicant in Great Lakes salmon? *Environmental Research, Section A, 80,* S38–S45.

Seligman, M. E. P. (1975). *Helplessness: On depression, development, and death.* New York: W. H. Freeman.

Senechal, M., & LeFevre, J. (2001). Storybook reading and parent teaching: Links to language and literacy development. In P. R. Britto & J. Brooks-Gunn (Eds.), *The role of family literacy environments in promoting young children's emerging literacy skills: New directions for child and adolescent development* (pp. 39–52). San Francisco: Jossey-Bass.

Senthilselvan, A., McDuffie, H. H., & Dosman, J. A. (1992). Association of asthma with use of pesticides. *American Review of Respiratory Disease, 146*(4), 884–887.

Shalit, A. G. (2000). The precautionary principle [Letter]. *Nature Biotechnology, 18,* 697.

Shcherbak, Y. M. (1989). *Chernobyl: A documentary story.* (Ian Press, Trans.). New York: St. Martin's Press.

Shih, R. A., Glass, T. A., Bandeen-Roche, K., Carlson, M. C., Bolla, K. I., Todd, A. C., & Schwartz, B. S. (2006). Environmental lead exposure and cognitive function in community-dwelling older adults. *Neurology, 67,* 1556–1562.

Shrader-Frechette, K. S. (1983). *Nuclear power and public policy* (2nd ed.). Boston: D. Reidel Publishing.

Shrader-Frechette, K. S. (1985). *Risk analysis and scientific method.* Boston: D. Reidel Publishing.

Shrader-Frechette, K. S. (1987). Parfit and mistakes in moral mathematics. *Ethics, 98,* 50–60.

Shrader-Frechette, K. S. (1991). *Risk and rationality: Philosophical foundations for populist reforms.* Berkeley: University of California Press.

Shrader-Frechette, K. S. (1994). *Ethics of scientific research.* Lanham, MD: Rowman & Littlefield.

Shrader-Frechette, K. S. (2002). Trading jobs for health: Ionizing radiation, occupational ethics, and the welfare argument. *Science and Engineering Ethics, 8*(2), 139–154.

Shrader-Frechette, K. (2007). Trimming exposure data, putting radiation workers at risk: Improving disclosure and consent through a national radiation dose-registry. *American Journal of Public Health, 97,* 1782–1786.

Shrivastava, P. (1987). *Bhopal: Anatomy of a crisis.* Cambridge, MA: Ballinger.

Siebein, G. W., Gold, M. A., Siebein, G. W., & Ermann, M. G. (2000). Ten ways to provide a high-quality acoustical environment in schools. *Language, Speech, and Hearing Services in Schools, 31*(4), 376–384.

Sierra Pacific Resources. (2000). *2000 Annual Report.* Reno, NV. Available online at www.sierrapacificresources.com.

Sjoberg, L. (2000). Factors in risk perception. *Risk Analysis, 20*(1), 1–11.

Slotkin, T. A. (1998). Fetal nicotine or cocaine exposure: Which one is worse? *Journal of Pharmacology and Experimental Therapeutics, 285,* 931–945.

Slotkin, T. A. (1999). Developmental cholinotoxicants: Nicotine and chlorpyrifos. *Environmental Health Perspectives, 107*(Supplement 1), 71–80.

Slovic, P. (1987). Perception of risk. *Science, 236,* 280–285.

Slovic, P., Flynn, J. H., & Layman, M. (1991). Perceived risk, trust, and the politics of nuclear waste. *Science, 254,* 1603–1607.

Smith, C. V. (1995). Vibroacoustic stimulation. *Clinical Obstetrics and Gynecology, 38*(1), 68–77.

Smith, M. (1985). Recent work on low level lead exposure and its impact on behaviour, intelligence and learning. *Journal of the American Academy of Child Psychiatry, 24,* 24–32.

Smith, R. N. (1969). *Chemistry: A quantitative approach.* New York: Ronald Press.

Sobrian, S. K., Vaughn, V. T., Ashe, W. K., Markovic, B., et al. (1997). Gestational exposure to loud noise alters the development of postnatal responsiveness of humoral and cellular components of the immune system in offspring. *Environmental Research, 73,* 227–241.

Sofer, S., Tal, A., & Shahak, E. (1989). Carbamate and organophosphate poisoning in early childhood. *Pediatric Emergency Care, 5,* 222–225.

Solomon, S. D., Smith, E. M., Robins, L. N., & Fischbach, R. L. (1987). Social involvement as a mediator of disaster-induced stress. *Journal of Applied Social Psychology, 17*(12), 1092–1112.

Sorkin, D. L. (2000). The classroom acoustical environment and the Americans With Disabilities Act. *Language, Speech, and Hearing Services in Schools, 31*(4), 385–388.

Soule, E. (2000). Assessing the precautionary principle. *Public Affairs Quarterly, 14*(4), 309–328.

Spears, G. (1994). Tests reveal dangerous levels of lead in crayons. *Wisconsin State Journal* (Madison, WI), April 6, 1994.

Spelt, D. K. (1948). The conditioning of the human fetus in utero. *Journal of Experimental Psychology, 38,* 338–346.

Stajich, G. V., Lopez, G. P., Harry, S. W., & Sexson, W. R. (2000). Iatrogenic exposure to mercury after hepatitis B vaccination in preterm infants. *Journal of Pediatrics, 136,* 679–681.

Stangle, D. E., Smith, D. R., Beaudin, S. A., Strawderman, M. S., Levitsky, D. A., & Strupp, B. J. (2007). Succimer chelation improves learning, attention, and arousal regulation in lead-exposed rats but produces lasting cognitive impairment in the absence of lead exposure. *Environmental Health Perspectives, 115*(2), 201–209.

Stanovich, K. E. (1986). Matthew effects in reading: Some consequences of individual differences in the acquisition of literacy. *Reading Research Quarterly, 21,* 360–406.

Stansfeld, S. A. (1992). Noise, noise sensitivity and psychiatric disorder: Epidemiological and psychophysiological studies. *Psychological Medicine* (Monograph Supplement 22).

Stansfeld, S.A., Berglund, B., Clark, C., Lopez-Barrio, I., Fischer, P. et al. (2005). Aircraft and road traffic noise and children's cognition and health: A cross-national study. *Lancet, 365*(9475), 1942–1949.

Statz, R. N. (1991). Chippewa treaty rights. *Transactions of the Wisconsin Academy of Sciences, Arts and Letters, 79*(1).

Steenland, K., Dick, R. B., Howell, R. J., Chrislip, D. W., et al. (2000). Neurologic function among termiticide applicators exposed to chlorpyrifos. *Environmental Health Perspectives, 108,* 293–300.

Stender, W., & Walker, E. (1974). The National Records Center fire: A study in disaster. *American Archivist, 37,* 521–549.

Stern, A. H., & Gochfield, M. (1998). Effects of methylmercury exposure on neurodevelopment [Letter]. *Journal of the American Medical Association, 281*(10), 896–897.

Stewart, A., Webb, J., & Hewitt, D. (1958). A survey of childhood malignancies. *British Medical Journal*, 1495–1508.

Stewart, P., Reihman, J., Lonky, E., Darvill, T., & Pagano, J. (2000). Prenatal PCB exposure and neonatal behavioral assessment scale (NBAS) performance. *Neurotoxicology and Teratology, 22*, 21–29.

Stewart, W. F., Schwartz, B. S., Simon, D., Bolla, K. I., et al. (1999). Neurobehavioral function and tibial and chelatable lead levels in 543 former organolead workers. *Neurology, 52*, 1610–1617.

Stewart, W. F., & Schwartz, B. S. (2007). Effects of lead on the adult brain: A 15-year exploration. *American Journal of Industrial Medicine, 50*, 729–739.

Stokstad, E. (2008). Stalled trial for autism highlights dilemma of alternative treatments (news). *Science, 321*(5887), 326.

Stone, R. A., & Levine, A. G. (1985). Reactions to collective stress: Correlates of active citizen participation at Love Canal. *Prevention in Human Services, 4*(1–2), 153–177.

Strain, J. J., Davidson, P. W., Bonham, M. P., Duffy, E. M., et al. (2008). Associations of maternal long-chain polyunsaturated fatty acids, methyl mercury, and infant development in the Seychelles Child Development Nutrition study. *NeuroToxicology*, doi:10.1016/j.neuro.2008.06.002.

Sulkowski, W. J., & Lipowczan, A. (1982). Impulse noise-induced hearing loss in drop forge operators and the energy concept. *Noise Control Engineering, 18*(1), 24–29.

Suppe, F. (Ed.). (1977). *The structure of scientific theories* (2nd ed.). Urbana: University of Illinois Press.

Susser, M. (1997). Consequences of the 1979 Three Mile Island accident continued: Further comment. *Environmental Health Perspectives, 105*(6), 566–567.

Sutherland, L., & Lubman, D. (2000). Overcoming acoustic barriers to learning: Report on the ASA/ANSI standard on classroom acoustics. Presentation at the Federal Interagency Committee on Aviation Noise symposium on the Effects of Noise on Children's Learning, San Diego, CA, February, 2000. Accessed online at www.fican.org/pages/sympos02.html on October 20, 2002.

Sutton, P. M., Vergara, X., Beckman, J., Nicas, M., & Das, R. (2007). Pesticide illness among flight attendants due to aircraft disinsection. *American Journal of Industrial Medicine, 50*, 345–356.

Stansfeld, S. A., Berglund, B., Clark, C. Lopez-Barrio, I., et al. (2005). Aircraft and road traffic noise and children's cognition and health: A cross-national study. *Lancet, 365*, 1942–1949.

Swanson, G. M., Ratcliffe, H. E., & Fischer, L. J. (1995). Human exposure to polychlorinated biphenyls (PCBs): A critical assessment of the evidence for adverse health effects. *Regulatory Toxicology and Pharmacy, 21*, 136–150.

Sydney Morning Herald (2008). Nursing a nuclear test hangover. (August 18, 2008). http://www.smh.com.au/news/world/nursing-a-nuclear-test-hangover/2008/08/17/1218911461036.html accessed on August 25, 2008.

Tainted geese can't go to pantries. (2001, June 27). *Capital Times.*

Talbott, E. O., Youk, A. O., McHugh, K. P., Shire, J. D., Zhang, A., et al. (2000a). Mortality among the residents of the Three Mile Island accident area: 1979– 1992. *Environmental Health Perspectives, 108*(6), 545–552.

Talbott, E. O., Zhang, A., Youk, A. O., McHugh-Pemu, K. P., & Zborowski, J. V. (2000b). Re: "Collision of evidence and assumptions: TMI deja view" [Letter]. *Environmental Health Perspectives, 108*(12), A547–A549.

Tarnopolsky, A., Barker, S. M., Wiggins, R. D., & McLean, E. K. (1978). The effects of aircraft noise on the mental health of a community sample: A pilot study. *Psychological Medicine, 8*(2), 219–233.

Taylor, S. M., Breston, B. E., & Hall, F. L. (1982). Effect of road traffic noise on house prices. *Journal of Sound and Vibration, 80*(4), 523–542.

Tharpe, A. M., & Bess, F. H. (1991). Identification and management of children with minimal hearing loss. *International Journal of Pediatric Otorhinolaryngology, 21*(1), 41–50.

Thompson, J. P., Casey, P. B., & Vale, J. A. (1994). Suspected paediatric pesticide poisoning in the UK. II—Home accident surveillance system 1989–1991. *Human & Experimental Toxicology, 13,* 534–536.

Thompson, L. A., Fagan, J. F., & Fulker, D. W. (1991). Longitudinal prediction of specific cognitive abilities from infant novelty preference. *Child Development, 62,* 530–538.

Tilden, J., Hanrahan, L. P., Anderson, H., Palit, C., et al. (1997). Health advisories for consumers of Great Lakes sport fish: Is the message being received? *Environmental Health Perspectives, 105,* 1360–1365.

Tilson, H. A., Jacobson, J. L., & Rogan, W. J. (1990). Polychlorinated biphenyls and the developing nervous system: Cross-species comparisons. *Neurotoxicology and Teratology, 12,* 239–248.

Trunk, A. D., & Trunk, E. V. (1981). Three Mile Island: A resident's perspective. *Annals of the New York Academy of Sciences, 365,* 175–185.

Tryphonas, H. (1995). Immunotoxicity of PCBs (Aroclors) in relation to Great Lakes. *Environmental Health Perspectives, 103*(Supplement 9), 35–45.

Tsuruta, D., Sowa, J., Higashi, N., Kobayashi, H., & Ishii, M. (2004). A red tattoo and a swordfish supper. *Lancet, 364,* 730.

Tyack, P. L. (2008). Implications for marine mammals of large-scale changes in the marine acoustic environment. *Journal of Mammalogy, 89,* 549–558.

Udall, S. L. (1964). The legacy of Rachel Carson. *Saturday Review,* May 16, pp. 23, 59.

United Nations. (1992). Rio Declaration on Environment and Development. Accessed online at www.unep.org/Documents/Default.asp?DocumentID=78 &ArticleID=1163 on October 20, 2002.

United Nations. (2002, February 6). Press release. Accessed online at www. un.org/apps/news/story.asp?NewsID=2817&Cr=Chernobyl&Cr1= on January 5, 2002.

United Nations Scientific Committee on the Effects of Atomic Radiation. (2000). *Sources and effects of ionizing radiation, Volume I: Sources.* United Nations Scientific Committee on the Effects of Ionizing Radiation. Accessed online at www.unscear.org/2000vol1.htm on July 5, 2002.

United Nations Scientific Committee on the Effects of Atomic Radiation. (2001a). *Annex J: Exposures and effects of the Chernobyl accident.* United Nations Scientific Committee on the Effects of Ionizing Radiation. Accessed online at www.unscear.org on January 22, 2002.

United Nations Scientific Committee on the Effects of Atomic Radiation. (2001b). *Conclusions: Third International Conference, health effects of the Chernobyl accident: Results of 15-year follow-up studies.* Kiev (Ukraine), June 4–8, 2001. Accessed online at www.iaea.org/worldatom/Press/Focus/Chernobyl-15/kiev-conference.shtml on January 22, 2002.

University of Minnesota (2008). The Raptor Center: Lead Poisoning. http://www.cvm.umn.edu/raptor/news/healthtopics/leadpoisoning/home.html accessed on August 30, 2008.

Upton, A. C. (1981). Health impact of the Three Mile Island accident. *Annals of the New York Academy of Sciences, 365,* 63–75.

U.S. Department of Agriculture (2008). National nutrition nutrient database for standard reference. http://www.ars.usda.gov/ba/bhnrc/ndl accessed on August 15, 2008.

U.S. Food and Drug Administration (2006). Mercury levels in commercial fish and shellfish. http://www.cfsan.fda.gov/~frf/sea-mehg.html accessed on August 15, 2008.

U.S. Food and Drug Administration (2007). 21 CFR700.13 Revised as of April 1, 2007. http://www.accessdata.fda.gov/scripts/cdrh/cfdocs/cfcfr/CFRSearch.cfm?fr=700.13 accessed on August 15, 2008.

Van Kempen, E. E., Kruize, H., Boshuizen, H. C., Ameling, C. B., et al. (2002). The association between noise exposure and blood pressure and ischemic heart disease: A meta-analysis. *Environmental Health Perspectives, 110*(3), 307–317.

Vernon, T. M. (1999). Vernon re Needleman. *American Journal of Public Health, 89,* 1128–1129.

Vianna, N. J., & Polan, A. D. (1984). Incidence of low birth weight among Love Canal residents. *Science, 226,* 1217–1219.

Viel, J-F., Curbakova, E., Dzerve, B., Eglite, M., et al. (1997). Risk factors for long-term mental and psychosomatic distress in Latvian Chernobyl liquidators. *Environmental Health Perspectives, 105*(Supplement 6), 1539–1544.

Voitsekhovitch, O., Prister, B., Nasvit, O., Los, I., & Berkovski, V. (1996). Present concept on current water protection and remediation activities for the areas contaminated by the 1986 Chernobyl accident. *Health Physics, 71*(1), 19–28.

Volante, J. A. (1998). Statement submitted for the record by Joseph A. Volante, national legislative director, the Disabled American Veterans. *Ionizing radiation, veterans health care, and related issues: Hearing before the Committee on Veterans' Affairs,* 105th Congress, 2nd sess. (April 21, 1998). Washington, D.C.: U.S. Government Printing Office.

von Gierke, H. E. (1982). All noise is noise [Editorial]. *Noise Control Engineering Journal, 18*(1), 3.

Vreugdenhil, H. J., Lanting, C. I., Mulder, P. G. H., Boersma, E. R., & Weisglas-Kuperus, N. (2002). Effects of prenatal PCB and dioxin background on cognitive and motor abilities in Dutch children at school age. *Journal of Pediatrics, 140*(1), 48–56.

Vreugdenhil,, H. J. I., Mulder, P. G. H., Emmen, H. H., & Weisglas-Kuperus, N. (2004a). Effects of perinatal exposure to PCBs on neuropsychological functions in the Rotterdam cohort at 9 years of age. *Neuropsychology, 18*(1), 185–193.

Vreugdenhil,, H. J. I., Van Zanten, G. A., Brocaar, M. P., Mulder, P. G. H., & Weisglas-Kuperus, N. (2004b). Prenatal exposure to polychlorinated biphenyls and breastfeeding: Opposing effects on auditory P300 latencies in Dutch children. *Developmental Medicine & Child Neurology, 46*, 398–405.

Vyas, N. B. (1999). Factors influencing estimation of pesticide-related wildlife mortality. *Toxicology and Industrial Health, 15*, 186–191.

Wachs, T. D., & Gruen, G. E. (1982). *Early experience and human development*. New York: Plenum Press.

Wadhwa, P. D. (1998). Prenatal stress and life-span development. *Encyclopedia of mental health* (Vol. 3). Orlando, FL: Academic Press.

Wagner, R., & Torgesen, J. (1987). The nature of phonological processing and its causal role in the acquisition of reading skills. *Psychological Bulletin, 101*, 192–212.

Wahlen, M., Kunz, C. O., Matuszek, J. M., Mahoney, W. E., & Thompson, R. C. (1980). Radioactive plume from the Three Mile Island accident: Xenon-133 in air at a distance of 375 kilometers. *Science, 207*, 639–640.

Walker, P. (2007). From Windscale to Sellafield: A history of controversy. *The Guardian*, April 18, 2007. http://www.guardian.co.uk/environment/2007/apr/18/energy.nuclearindustry accessed on August 25, 2008.

Walker, R. B., Gessel, S. P., & Held, E. E. (1997). The ecosystem study on Rongelap atoll. *Health Physics, 73*(1), 223–233.

Walkowiak, J., Wiener, J-A., Fastabend, A., Heinzow, B., et al. (2001). Environmental exposure to polychlorinated biphenyls and the home environment: Effects on psychodevelopment in childhood. *Lancet, 358*(9293), 1602–1607.

Wampold, B. E., & Serlin, R. C. (2000). The consequence of ignoring a nested factor on measures of effect size in analysis of variance. *Psychological Methods, 5*(4), 425–433.

Wang, C-L., Chuang, H-Y., Chang, C-Y., Liu, S-T., et al. (2000). An unusual case of organophosphate intoxication of a worker in a plastic bottle recycling plant: An important reminder. *Environmental Health Perspectives, 108*, 1103–1105.

Ward, D. H., Stehn, R. A., Erickson, W. P., & Derksen, D. V. (1999). Response of fall-staging brant and Canada geese to aircraft overflights in southwestern Alaska. *Journal of Wildlife Management, 63*(1), 373–381.

Wargo, J. (1996). *Our children's toxic legacy*. New Haven, CT: Yale University Press.

Warner, F., & Kirchmann, R. J. C. (Eds.). (2000). *Nuclear test explosions: Environmental and human impacts (SCOPE 59)*. West Sussex, England: John Wiley and Sons.

Wasserman, H., & Solomon, N. (1982). *Killing our own: The disaster of America's experience with atomic radiation*. Delta. Available online at www.ratical.org/radiation/KillingOurOwn/.

Wassermann, M., Wassermann, D., Cucos, S., & Miller, H. J. (1979). World PCBs map: Storage and effects in man and his biologic environment in the 1970s. *Annals of the New York Academy of Sciences, 320*, 69–110.

WDNR. (2002). Fish consumption advisory. Accessed online at www.dnr.state.wi.us/org/water/fhp/fish/advisories/ on October 20, 2002.

Weaver, K. L, Ivester, P., Chilton, J. A., Wilson, M. D., Pandey, P., & Chilton, F. H. (2008). The content of favorable and unfavorable polyunsaturated fatty acids found in commonly eaten fish. *Journal of the American Dietetic Association, 108*, 1178–1185.

Weinstein, N. D. (1978). Individual differences in reactions to noise: A longitudinal study in a college dormitory. *Journal of Applied Psychology, 63*, 458–466.

Weinstein, N. D. (1982). Community noise problems: Evidence against adaptation. *Journal of Environmental Psychology, 2*, 87–97.

Weisglas-Kuperus, N., Patandin, S., Berbers, G. A., Sas, T. C., et al. (2000). Immunologic effects of background exposure to polychlorinated biphenyls and dioxins in Dutch preschool children. *Environmental Health Perspectives, 108*, 1203–1207.

Weiss, B. (1983). Behavioral toxicology and environmental health science: Opportunity and challenge for psychology. *American Psychologist, 38*, 1174–1187.

Weiss, B. (2000). Vulnerability of children and the developing brain to neurotoxic hazards. *Environmental Health Perspectives Supplements, 108*(3), 375–381.

Welch, J. (2001). Remarks by Jack Welch. GE Annual Share Owner's Meeting, Atlanta, April 25, 2001. Available online at www.ge.com.

Weppner, S., Elgethun, K., Lu, C., Herbert, V., Yost, M. G., & Fenske, R. A. (2006). The Washington aerial spray drift study: Children's exposure to methamidophos in an agricultural community following fixed-wing aircraft applications. *Journal of Exposure Science and Environmental Epidemiology, 16*, 387–396.

White, K. R. (1982). The relation between socioeconomic status and academic achievement. *Psychological Bulletin, 91*, 461–481.

Whitmore, R. W., Immerman, F. W., Camann, D. E., Bond, A. E., Lewis, R. E., & Schaum, J. L. (1994). Non-occupational exposures to pesticides for residents of two U.S. cities. *Archives of Environmental Contamination and Toxicology, 26*, 47–59.

Wightman, F. L., & Kistler, D. J. (2005). Informational masking of speech in children: Effects of ipsilateral and contralateral distracters. *Journal of the Acoustical Society of America, 118*, 3164–3176.

Wilgoren, J. (2002, January 19). 100 families leaving tainted town for cleanup. *New York Times.* Accessed online at www.nytimes.com/2002/01/19/national/19LEAD.html?ex=1012582259&ei=1&en=db9aca9add0406dd on January 20, 2002.

Williams, M. K., Barr, D. B., Camann, D. E., Cruz, L. A., et al. (2006). An intervention to reduce residential insecticide exposure during pregnancy among an inner-city cohort. *Environmental Health Perspectives, 114*, 1684–1689.

Williams, N. (1995). Chernobyl: Life abounds without people. *Science, 269*, 304.

Wing, S. (2003). Objectivity and ethics in environmental health science. *Environmental Health Perspectives, 111*, 1809–1818.

Wing, S., & Richardson, D. (2000). Collision of evidence and assumptions: TMI deja view [Letter]. *Environmental Health Perspectives, 108*(12), A546–A547.

Wing, S., Richardson, D., & Armstrong, D. (1997a). Response: Science, public health, and objectivity: Research into the accident at Three Mile Island. *Environmental Health Perspectives*, *105*(6), 567–570.

Wing, S., Richardson, D., Armstrong, D., & Crawford-Brown, D. (1997b). A reevaluation of the cancer incidence near the Three Mile Island nuclear plant: The collision of evidence and assumptions. *Environmental Health Perspectives*, *105*(1), 52–57.

Wing, S., Shy, C. M., Wood, J. L., Wolf, S., Crangle, D. L., & Frome, E. L. (1991). Mortality among workers at Oak Ridge National Laboratory. *Journal of the American Medical Association*, *265*, 1397–1402.

Winneke, G., Bucholski, A., Heinzow, B., Kramer, U., Schmidt, E., et al. (1998). Developmental neurotoxicity of polychlorinated biphenyls (PCBs): Cognitive and psychomotor functions in 7-month-old children. *Toxicology Letters*, *102–103*, 423–428.

Wisconsin Department of Natural Resources. (2000). *Important health information for people eating fish from Wisconsin waters, 2000*. Pub no FH824 00Rev. Madison, WI, 53707.

Wisconsin Department of Natural Resources (2008). http://www.dnr.state.wi.us/fish/consumption/FishAdv08WebList.pdf accessed on November 19, 2008.

Wisconsin Stewardship Network. (2002). *Mining issues page*. Accessed online at www.wsn.org on July 7, 2002.

Wolfe, H. R., Durham, W. F., & Walker, K. C., et al. (1961). Health hazards of discarded pesticide containers. *Archives of Environmental Health*, *3*, 45–51.

Wood, D. F. (2001). Decision in 99–2618, U.S. Court of Appeals, State of Wisconsin v EPA and Sakaogon Chippewa Community. Accessed online at laws.findlaw.com/7th/992618.htm on July 7, 2002.

Wood, J. M., Kennedy, F. S., & Rosen, C. G. (1968). Synthesis of methyl-mercury by extracts of a methanogenic bacterium. *Nature*, *220*, 173–174.

World Health Organization (2006). Health effects of the Chernobyl accident and special health care programmes. Geneva. http://www.who.int/ionizing_radiation/chernobyl/WHO%20Report%20on%20Chernobyl%20Health%20Effects%20July%2006.pdf accessed on August 25, 2008.

Wright, R. J. (2005). Stress and atopic disorders. *Journal of Allergy and Clinical Immunology*, *116*, 1301–1306.

Yamazaki, J. N., & Schull, W. J. (1990). Perinatal loss and neurological abnormalities among children of the atomic bomb: Nagasaki and Hiroshima revisited, 1949 to 1989. *Journal of the American Medical Association*, *264*(5), 605–609.

Yamins, J. G. (1977). The relationship of subclinical lead intoxication to cognitive and language functioning in preschool children. *Dissertation Abstracts International*, *37*(8-B), 4176-B.

Yevelson, I. I., Abdelgani, A., Cwikel, J., & Yevelson, I. S. (1997). Bridging the gap in mental health approaches between East and West: The psychosocial consequences of radiation exposure. *Environmental Health Perspectives*, *105*(Supplement 6), 1551–1556.

Yorifuji, T., Tsuda, T., Takao, S., & Harada, M. (2008). Long-term exposure to methylmercury and neurologic signs in Minamata and neighboring communities. *Epidemiology, 19*, 3–9.

Young, J. G., Eskenazi, B., Gladstone, E. A., Bradman, A., et al. (2005). Association between in utero pesticide exposure and abnormal reflexes in neonates. *NeuroToxicology, 26*, 199–209.

Yu, M., Hsu, C., Gladen, B. C., & Rogan, W. J. (1991). In utero PCB/PCDF exposure: Relation of developmental delay to dysmorphology and dose. *Neurotoxicology and Teratology, 13*, 195–202.

Zayas, L. H., & Ozuah, P. O. (1996). Mercury use in Espiritismo: A survey of botanicas. *American Journal of Public Health, 86*, 111–112.

Zeckhauser, R. J., & Viscusi, W. K. (1990). Risk within reason. *Science, 248*, 559–564.

Zelman, M., Camfield, P. R., Moss, M., Camfield, C., et al. (1991). Toxicity from vacuumed mercury: A household hazard. *Clinical Pediatrics, 30*(2), 121–123.

Zimbrick, J. D. (1998). The ultimate questions: Future research at RERF. In L. E. Peterson & S. Abrahamson (Eds.), *Effects of ionizing radiation: Atomic bomb survivors and their children (1945–1995)*. Washington, D.C.: John Henry Press.

Zimmer, K., & Ellermeier, W. (1999). Psychometric properties of four measures of noise sensitivity: A comparison. *Journal of Environmental Psychology, 19*, 295–302.

Index

Japan:
  mercury poisoning, 41–43, 46
  PCB contamination, 71–72, 83
  PCB poisoning, 71–72, 175
  radiation exposure, health effects
    of, 198
Jensen, Soren:
  PCBs, 70
  University of Stockholm, 70
Johns Hopkins University, 107

Kaimer, F. R., 93
Kaplan, A., 287 n. 21
Kazakhstan, 205
Kehoe, Robert:
  Kettering Laboratory, 7, 8
  Lead Industry Association
    (LIA), 10
  lead poisoning, 9–10
  lead pollution, 10–14
  tetraethyl lead, 8–9, 207, 230
Kehoe Paradigm, 8, 231, 238
  lead poisoning, 16
  tetraethyl lead, 7
Kelly, R. E., 93
Kennedy, John F., 107
Kennedy, Robert F., 147
Kettering Laboratory, 7, 8
Kiev (Ukraine), 189
Kiste, Robert, 201
Kryter, Karl, 165–166, 169,
  272 n. 5
Kuhn, Thomas, 287 n. 21
Kunkel School (Pa.), 176
Kuratsune, M., 97
Kyushu University (Japan), 72

Lake Michigan:
  PCB contamination, 74–76, 80–81,
    83–87, 90, 92
Lane, S. R., 273 n. 18
Lead, 240–243
  atmospheric, 11–12
  in blood, 14
  exposure, behavioral effects of, 22,
    29, 225

  exposure, developmental effects of, 3,
    15, 20–22, 29–33, 105, 225
  exposure, health effects of, 34–35
  exposure, monitoring, 11–12, 13,
    240, 260 n. 19
  exposure, prenatal, 76, 225
  in gasoline. See Tetraethyl lead
  in paint. See Lead paint
  sources, 25
  tetraethyl. See Tetraethyl lead
Lead abatement:
  Alliance to End Childhood Lead
    Poisoning, 28
  Harvard University, 27–28
  lead paint, 27–29
  Massachusetts Department of Public
    Health, 27–28
  Needleman, Herbert, 28
  U.S. Department of Housing and
    Urban Development (HUD), 28
  U.S. Public Health Service, 28
Lead exposure:
  African Americans, 4, 23, 25,
    26–27, 170
  Australia, 22–23
  Christchurch (New Zealand),
    20–22, 30
  Cincinnati (Ohio), 23, 30
  Cleveland (Ohio), 30
  Hispanics, 170
  and Intelligence Quotient (IQ),
    29–30, 34
  New Zealand, 20–22, 30
  Port Pirie (Australia), 22–23
Lead exposure, behavioral effects:
  African Americans, 23
  Australia, 22–23
  Cincinnati (Ohio), 23
  Port Pirie (Australia), 22–23
Lead exposure, developmental
    effects:
  African Americans, 23, 26–27
  Christchurch (New Zealand),
    20–22, 30
  Cleveland (Ohio), 30
  New Zealand, 20–22, 30

New Hampshire, 34
New Orleans (La.)
    Agricultural Street Dump, 222
    chemical waste, 221–222
New York, 221
New York City (N.Y.), 14
New York City Metropolitan
    Achievement Test, 145
New York City Prenatal Pesticide Study,
    120–121
New York State Department of
    Health, 211
    Love Canal (N.Y.), 218, 221, 223
    Roswell Park Memorial Institute, 285
    n. 256
    Seegal, Richard, 88
New York State Office of Mental
    Health, 217
New York State Supreme Court, 217
New York Times:
    Carson, Rachel, 107
    Herculaneum (Mo.), 220
    tetraethyl lead, 5–6
New York University, 151, 169
New Zealand:
    lead exposure, developmental effects
    of, 20, 31
    mercury, exposure to, 39
Niagara County (N.Y.), 218
Niagara Falls, City of, 219
Nicotine:
    animal research, 114
    and apoptosis, 114
    epidemiological studies, 113
    exposure, behavioral effects of, 113
    exposure, developmental effects of,
    114–115
    exposure, prenatal, 120, 141, 225
    exposure, secondhand, 248
Niigata (Japan), 44
Nixon, Richard M.:
    Environmental Protection Agency
    (EPA), 107
    Lead Poisoning Prevention Act of
    1970, 14
Noise, xi. *See also* Sound

24 Leq, 143
abatement, 148, 152, 162–163
ambient, 142
animal research, 157, 159–160
annoyance, 162–164
and auditory discrimination,
    145–146, 147, 148
auditory effects, 140, 141
and blood pressure, 148–149, 161, 168
country road and airport study, 154
decision criteria, 162–166
DNL (day and night sound level), 142
effects on reading, 154–156
Environmental Protection Agency
    (EPA), 164
exposure, neonatal, 158–159
exposure, postnatal, 157
exposure, prenatal, 141, 156–157,
    159–160, 225
and hearing loss, 170, 171
heating and air conditioning, 156
and house prices, 169
impulsive, 272 n. 4
nonauditory effects, 140, 141
and reading test scores, 145–146,
    147, 151–153, 157–158, 273 n. 26
regulation of, 162, 249
sensitivity to, 172
and socioeconomic status, 169
and speech perception, 154–155
steady-state, 272 n. 4
and stress, 156, 157, 160, 274 n. 73
and wildlife, 170–173
work safety rules, 289 n. 24
Noise, exposure to:
    African Americans, 150, 170
    Hispanics, 169
    Mexican Americans, 150
North Carolina:
    PCB contamination, 77–78, 95
    pesticide exposure, neurobehavioral
    effects of, 134–135, 247
Nuclear Claims Tribunal, 201
Nuclear power, 188–189, 228
Nuclear Regulatory Commission, 177,
    178, 212

Nuclear weapons testing:
  Alamagordo (N.M.), 201
  Algeria, 206
  Australia, 206
  China, 206
  Christmas Island, 206
  Fangataufa, 206
  France, 206
  Great Britain, 206
  India, 206
  Kazakhstan, 205
  Marshall Islands, Republic of, 176
  military personnel, 202–203
  Monte Bello Islands, 206
  Mururoa, 206
  Nevada Test Site, 201–202
  Pakistan, 206
  Soviet Union, 205, 206
  United States, 206

Occidental Chemical Company,
    211, 218
Odense University (Denmark):
  Faroe Islands, 45
  Grandjean, Phillippe, 45
Oklahoma, 205
Omega-3 fatty acids, in fish, 62–63
OPIDN (organophosphate-induced
    delayed polyneuropathy), 109,
    110–111
OPIDP. See OPIDN (organophosphate-
    induced delayed polyneuropathy)
Oswego (N.Y.), 85, 87

P300, 79
Paigen, Beverly:
  Love Canal (N.Y.), 212, 213–214
  Roswell Park Memorial Institute, 213
Pakistan:
  mercury poisoning, 41
  nuclear weapons testing, 206
Parade, 14
Patel, Marilyn, 203
Patterson, Clair "Pat":
  California Institute of Technology, 9
  lead pollution, 9–10, 30

Pauling, Linus, 202
PBBs (polybrominated biphenyls),
    77, 86
PBDE contamination, 94–95, 101
PBDEs (polybrominated diphenyl
    ethers), 101
PCB contamination animal research, 87
  arctic regions, 95–96
  Belgium, 98
  in breast milk, 79–82
  Environmental Protection Agency
    (EPA), 91
  in fish, 74, 86
  Fox River Valley (Wisc.), 101
  Great Lakes, 100, 247
  Hudson River (N.Y.), 71, 99, 102–103
  Inuit people, 95–96
  Japan, 72, 83
  Lake Michigan, 74, 84, 91, 101
  Lobelville (Tenn.), 101–102
  Love Canal (N.Y.), 209
  Netherlands, 78
  North Carolina, 77–78, 95
  oil disease, 72, 73
  Oswego (N.Y.), 85, 87
  Perry County (Tenn.), 101–102
  St. Lawrence River, 221
  Taiwan, 73–74, 87, 175
  Tenneco, 101–102
PCBs, xi, xiii, 232
  in animal feed, 98
  in breast milk, 79–82, 95
  congeners, 71, 89
  disposal, 71
  exposure, health effects of, 94
  exposure, neurodevelopmental effects
    of, 88–89
  exposure, occupational, 90, 91
  exposure, postnatal, 86
  exposure, prenatal, xiii, 45, 68, 77,
    80–81, 89, 225
  exposure, sources of, 97–99
  in food chain, 71, 95, 100
  industrial uses, 69–70
  regulation of, 96
  in wildlife, 98–99, 103

burden of proof, 230, 234–235
and PCB exposure, 93–94
public participation, 236–237
Rio Declaration, 229
and uncertainty, 230–231
President's Commission on the Accident
at Three Mile Island:
Fabrikant, Jacob, 180
Public Health Safety Task Force, 180
Pripyat (Ukraine), 189, 190, 191
PubMed, 84

Quality of life, xviii
Quine, W. V., 287 n. 21

Radiation exposure, cancer effects:
Chernobyl (Ukraine), 206, 207, 218
Japan, 200
Marshall Islands, Republic of,
200, 201
Radiation exposure, developmental
effects:
Belarus, 198
Chernobyl (Ukraine), 198
Radiation exposure, health effects:
Chernobyl (Ukraine), 199
Japan, 200
Marshall Islands, Republic of, 200
Radiation exposure, prenatal:
Belarus, 198
Chernobyl (Ukraine), 197, 198
Ukraine, 194–195
Radiation exposure, psychological
effects:
Chernobyl (Ukraine), 197–198
Gomel (Belarus), 196
Ukraine, 194–195
Radiation exposure, reproductive
effects, 198–199
Radiation Exposure Compensation
Act, 203
Radioactivity, xi, 174, 179
background radiation, 278 n. 25
cancer effects, 180, 198, 200, 202, 203
depleted uranium, 197, 282 n. 142
ecosystems, effects on, 206–207

exposure, developmental effects of,
193–194, 225
exposure, occupational, 205
exposure, prenatal, 193, 225
exposure to, 193–194
fallout, 201
farm animal deaths, 180
health effects, 197
medical and dental uses, 252
and natural resources, 200
psychological effects, 182–184, 193
radiation poisoning, symptoms
of, 180
radioactive cesium, 204
radioactive iodine, 177, 206
radioactive krypton, 177, 178
radioactive xenon, 177, 178
radon, 252
reproductive effects, 180,
198–199, 203
risk perception and coping, 188–189
tree kills, 180
waste disposal, 201, 223
and wildlife, 199, 206–207
Radioactivity, exposure to, 193–194
Reading test scores:
and African Americans, 150
and Hispanics, 150
and noise, 145–146, 148,
150–154, 156
Reference dose (RfD), 57
Renco Group, Inc.:
Doe Run, 220
Herculaneum (Mo.), 220
Reproductive effects:
chemical waste, 208
Chernobyl (Ukraine), 198–199
radioactivity, 180, 198, 203
Three Mile Island (TMI), 182
Research, objectivity in, xv–xvi
Retinoids, 88
Rice, Deborah, 94
Rio Tapajos (Brazil), 49–50
Risk assessment, xv, 54–61, 64,
226–227, 264 n. 63
adverse event definition, 54–55

Sternglass, Ernest, 188
Stewart, Alice, 188
Stress:
    and community pollution,_175
    and coping, 184, 250
    and noise, 153, 156, 160, 274 n. 73
Superfund, 19, 25–261 n. 43
    Agricultural Street Dump (New
        Orleans, La.), 222
    Carter, Jimmy, 207
    Fox River Valley (Wisc.), 101
    Hudson River (N.Y.), 102–103
    Lobelville (Tenn.), 101–102
    PCB contamination, 99
    St. Lawrence River, 221
    waste sites, 221, 251
Suppe, F., 260 n. 19, 287 n. 21
Susquehanna River, 176, 178
Swanson, Mary:
    decision criteria, 92–93
    Michigan State University, 89
Symptoms:
    insecticides, carbamate, 128
    insecticides, organophosphorus (OP),
        109, 128, 131, 134
    lead poisoning, 13
    mercury poisoning, 37–38
    pink disease, 40

Taiwan:
    PCB contamination, 73–74, 83, 175
Tamplin, Arthur, 188
TCB (2,4,4',4'-TCB), 88–89
Telone, 122
Tenneco, 101–102
Tetraethyl lead, 5–8, 11–12, 224
    decision standards, 8
    DuPont, 5–6
    Edsall, David, 6
    Hamilton, Alice, 5–6
    Henderson, Yandell, 6
    Kehoe, Robert, 7–8, 207, 229
    Kehoe Paradigm, 6–7
    Standard Oil Company, 6
Texas, 133
Thermoluminescence dosimeter
    (TLD), 178

Thimersol, 51–52, 53
Thomas Panel, 213–214, 219
Thornburgh, Richard, 177
Three Mile Island (TMI), 175, 176–177
    background radiation, 278 n. 25
    cancer effects, 181, 206–207
    psychological effects, 182–187
    radioactive cesium, 204
    reproductive effects, 180
    risk perception and coping, 184–186
    and stress, 184–187
Threshold view, xiii, 13
Thyroid-secreting hormone (TSH), 88
Tilson, H. A., 88
Time, 13
Times Beach (Mo.), 220
TLD (thermoluminescence
    dosimeter), 178
Tobacco:
exposure, developmental effects
    of, xvii, 225
TOCP (tri-o-cresyl phosphate), 110
Tolerable daily intake (TDI), 124
Toxic equivalency (TEQ), 98
Toxicity:
    acute, 70, 110, 128
Tragedy of the commons, 239
Trait anger, 275 n. 84
Traumeel, 245
Tri-o-cresyl phosphate (TOCP), 110
TSH (thyroid-secreting hormone), 88

Udall, Stewart, 107
Ukraine:
    Chernobyl reactor accident, 178
    radiation exposure, 194–196
Ullian, J. S., 287 n. 21
Uncertainty, scientific, xv, 230–234,
    287 n. 21
    uncertainty factor, 58, 63–64
Uniformed Services University of the
    Health Sciences, 186, 187, 212
Union Carbide:
    Bhopal (India), 219
    chemical waste, 219
United Nations:
    Chernobyl reactor accident, 178